INTRODUCTION TO CONTROLLED-SOURCE ELECTROMAGNETIC METHODS

This volume describes how controlled-source electromagnetic methods are used to determine the electrical conductivity and hydrocarbon content of the upper few kilometres of the earth, on land and at sea. The authors show how the signal-to-noise ratio of the measured data may be maximised via suitable choice of acquisition and processing parameters and selection of subsequent data analysis procedures. Complete impulse responses for every electric and magnetic source and receiver configuration are derived, providing a guide to the expected response for real data. One-, two- and three-dimensional modelling and inversion procedures for recovery of earth conductivity are presented, emphasising the importance of updating model parameters using complementary geophysical data and rock physics relations. Requiring no specialist prior knowledge of electromagnetic theory, and providing a step-by-step guide through the necessary mathematics, this book provides an accessible introduction for advanced students, researchers and industry practitioners in exploration geoscience and petroleum engineering.

ANTON ZIOLKOWSKI is Professor of Petroleum Geoscience at the University of Edinburgh. He co-invented the multichannel transient electromagnetic (MTEM) surveying method for hydrocarbon reservoir detection and co-led the technology spin-out. He is a member of the Institute of Electrical and Electronics Engineers and the American Geophysical Union, an honorary member of the Society of Exploration Geophysicists, a Fellow of the Royal Academy of Engineering and a fellow of the Royal Society of Edinburgh. He has received the Conrad Schlumberger and Desiderius Erasmus awards of the European Association of Geoscientists and Engineers.

EVERT SLOB is Professor of Geophysical Electromagnetic Methods at the Delft University of Technology, where he teaches undergraduate and graduate classes. He was Editor-in-Chief of the journal *Geophysics* and on the board of directors of the Society of Exploration Geophysicists from 2013 to 2015, of which he remains a member. He is also a member of the European Association of Geoscientists and Engineers and the American Geophysical Union.

INTRODUCTION TO CONTROLLED-SOURCE ELECTROMAGNETIC METHODS

Detecting Subsurface Fluids

ANTON ZIOLKOWSKI
The University of Edinburgh

EVERT SLOB
Delft University of Technology

CAMBRIDGE
UNIVERSITY PRESS

University Printing House, Cambridge CB2 8BS, United Kingdom

One Liberty Plaza, 20th Floor, New York, NY 10006, USA

477 Williamstown Road, Port Melbourne, VIC 3207, Australia

314–321, 3rd Floor, Plot 3, Splendor Forum, Jasola District Centre, New Delhi – 110025, India

79 Anson Road, #06–04/06, Singapore 079906

Cambridge University Press is part of the University of Cambridge.

It furthers the University's mission by disseminating knowledge in the pursuit of education, learning, and research at the highest international levels of excellence.

www.cambridge.org
Information on this title: www.cambridge.org/9781107058620
DOI: 10.1017/9781107415904

First published 2019

Printed in the United Kingdom by TJ International Ltd. Padstow Cornwall

A catalogue record for this publication is available from the British Library.

Library of Congress Cataloging-in-Publication Data
Names: Ziolkowski, Anton, 1946– author. | Slob, Evert C. (Evert Cornelis), 1962– author.
Title: Introduction to controlled-source electromagnetic methods : detecting subsurface fluids / Anton Ziolkowski (The University of Edinburgh), Evert Slob (Delft University of Technology).
Description: Cambridge ; New York, NY : Cambridge University Press, 2019. | Includes bibliographical references and index.
Identifiers: LCCN 2018034518 | ISBN 9781107058620 (hardback) | ISBN 9781107634855 (pbk.)
Subjects: LCSH: Earth (Planet)–Electric properties. | Earth (Planet)–Magnetic properties. | Earth (Planet)–Crust. | Electromagnetic fields. | Electric prospecting.
Classification: LCC QE501.3 .Z56 2019 | DDC 551–dc23
LC record available at https://lccn.loc.gov/2018034518

ISBN 978-1-107-05862-0 Hardback

Additional resources for this publication at www.cambridge.org/csem.

Contents

Preface

The aim of this book is to make the benefits of controlled-source electromagnetic (CSEM) methods more widely appreciated by geoscientists and engineers, and to provide an approach that has sound theoretical foundations and a clear description of the practical aspects of CSEM data acquisition, processing and interpretation.

CSEM methods are used to explore for contrasts in subsurface electrical conductivity and are especially useful to search for subsurface fluids, including resistive hydrocarbons and conductive saline water. For example, CSEM methods have the potential to detect hydrocarbons before drilling. Since three out of four exploration wells contain no hydrocarbons, it may pay to carry out CSEM exploration before drilling to increase the likelihood of finding oil or gas. Saline water at depths of 2–4 km is usually hot enough to provide heat for buildings. In many countries, heating consumes more energy than transport and electricity generation combined. CSEM has the potential to find the geothermal resources that can reduce our dependence on fossil fuels.

Theoretical work on the concept of CSEM methods and the use of loops and antennas for exploration dates back to the 1950s. Onshore techniques were developed commercially and by the academic community. Offshore techniques were developed initially by academics. By 1991, Misac N. Nabighian was able to bring all this work together in the two-volume book *Electromagnetic Methods in Applied Geophysics*, published by the Society of Exploration Geophysics. In the first decade of the twenty-first century, CSEM became a tool for de-risking exploration drilling for deep-water prospects. Compared with seismic exploration, however, CSEM is still in its infancy and is still expensive per data point. There is clearly room for development.

It is now well understood in seismic exploration that broad bandwidth data are essential for good imaging of subsurface structures, whether the data are processed in the time domain or the frequency domain. A key concept is the idea of an impulsive source and the resulting impulse response of the earth. This concept is equally applicable to CSEM and is at the heart of our description of the method. For the source time function, CSEM has a big advantage over seismic exploration methods: it is very easy to reverse the polarity of current flow and create source time functions that have desirable properties. Furthermore, the source time function is easily measured and recovery of the resulting impulse responses from the measured data by deconvolution is straightforward. The impulse responses may be processed in the time domain or the frequency domain to determine subsurface resistivities.

There are some similarities with seismic exploration, but there are major differences. The most important difference is, of course, the physics. Seismic data obey the wave equation; electromagnetic (EM) data in conducting media such as fluid-filled rocks obey the diffusion equation. Seismologists often use ray theory to describe what happens to the waves – how they reflect, refract and diffract. Unfortunately, ray theory does not apply to diffusive data. Seismologists are accustomed to lining up seismic arrivals that have the same shape and estimating seismic velocities as a result – the velocities are determined from the data themselves. Such techniques cannot be used to estimate resistivities from EM data, because the shape of the wave changes as it propagates. Instead, the resistivities are normally estimated from the data by inversion, which is a kind of modelling. For a seismologist this can be frustrating. This book is written partially for seismologists who would like an easy way 'in' to understanding electromagnetics.

The book is written for students, researchers and practitioners. Much of the material has been presented as courses for undergraduate and graduate geophysics students at the University of Edinburgh and at Delft University of Technology. The mathematical background required is partial differential equations, vector algebra, Fourier transforms and Laplace transforms.

We have had discussions with many friends and colleagues, and thank in particular, Bruce Hobbs, Paul Stoffa, David Wright, David Taylor, Dieter Werthmüller and our students for all their help and comments. We thank Cambridge University Press for agreeing to publish the book. Susan Francis has been especially kind, helpful, encouraging and patient.

Anton thanks his lovely wife Kate for constant support.

Notation and Conventions

Symbols

Symbol	Description	SI units
a	tortuosity factor in Archie's law	–
A	area	m^2
$\mathbf{B}$	vector magnetic induction	$V\,s\,m^{-2}$
c	propagation velocity	$m{\cdot}s^{-1}$
c_w	speed of sound in water	$m{\cdot}s^{-1}$
$c_0 = 299,792,458$	electromagnetic wave propagation velocity	$m{\cdot}s^{-1}$
$\mathbf{D}$	electric flux density	$C\,m^{-2}$
∇	del, or nabla, vector operator	m^{-1}
$\mathbf{E}$	vector electric field intensity	$V{\cdot}m^{-1}$
ε	electrical permittivity	$C^2{\cdot}N^{-1}{\cdot}m^{-2}$
$\varepsilon_0 \approx 8.85 \times 10^{-12}$	electrical permittivity of free space	$C^2{\cdot}N^{-1}{\cdot}m^{-2}$
δ	skin depth	m
$\delta(t)$	impulse function	s^{-1}
Δx_s	distance between source electrodes	m
Δx_r	distance between receiver electrodes	m
f	frequency	Hz
F	formation factor	–
G	Green's function (units depend on problem)	–
$\gamma = \sqrt{\zeta\sigma}$	horizontal wavenumber	m^{-1}
$\gamma_v = \sqrt{\zeta\sigma_v}$	vertical wavenumber	m^{-1}
$\Gamma = \sqrt{\kappa^2 + \gamma^2}$	vertical wavenumber	m^{-1}
$\Gamma_v = \sqrt{\lambda^2\kappa^2 + \gamma^2}$	vertical wavenumber	m^{-1}
h_j	thickness of *j*th layer (Chapter 1)	m
$\mathbf{H}$	vector magnetic field intensity	$A\,m^{-1}$

I	electric current	A
$\mathbf{I}^{\mathrm{m}}$	magnetic source current dipole moment	V m
$\mathbf{I}^{\mathrm{e}}$	electric source current dipole moment	A m
I_n	modified Bessel function of first kind and order n	–
i, j, k, l	indices	–
$\mathbf{J}$	volume density of induced current vector	A m^{-2}
$\mathbf{J}^{\mathrm{e}}$	volume density of external electric current	A m^{-2}
$\mathbf{J}^{\mathrm{m}}$	volume density of external magnetic current	V m^{-2}
J_n	ordinary Bessel function of first kind and order n	–
$k = \omega/c$	wavenumber	m^{-1}
k_x, k_y, k_z	wavenumber components	m^{-1}
K	bulk modulus	Pa
K_n	modified Bessel function of second kind and order n	–
$\kappa = \sqrt{k_x^2 + k_y^2}$	horizontal wavenumber	m^{-1}
l	length	m
L	length	m
$\lambda = \sqrt{\sigma/\sigma_v}$	coefficient of anisotropy	–
m	cementation factor in Archie's law	–
μ	magnetic permeability	H·m^{-1}
$\mu_0 = 4\pi \times 10^{-7}$	magnetic permeability of free space	H·m^{-1}
n	saturation exponent in Archie's law	–
p	pressure	Pa
$\Pi(t)$	rectangle function	–
ϕ	porosity	–
q	fluid monopole source time function	Pa·m
r	distance	m
R	electrical resistance	Ω
R_h	electrical resistance in horizontal direction	Ω
R_v	electrical resistance in vertical direction	Ω
ρ	electrical resistivity	Ω·m
ρ_0	electrical resistivity of a rock saturated with salt water	Ω·m
ρ_t	resistivity of a rock	Ω·m
ρ_w	electrical resistivity of salt water	Ω·m
ϱ	density	kg·m^{-3}
ϱ_f	volume density of free charge	C m^{-3}

$s = -i\omega$	Laplace variable	Hz
S_{hc}	hydrocarbon saturation	–
S_w	water saturation	–
σ	electrical conductivity, horizontal conductivity	S·m^{-1}
σ_v	vertical conductivity	S·m^{-1}
t	time	s
T	period	s
T_g	duration of impulse response	s
T_s	duration of source time function	s
V	voltage	V
$\omega = 2\pi f$	angular frequency	rad s^{-1}
$\mathcal{V}$	volume	m^3
x, y, z	Cartesian coordinates	m
z_s	source depth	m
$\zeta = s\mu_0$	zeta	H·m^{-1}s^{-1}

Cartesian Coordinates

We use a right-handed Cartesian coordinate system with the z-axis positive downwards and the air–earth interface at $z = 0$, as shown in Figure 1.

Special Functions

The **ordinary Bessel function of the first kind** and order n is defined as

$$\mathrm{J}_n(\xi) = \frac{\mathrm{i}^{-n}}{\pi} \int_{\psi=0}^{\pi} \exp[-\mathrm{i}\xi \cos(\psi)] \cos(n\psi) d\psi. \tag{1}$$

The **modified Bessel functions of the first and second kinds** and order n are given by

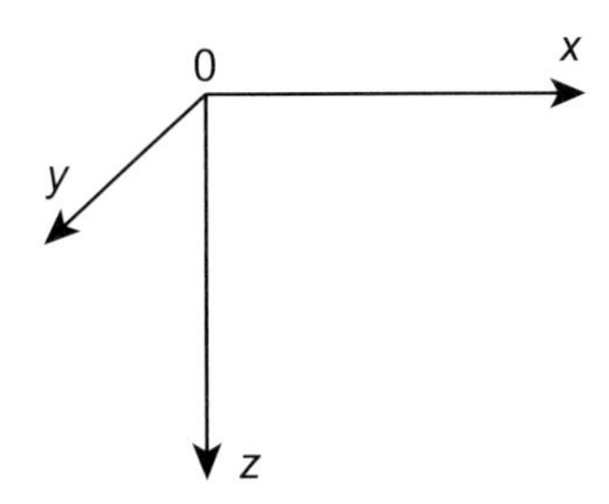

Figure 1 Cartesian coordinates.

$$\mathrm{I}_n(\xi) = \frac{1}{\pi}\int_{\psi=0}^{\pi} \exp[\xi \cos(\psi)]\cos(n\psi)d\psi, \tag{2}$$

$$\mathrm{K}_n(\xi) = \int_{\psi=0}^{\infty} \exp[-\xi \cosh(\psi)]\cosh(n\psi)d\psi. \tag{3}$$

The **error function** is defined as

$$\mathrm{erf}(x) = \frac{2}{\sqrt{\pi}}\int_{u=0}^{x} \exp(-u^2)du. \tag{4}$$

The **complementary error function** is defined as

$$\mathrm{erfc}(x) = 1 - \mathrm{erf}(x) = \frac{2}{\sqrt{\pi}}\int_{x}^{\infty} \exp(-u^2)du. \tag{5}$$

The **gamma function** is defined for all complex numbers except the non-positive integers. For complex numbers with a positive real part, it is defined via a convergent improper integral:

$$\Gamma(z) = \int_{0}^{\infty} x^{z-1}e^{-x}dx. \tag{6}$$

The **Heaviside function**, also known as the **Step function**, is defined as

$$H(t) = \begin{cases} 0, & t < 0 \\ \frac{1}{2}, & t = 0. \\ 1, & t > 0 \end{cases} \tag{7}$$

An **impulse**, also known as the **Dirac delta function**, is an infinitely strong pulse of unit area that can be defined as the two conditions:

$$\delta(t) = 0, t \neq 0;$$

$$\int_{-\infty}^{\infty} \delta(t)dt = 1. \tag{8}$$

The **rectangle function** is defined as

$$\Pi(t) = \begin{cases} 0, & |t| > \frac{1}{2} \\ \frac{1}{2}, & |t| = \frac{1}{2}. \\ 1, & |t| < \frac{1}{2} \end{cases} \tag{9}$$

The **normalised sinc function** is defined as

$$\mathrm{sinc}(f) = \frac{\sin(\pi f)}{\pi f}. \tag{10}$$

Transforms

A wavefield may be described as a function $a(x, y, z, t)$ that varies with both position (x, y, z) and time (t). We define the **temporal Fourier transform** as

$$\hat{a}(x, y, z, \omega) = \int_{-\infty}^{\infty} a(x, y, z, t)e^{i\omega t}dt, \tag{11}$$

with inverse

$$a(x.y, z, t) = \frac{1}{2\pi}\int_{-\infty}^{\infty} \hat{a}(x, y, z, \omega)e^{-i\omega t}d\omega. \tag{12}$$

The **double spatial Fourier transform** of the space–frequency domain function $\hat{a}(x, y, z, \omega)$ is defined as

$$\tilde{a}(k_x, k_y, z, \omega) = \int_{-\infty}^{\infty}\int_{-\infty}^{\infty} \hat{a}(x, y, z, \omega)e^{-i(k_x x+k_y y)}dxdy, \tag{13}$$

with inverse

$$\hat{a}(x, y, z, \omega,) = \frac{1}{4\pi^2}\int_{-\infty}^{\infty}\int_{-\infty}^{\infty} \tilde{a}(k_x, k_y, z, \omega)e^{i(k_x x+k_y y)}dk_x dk_y, \tag{14}$$

where k_x and k_y are the *horizontal wavenumbers* and the tilde $\tilde{}$ indicates the further change of domain. Here we have chosen the negative sign for the exponential for transformation from space to wavenumber and therefore the positive sign for the inverse transform.

The forward temporal and spatial Fourier transforms can be combined to give the **forward triple Fourier transform**

$$\tilde{a}(k_x, k_y, z, \omega) = \int_{-\infty}^{\infty}\int_{-\infty}^{\infty}\int_{-\infty}^{\infty} a(x, y, z, t)e^{i(\omega t-k_x x-k_y y)}dxdydt. \tag{15}$$

Similarly, the inverse transforms 12 and 14 can be combined to give the **inverse triple Fourier transform**

$$a(x, y, z, t) = \frac{1}{8\pi^3}\int_{-\infty}^{\infty}\int_{-\infty}^{\infty}\int_{-\infty}^{\infty} \tilde{a}(k_x, k_y, z, \omega)e^{-i(\omega t-k_x x-k_y y)}dk_x dk_y d\omega. \tag{16}$$

The **two-sided time-Laplace transform** of $a(x, y, z, t)$ is defined as

$$\hat{a}(x, y, z, s) = \int_{-\infty}^{\infty} a(x, y, z, t)e^{-st}dt, \tag{17}$$

and is the same as the temporal Fourier transform for the substitution $s = -i\omega$, where s is complex. When the real part of s is zero, it becomes identical with the Fourier transform. If $a(x, y, z, t) = 0$ for $t < 0$, only half the integral is required.

The **one-sided time-Laplace transform** is defined as

$$\hat{a}(x, y, z, s) = \int_{0+}^{\infty} a(x, y, z, t)e^{-st}dt. \tag{18}$$

Because s is complex, the inverse transform is a contour integration in the complex plane

$$a(x, y, z, t)H(t) = \frac{1}{2\pi i}\int_{c-i\infty}^{c+i\infty} \hat{a}(x, y, z, s)e^{st}ds, \tag{19}$$

where c is a positive constant.

The **time-Laplace and three-dimensional spatial Fourier transform** of $a(x, y, z, t)$ is

$$\breve{a}(k_x, k_y, k_z, s) = \int_{0}^{\infty}\int_{-\infty}^{\infty}\int_{-\infty}^{\infty}\int_{-\infty}^{\infty} a(x, y, z, t)e^{-st}e^{-i[k_x x+k_y y+k_z z]}dxdydzdt. \tag{20}$$

1
Introduction

The aim of this book is to explain how controlled-source electromagnetic (CSEM) methods can be used to locate resistivity variations in the top few kilometres of the Earth's crust. Applications include the search for hydrocarbons and the search for hot brine. Hydrocarbons increase the resistivity of reservoir rocks; hot brine, which is useful for geothermal purposes, reduces the resistivity of an aquifer relative to the rocks above and below. This chapter begins with Ohm's law and resistivity and proceeds to a brief discussion of resistivity of rocks, how layering introduces anisotropy, and the effect of replacing normal pore fluid with hydrocarbons in sandstone reservoirs. Hydrocarbons can increase the rock resistivity by orders of magnitude, while the P-wave velocity is hardly affected. This is demonstrated with laboratory measurements and logs from a North Sea well. As an introduction to CSEM data, acoustic propagation in water from an impulsive monopole seismic source is compared with electromagnetic propagation from an impulsive point dipole current source also in water. The effect of a buried resistor on the response is then illustrated for the simple case of a dipole source and a line of dipole receivers over a one-dimensional Earth. The effect of a buried conductor is illustrated using an identical source–receiver configuration. How subsurface resistivities may be obtained from CSEM data is not obvious. An outline of the procedure for finding the resistivities by inversion is presented, including constraints imposed by borehole and seismic data. This is followed by an outline of the book.

1.1 Ohm's Law and Resistivity

If a potential difference V volts is maintained across the ends of an electrical conductor by an external source, such as a battery, a current I amps flows in the conductor. The ratio of the voltage V to the current I is a constant R, known as the resistance of the conductor, which has units of ohms. This relationship is known as Ohm's law and is usually expressed as

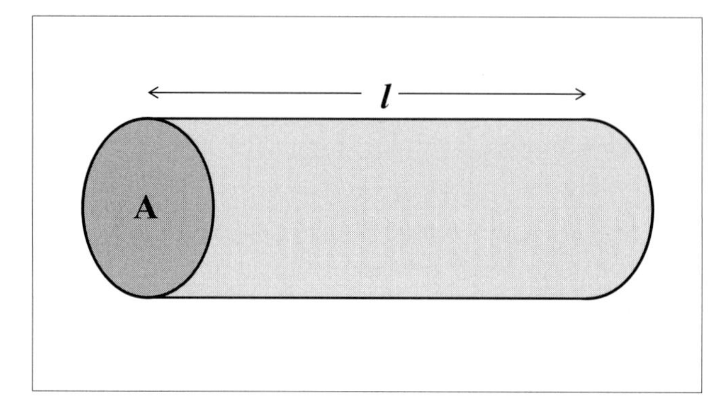

Figure 1.1 Cylindrical conductor of length l and cross-sectional area A.

$$V = IR \tag{1.1}$$

Figure 1.1 shows a cylindrical conductor of length l m and cross-sectional area A m^2. The resistance R of the conductor is proportional to its length and inversely proportional to its cross-sectional area, expressed as

$$R = \rho \frac{l}{A}. \tag{1.2}$$

The constant of proportionality ρ in equation 1.2 is a physical property of the material of the conductor, known as its resistivity, which has units of ohm-m. The reciprocal of resistivity is conductivity σ,

$$\sigma = \frac{1}{\rho}, \tag{1.3}$$

which has SI units of S m^{-1}.

1.2 Resistivity of Rocks

In metallic conductors the current flows by means of moving electrons. In other conductors the flow is by the movement of charged objects or ions. Positive ions move towards the negative potential and negative ions move towards the positive potential. By convention, the direction of current flow is taken to be the direction of flow of positively charged objects. Electrons are negatively charged, so they move towards the positive potential and thus in the opposite direction to the flow of current.

Rocks are composed of minerals that form a solid matrix that contains pores. The fraction of rock volume occupied by pore space is the porosity ϕ. The pores are full of fluids. The solid matrix is normally extremely resistive, as there are very few charged objects or ions free to move and conduct electricity. The fluid in the pores, on the other hand, contains ions and can therefore conduct electricity. Normally the fluid is salt water and the conductivity of the rock depends on the concentration of salt in the water, the fraction of the pore space that contains salt water and the freedom of movement of the ions between pores.

Sometimes the pores also contain hydrocarbons as solid, liquid, gas or a combination of phases. When hydrocarbons are present there are usually three fluid phases: salt water, hydrocarbon liquid and hydrocarbon gas – normally methane. The hydrocarbons are not ionised and so they are not conductors of electricity. It follows that the presence of hydrocarbons increases the resistivity of the rock. The greater the fluid fraction, or saturation, of hydrocarbons, the greater is the resistivity of the rock. As shown in the following, the effect of replacing salt water by hydrocarbons can increase the resistivity by orders of magnitude.

1.3 Resistivity Anisotropy

Very often rocks are layered, as indicated in Figure 1.2. A current flowing vertically through the sequence of layers sees resistances in series, with the resistance of the stack of layers being

$$R_v = \frac{l}{L^2} \sum_{j=1}^{n} \rho_j h_j. \tag{1.4}$$

In the horizontal direction the resistance of the jth layer is $R_j = \rho_j / h_j$ and the horizontal resistance R_h is given by

$$\frac{1}{R_h} = \sum_{j=1}^{n} \frac{1}{R_j}. \tag{1.5}$$

Resistors in parallel offer less resistance than the same resistors in series, so $R_h < R_v$. The scale of Figure 1.2 is arbitrary. It could be metres or millimetres. The point is that layering, and rock heterogeneity in general, gives rise to resistive anisotropy which is normally very important. The ratio R_v/R_h is often greater than 2, sometimes much greater.

1.4 Effect of Hydrocarbons on Resistivity: Archie's Law

In 1942 Gustavus E. Archie of Shell published results of laboratory experiments on Gulf Coast reservoir rock core samples in what has become a classic paper in rock physics. Archie (1942) found that the resistivity ρ_0 of a reservoir rock sample

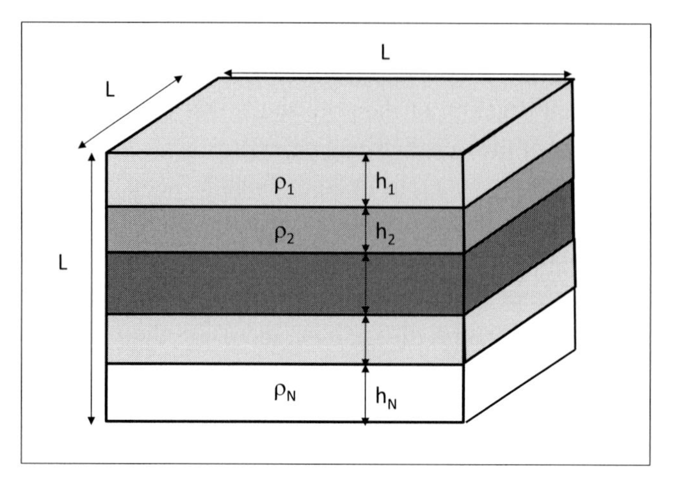

Figure 1.2 Horizontally layered cube of rock: resistivity varying with depth.

saturated with salt water was always related to the resistivity of the water ρ_w by a constant factor, which he called the formation factor F:

$$F = \frac{\rho_0}{\rho_w}. \tag{1.6}$$

This formula implies that the electricity flows only through the salt water. It is a consequence that the rock matrix has zero conductivity. Archie also found a power-law relation between this formation factor and the porosity ϕ:

$$F = \frac{1}{\phi^m}, \tag{1.7}$$

where the exponent m depends on the rock. Archie then combined these findings with work that had been published by Wyckoff and Botset (1936), Jakosky and Hopper (1937), Martin et al. (1938) and Leverett (1939). These researchers had established that displacing varying amounts of conducting water from water-saturated sand with non-conducting oil or carbon dioxide increases the resistivity of the rock. Specifically, the water saturation S_w, the fraction of pore space filled with water, is related to the rock resistivity ρ_t as

$$S_w^n = \frac{\rho_0}{\rho_t}, \tag{1.8}$$

in which n is known as the saturation exponent. Eliminating the formation factor F from equations 1.6 and 1.7 results in an expression for ρ_0, which may be substituted in equation 1.8 to give:

$$S_w^n = \frac{1}{\phi^m}\frac{\rho_w}{\rho_t}. \tag{1.9}$$

This is Archie's law. The formation factor in equation 1.7 has subsequently been modified by multiplying the right-hand side by a factor a, known as the 'tortuosity factor'. This then leads to the generalised form of Archie's law:

$$S_w^n = \frac{a}{\phi^m} \frac{\rho_w}{\rho_t}. \tag{1.10}$$

As Rider (1996: 56) puts it, 'When S_w is not 100% there are hydrocarbons present.' Water saturation S_w and hydrocarbon saturation S_{hc} are related as

$$S_w = 1 - S_{hc}. \tag{1.11}$$

The saturation exponent n is normally 2; m, known as the cementation factor, is closely related to the shape of the grains, or texture, of the rock (Rider, 1996), and is normally about 2. Rider (1996: 57) says the most frequently used formula for the formation factor F is with $a = 0.62$ and $m = 2.15$, which is the best average for sandstones.

Archie's law works well on clean, uniform sandstones. It works less well when clay is present in the sandstone, as Archie was well aware. Clay minerals can choke the narrowest pore throats in the rock matrix; they are also electrically conductive. The conductivity of the clay invalidates equation 1.6, because the rock matrix is no longer a perfect insulator. Further, the choking effect at pore throats decreases the permeability and impedes the flow of charged objects. Many attempts have been made to find formulae to include the presence of clay (e.g. De Witte, 1957; Bussian, 1983) and to account for the variations in connectivity between pores (e.g. Wyllie and Rose, 1950).

For the interpretation of well logs, Archie's law is indispensable. For the interpretation of electromagnetic data, which sample much larger rock volumes and thus a large range of heterogeneities, Archie's law should be regarded as a guide.

Figure 1.3, redrawn from Wilt and Alumbaugh (1998), shows the variation of resistivity with brine saturation for a real sandstone with porosity 0.3. As the brine saturation decreases, and hydrocarbon saturation increases, the resistivity increases exponentially. These are real measurements and the variation of resistivity with saturation is according to Archie's law. On the same graph is shown the corresponding effect of variation in P-wave velocity with brine saturation. There is a small decrease in P-wave velocity for substantial decrease in brine saturation. Comparing the two curves, it is clear that the resistivity is much more sensitive than the P-wave velocity to variations in brine saturation and thus to the presence of hydrocarbons.

1.5 Example Well Logs: P-Wave Velocity and Resistivity

Figure 1.4 shows a resistivity log (in black) and a sonic log (in grey) from well 9/23B-7 in the North Sea Harding field. Harding is a medium-size oil and gas field

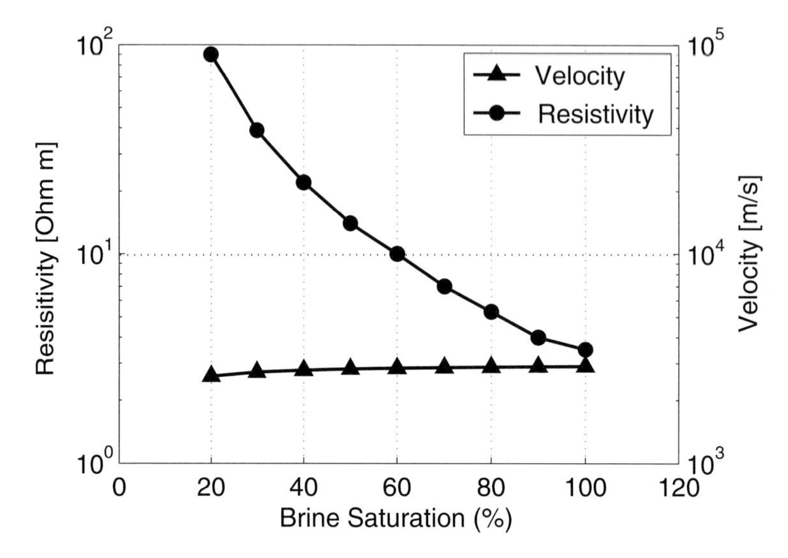

Figure 1.3 Resistivity and P-wave velocity as a function of brine saturation for a porous sandstone (redrawn from Wilt and Alumbaugh, 1998).

at a depth of about 1700 m below the sea floor in block 9/23B, about 320 km north-east of Aberdeen. The field has a high-quality Eocene Balder sandstone reservoir. Original oil in place was 300 million barrels. First oil production was in 1996, with gas being re-injected into the reservoir.

The measured resistivity is approximately 1 ohm-m for most of the 800–1800 m logged interval. At 1100–1150 m there are thin resistive beds with resistivities up to 200 ohm-m; the sonic log shows sharp fluctuations in the same interval. At 1570 m and 1600 m there are three more thin resistive beds with resistivities of 200 and 300 ohm-m which are correlated with slight increases in sonic velocity. Between 1630 and 1760 m the resistivity increases dramatically to as high as 1000 ohm-m. This is the Balder sandstone layer. In the same interval the sonic log shows two layers, with the upper one 1630–1700 m having a slightly lower velocity than the lower layer, 1700–1760 m. The huge increase in resistivity in this interval is caused by the replacement of brine in the sandstone by hydrocarbons.

From the sonic log there is very little indication of this hydrocarbon potential. It is the resistivity log that reveals this. This is a clear demonstration of the motive for searching for resistive reservoirs.

Suppose a potential sandstone reservoir of volume $\mathcal{V}$ has been identified using seismic data. Suppose also that the porosity ϕ of the reservoir has been estimated from an analysis of seismic attributes. The total volume of the pore space is then $\phi\mathcal{V}$. If now the resistivity ρ_t of the reservoir is known from electromagnetic survey data, the water saturation S_w can be estimated from Archie's law, equation 1.10,

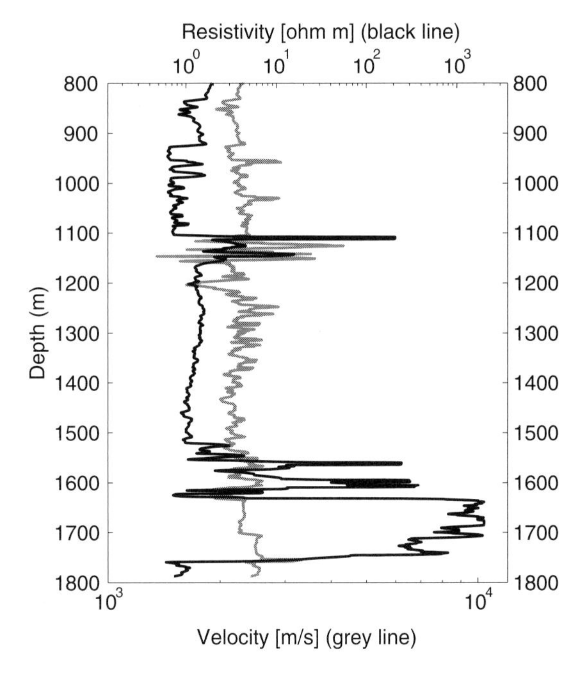

Figure 1.4 Logs from North Sea well 9/23b-7; resistivity is black; P-wave velocity is grey (redrawn from Ziolkowski et al., 2010).

and the volume of hydrocarbons is then $(1 - S_w)\phi\mathcal{V}$. Calculations like this are the motivation for using electromagnetic surveys to determine subsurface resistivity variations and rank drilling prospects already identified with seismic data.

1.6 Controlled-Source Electromagnetic Surveys

Conductivity variations in the Earth's crust and upper mantle have been investigated for decades using passive measurements of the electromagnetic field at the Earth's surface induced by natural variations in the Earth's magnetic field caused by ionospheric signals. This is the magnetotelluric (MT) method (Cagniard, 1953). MT measurements on the ocean floor suffer from attenuation of the ionospheric signals by the conducting sea water, the attenuation increasing with frequency and with the depth of the water.

It is not clear who first had the idea of using an active electromagnetic marine source to solve this problem. Certainly Bannister (1968), working in the US Navy Underwater Sound Laboratory at Fort Trumbull, New London, Connecticut, provided an early key step with the analysis of the responses of horizontal electric and magnetic dipoles in water. He argued that electric field measurements are preferable to magnetic measurements because the induced noise component is smaller. He showed that the sea bed conductivity may be determined by measuring only the horizontal electric field components produced by a subsurface horizontal magnetic dipole antenna or an electric dipole antenna, the configuration used today. Constable (2010) describes parallel work by the Scripps Institution of Oceanography, where Charles Cox and Jean Filloux developed the first equipment suitable for deep-water MT and CSEM soundings. In 1980, Cox proposed the use of an active man-made electromagnetic source at the sea bottom to overcome the problem caused by the attenuation of the magnetotelluric signal in deep water. According to Constable (2010), Cox appears to have been unaware of Bannister's 1968 paper and proposed the method independently. In 1981, Edwards et al. (1981) stated: 'Controlled source electromagnetic techniques are the obvious solution to this problem.' They showed theoretically that a vertical current bipole source in the water would produce detectable signals from below the sea floor in a horizontal magnetic receiver on the sea floor. The same group at Toronto had already pioneered the use of periodic pseudo-random binary sequences for CSEM survey on land (Duncan et al., 1980). Chave and Cox (1982) at Scripps showed theoretically for the marine case that 'Horizontal electric dipole sources produce much larger field amplitudes than their vertical counterparts for a given frequency range, and the horizontal electric field offers superior received signal performance.' By 1982, the horizontal electric dipole (HED) had become accepted as the preferred source.

Edwards et al. (1981) use the word *bipole*, while Chave and Cox (1982) use the word *dipole* for the same thing. The two poles of a dipole or a bipole are separated by a distance. In this book we use *dipole* whatever the distance between the poles. The meaning should be clear from the context.

1.7 Seismic and Electromagnetic Propagation

Here, we now compare seismic wave propagation with electromagnetic propagation in conducting media.

Some seismic sources, such as underground explosions and earthquakes, produce permanent displacements. That is, the displacement source time function contains a zero-frequency, or DC, component. To first order, seismic wave propagation is elastic; that is, there are no losses. The elastic waves from the source deform the media in which they propagate very slightly: the strains are usually smaller than 10^{-5}. After a seismic wave has passed through a medium, the medium returns to its original state, apart from any permanent displacement. Most man-made seismic sources, apart from explosives, do not generate permanent displacements. In fact, it is very difficult to generate very low-frequency energy (<1 Hz) with man-made seismic sources, especially in the marine environment. Measurements of seismic waves from such sources contain no DC component.

Electromagnetic waves are different from seismic waves. Electromagnetic waves in conducting media – fluids and solids – exhibit both electric and magnetic fields. The electric fields are associated with currents according to Ohm's law. The currents generate magnetic fields according to the Biot–Savart law. Changes in the magnetic fields cause changes in the electric field by Faraday's law of electromagnetic induction. Maxwell developed his famous theory of electromagnetism by starting with the experimental evidence presented by Faraday. The important point for the exploration geophysicist is that the electric and magnetic fields are related, and whenever there is current, as there must be in a conducting medium, there are losses. These losses are the principal difference between seismic and EM propagation. Another major difference is that it is very easy to create DC electromagnetic energy – for instance, simply by switching on a DC current. Whether the electromagnetic data contain DC energy or not depends only on the source time function.

Seismic waves propagate according to the wave equation, which is derived from two more fundamental equations: Newton's second law of mechanics (force equals mass times acceleration) and Hooke's law of elasticity (stress is proportional to strain). In solids there are two kinds of elastic waves: longitudinal, or P-waves, in which the particle vibration is parallel to the direction of wave propagation; and shear, or S-waves, in which the particle vibration is perpendicular to the direction of propagation. Fluids have no shear strength and therefore do not support shear waves. P-waves propagate in fluids and are known as acoustic waves.

Consider the simple case of a monopole source in water generating a spherical pressure wave $p(r, t)$ with the form

$$p(r,t) = \frac{\rho}{4\pi r} q'\left(t - \frac{r}{c}\right), \tag{1.12}$$

in which r is the distance from the source, t is time, $q(t)$ is the source time function with dimensions of volume divided by time and c is the speed of sound in water

$$c = \sqrt{\frac{K}{\varrho}}, \tag{1.13}$$

with K and ϱ the bulk modulus and density of the water. Consider also a similar simple case of an x-directed electric current impulsive dipole source in water generating a spherically spreading diffusive electric field $E_x(x, y, z, t)$ with the form

$$E_x(x, y = 0, z, t) = \frac{\mu I_x \exp[-\sigma\mu r^2/(4t)]}{4\pi t^{5/2}} \sqrt{\frac{\sigma\mu}{4\pi}} \left[1 - \frac{\sigma\mu z^2}{4t}\right]. \tag{1.14}$$

The derivation of this result is given in Chapter 4 and the corresponding Green's function is given in equation 4.95. Figure 1.5(a) shows the configuration used to

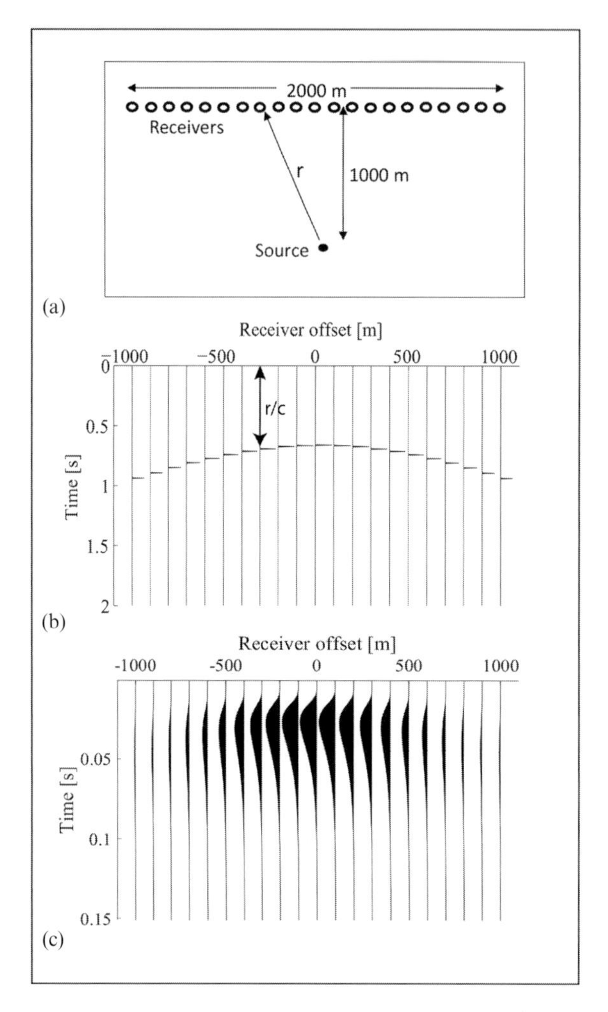

Figure 1.5 (a) Configuration of source and receivers in water of velocity 1500 m/s and resistivity 0.33 ohm-m; (b) pressure response at the receivers to an impulsive acoustic monopole at the origin; (c) x-component of electric field response to a x-directed impulsive current dipole source at the origin.

compute the whole space responses for the acoustic and electromagnetic situations. Figure 1.5(b) shows a representation of the pressure response to an impulsive compressional monopole source in an infinite body of water. The impulse is band-limited by the sampling and therefore has finite amplitude. The arrival time of the impulsive wave at a receiver is proportional to the distance r from the source, while the amplitude decays as $1/r$, as expressed in equation 1.12. Figure 1.5(c) shows the corresponding electric field response to an impulsive x-directed electric dipole source in an infinite body of water. The arrivals are not impulsive. The electric field of the dipole source is directional and happens to be predominantly negative for the receivers in this configuration, as can be seen from equation 1.14. This is dramatically different from the purely geometrical decay in the acoustic case.

The electric field at the point (x, y, z) and time t is the vector $\mathbf{E}(x, y, z, t)$, which has units of V m^{-1} and, except at the source point, satisfies the following wave equation:

$$\left(\nabla^2 + \mu\varepsilon\frac{\partial^2}{\partial t^2} + \mu\sigma\frac{\partial}{\partial t}\right)\mathbf{E} = 0, \tag{1.15}$$

in which σ is the electrical conductivity, μ is the magnetic permeability, which for most non-magnetic materials is normally taken to be equal to the magnetic permeability of free space $\mu_0 = 4\pi \times 10^{-7}$ H m^{-1}; ε is the electrical permittivity, which is defined in free space as $\varepsilon_0 = 8.85 \times 10^{-12}$ F m^{-1}. As we explain in Chapter 4, the term involving the electric permittivity can be ignored, thus reducing the electromagnetic problem to a purely diffusive problem. In any homogeneous region the equation is scalar and acts on every vector component separately. The field vector components interact due to the source vector and at boundaries across which the electric conductivity changes.

1.8 One-Dimensional Example of a Buried Resistive Layer

The effect of a buried resistive layer, for instance a hydrocarbon-saturated sandstone, is illustrated for the land case by modelling. Figure 1.6 illustrates a one-dimensional earth model consisting of a 20 ohm-m half-space with a layer 100 m thick and resistivity 400 ohm-m at a depth of 2 km. The source is a 1 A-m dipole and the receiver is in line with the source electrodes at an offset of 5 km.

The response at the surface of a half-space to a switch-off step in current in a surface dipole source was calculated by Weir (1980). Edwards (1997) showed how to calculate a layered-earth response to a switch-on source. The grey curve of Figure 1.7(a) shows the electric field response of the 20 ohm-m half-space to a switch-on step, computed using the theory of Edwards (1997). There is an instantaneous rise at $t = 0$, followed by a slower rise to an asymptotic value equal to twice the value of the initial step. The initial step is caused by propagation through the air at close to the speed of light. This is known as the air wave. The black curve in Figure 1.7(a) shows the effect of a 100 m thick resistive layer of 400 ohm-m at a depth of 2 km: the amplitude of the response after the initial step is greater. Figure 1.7(b) shows the corresponding impulse responses, which may be obtained by differentiating the curves shown in Figure 1.7(a). The initial step in Figure 1.7(a), the air wave, becomes an impulse at $t = 0$ in 1.7(b), and the earth impulse response of the half-space shows a rise to a peak, followed by an asymptotic decay to zero. The influence

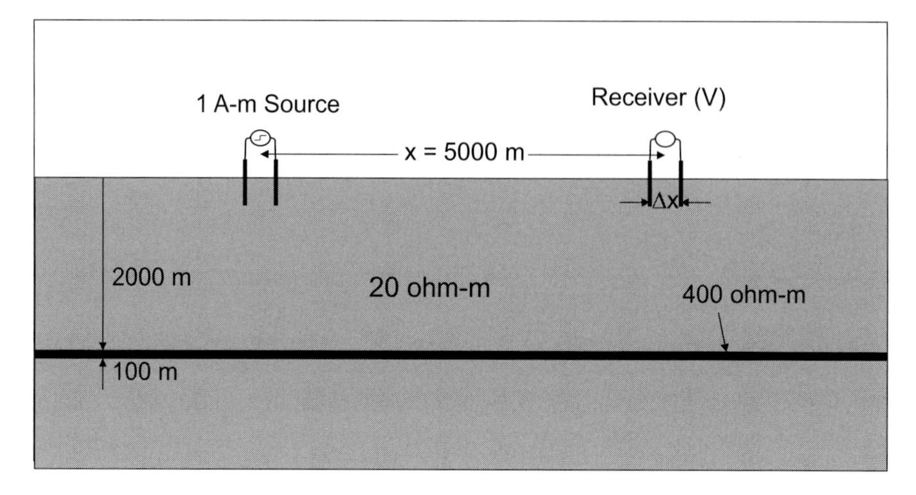

Figure 1.6 Model of a source and receiver 5000 m apart at the surface of a 20 ohm-m half-space. At a depth of 2000 m is a layer of thickness 100 m with a resistivity of 400 ohm-m.

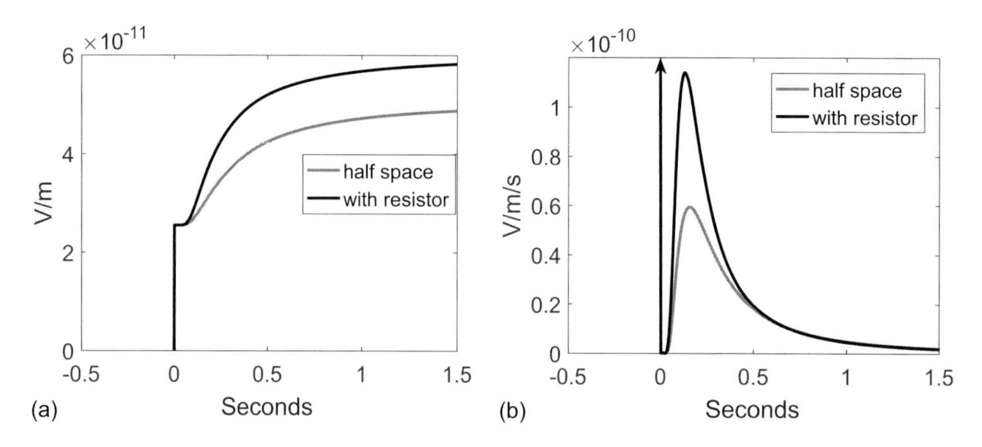

Figure 1.7 (a) Electric field response of a 20 ohm-m half-space at an offset of 5 km to a 1 A-m step at the source dipole (grey curve); and with a 100 m thick 400 ohm-m resistive layer at a depth of 2 km (black curve). (b) impulse response for a 20 ohm-m half-space (grey curve), ; and with a 100 m thick 400 ohm-m resistive layer at a depth of 2 km (black curve). The black vertical arrow at time $t = 0.0$ represents the arrival through the air of the impulse at the source.

of the thin resistive layer on the response is shown as the black curve, which has a significantly higher peak arriving slightly earlier than the peak of the half-space response.

The resistor anomaly is the response with the resistor present minus the half-space response. The first second of the impulse response anomaly is shown in Figure 1.8 as a function of source–receiver offset at intervals of 100 m, with a

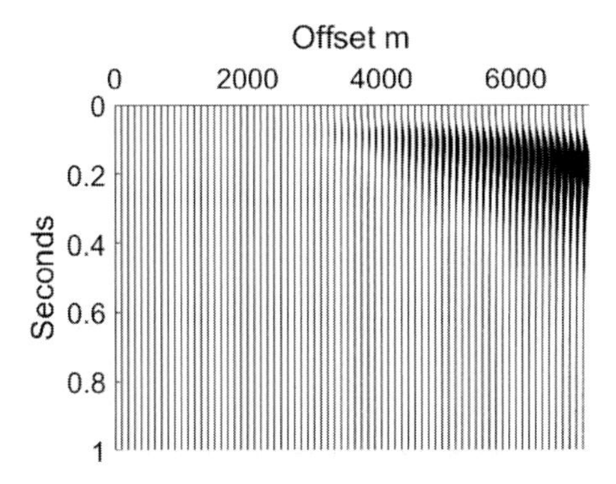

Figure 1.8 One-dimensional resistor anomaly as a function of time, vertically, and source–receiver offset, horizontally, in 100 m intervals.

positive anomaly as an excursion to the right with the area between zero and the excursion filled in black. The anomaly becomes clear at offsets greater than about 4000 m, or twice the depth of the resistor. The anomaly appears to be an approximately horizontal event, beginning almost immediately, with the time of the peak increasing with offset.

Thus the effect of the resistor can be seen in both the amplitude of the earth impulse response and the arrival time of the peak. These effects are seen at offsets greater than about twice the target depth. This is important information, but it does not lead directly to the determination of the depth, the thickness, or the resistivity contrast of the resistive layer. In fact, it is not obvious how these parameters may be determined from the CSEM data.

1.9 One-dimensional Example of a Buried More-Conducting Layer

Figure 1.9 shows a simple model of a 1 ohm-m layer 100 m thick, representing a saline aquifer, at a depth of 2 km in a 20 ohm-m half-space. Figure 1.10(a) shows the electric field response at the receiver for the model of Figure 1.9 for a 1 A-m switch-on step in current applied at time $t = 0$ between the two source current electrodes for the simple half-space in grey and for the half-space with buried conductor in black. The half-space step response is identical with that shown in Figure 1.7(a); the black curve shows the same initial step, followed by a slight dip and a rise to a lower final value. Figure 1.10(b) shows the time derivative of the two curves shown in Figure 1.10(a). These are the impulse responses. As for the resistor model, the vertical arrow at $t = 0$ represents the impulse that is the time derivative of the initial step. The grey curve is the impulse response of the 20 ohm-m half-space and is identical to the grey curve in Figure 1.7. The black curve is the impulse response of the half-space with buried conductor. It shows a negative dip before a rise to a

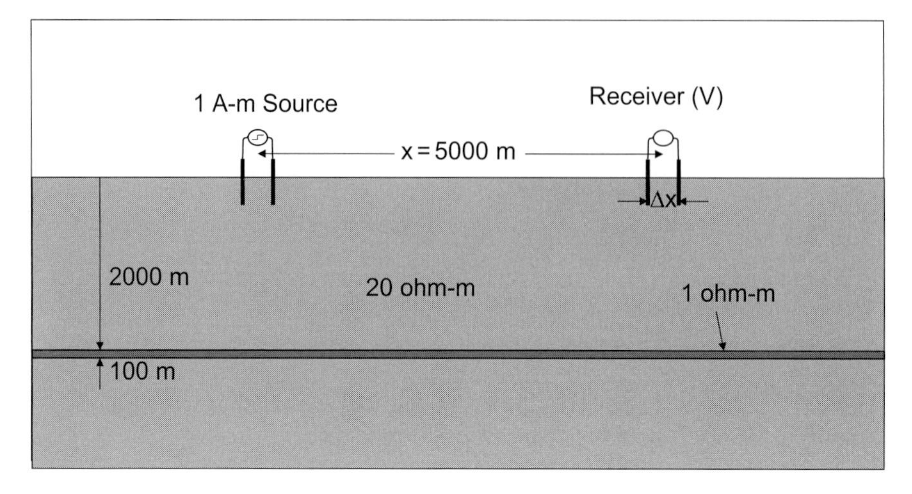

Figure 1.9 Model of a source and receiver 5000 m apart at the surface of a 20 ohm-m half-space. At a depth of 2000 m is a layer of thickness 100 m with a resistivity of 1 ohm-m.

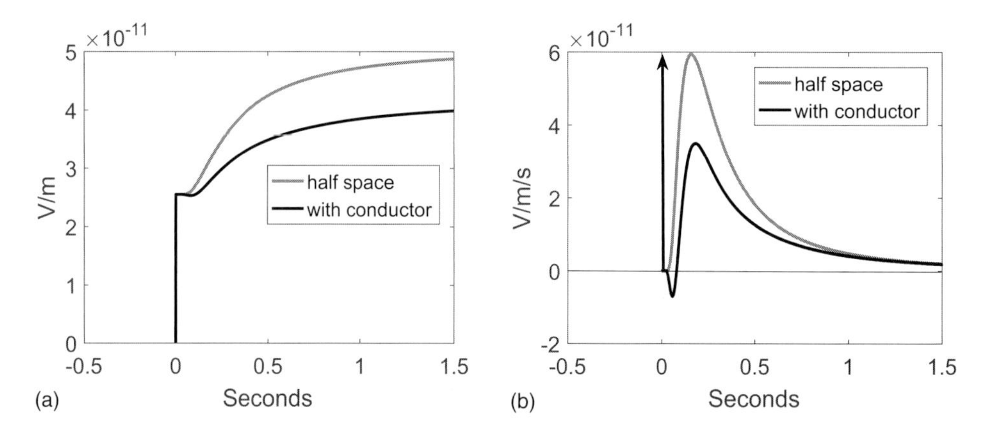

Figure 1.10 (a) Electric field response of 20 ohm-m half-space at an offset of 5000 m to a 1 A-m step (black curve); (b) with a buried 1 ohm-m conductor (grey curve).

later peak that is lower than the half-space response, followed by an asymptotic decay to zero.

Figure 1.11 shows the conductor anomaly, calculated as a function of time and source–receiver offset using the same procedure as for the previous resistor anomaly. That is, the half-space impulse response is subtracted from the impulse response for the model with the conductor.

The anomaly is negative, a trough, plotted as an excursion to the left and filled between zero and the excursion in grey. The anomaly is detectable at offsets greater than 4000 m, or about twice the depth to the conductor. It appears to be an

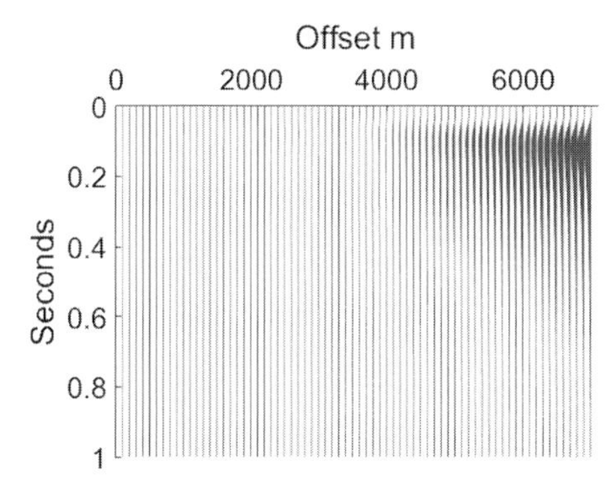

Figure 1.11 One-Dimensional conductor anomaly as a function of time, vertically, and source–receiver offset, horizontally, in 100 m intervals.

approximately horizontal event, beginning almost immediately, with the time to the trough minimum virtually constant with offset and indicative for depth to layer.

1.10 Extraction of Resistivities from CSEM Data: The Problem of Inversion

Electromagnetic propagation in conducting media satisfies a diffusion equation in which the shape of the pulse changes as it propagates, progressively converting electrical energy to heat, with high-frequency electrical energy being converted more rapidly than low-frequency energy. This differs from seismic wave propagation in which energy is contained in the propagating waves, which satisfy the wave equation, and the amplitude of the propagating pulse decreases with the distance travelled, but without change of shape.

Processing of seismic data exploits the invariance of the pulse shape with propagation distance to line up arrivals and determine the fundamental medium property: the velocity of propagation, which is, to first order, frequency-independent. Normal-moveout correction in seismic data processing is an example of this. Seismic velocity analysis is the process of finding the velocity model that allows the alignments of arrivals to be made. In seismic data processing the subsurface propagation velocity distribution is determined from the seismic data themselves.

Hitherto, this approach has not been open to the processing and analysis of CSEM data because the propagation velocity is a function of both the fundamental medium property – the resistivity – and the frequency. Therefore the pulse shape changes with propagation distance.

The normal method to determine the subsurface resistivity distribution from CSEM data is by inversion (e.g. Zhdanov, 2002; Constable and Srnka, 2007; Newman et al., 2010). The scheme is illustrated in Figure 1.12. Inversion is an iterative

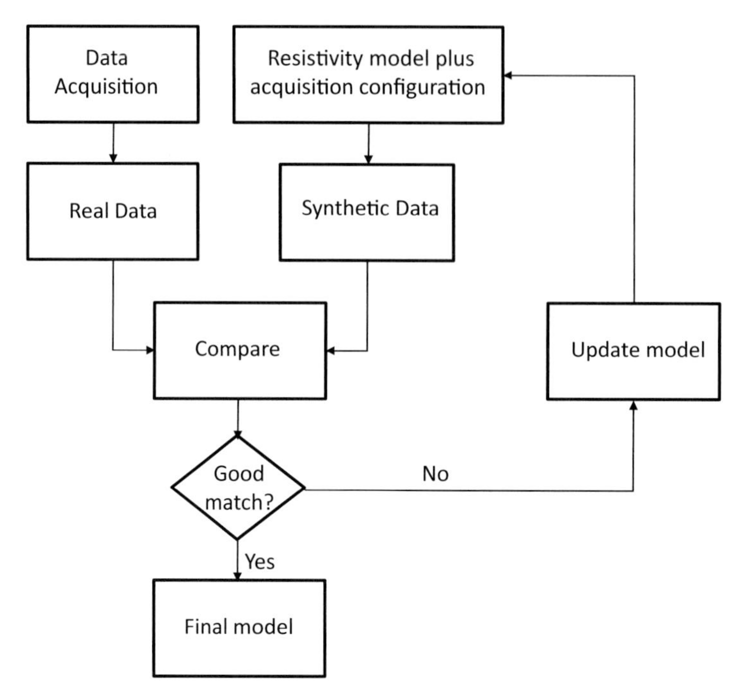

Figure 1.12 Present procedure for finding a resistivity model by inversion.

modelling procedure in which synthetic data are computed using a model of the subsurface resistivity distribution and the known data acquisition parameters. The model that produces synthetic data that are in some sense a best fit to the real CSEM data is the model that best represents the subsurface. The model is not derived directly from the CSEM data themselves. Since the inversion process operates primarily on model updates and not on the data, the key issues are the origin of the model and the criteria for updating it.

1.11 Outline of the Book

Chapter 2 gives an overview of the most common CSEM methods used for hydrocarbon exploration, and includes a description of sources and receivers, source time functions and sources of noise.

Chapters 3 and 4 present theory. Chapter 3 covers elements of Fourier transforms and linear filters, and emphasises that the Earth is a linear filter for the propagation of seismic and electromagnetic waves. Chapter 4 provides the essential elements of electromagnetic wave theory in layered media, beginning with Maxwell's equations and using Fourier and Laplace transforms. The principal results of this chapter are analytical expressions for the impulse response, or Green's function, of a current dipole source in any layer. Chapter 5 presents numerical examples of various layered media configurations to help the understanding of the theory presented in Chapter 4.

Chapter 6 discusses source control; that is, various source time functions and their properties in the frequency domain. Analysis of the CSEM data might be performed in the time domain or the frequency domain; some signals have broader application than others. This chapter also includes signature deconvolution, to recover the earth impulse response from the CSEM data.

The next three chapters discuss three CSEM methods. Chapter 7 is devoted to current practice of marine CSEM with ocean bottom receiver nodes and a current dipole source towed near the sea floor with a continuous source time function. Chapter 8 discusses land CSEM with a transient source signal. Chapter 9 discusses marine CSEM with a transient source signal and looks at ocean bottom cable and towed sensor streamer data. Each chapter is illustrated with examples to show how the data are interpreted.

The goal of all the methods is to determine subsurface resistivities. Current practice does this by finding models that fit the data. Chapter 10 discusses forward modelling and inversion of CSEM data, with a focus on 3D methods.

Chapter 11 is devoted to recovery of resistivities from CSEM data by inversion, in which an initial model is updated to find a model that is a best fit to the data. The chapter examines two aspects of this procedure: the determination of model parameters from other geophysical data, and the manipulation of the data to determine the goodness of fit.

Finally, Chapter 12 discusses some practical issues of how to obtain the necessary 2D and 3D data for a reasonable cost. CSEM is in its infancy. There is a long way to go before the cost per data point is comparable with seismic data acquisition costs. We recognise that unless the data acquisition costs are reduced, however, CSEM will not be widely used. Therefore we offer some suggestions on the directions in which it might be developed, and topics for further research.

2

Sources, Receivers, Acquisition Configurations and Source Time Functions

This chapter describes the four basic elements in the acquisition of controlled-source electromagnetic (CSEM) data: (1) the source, (2) the receiver, (3) the source-receiver configuration and (4) the source time function. The Earth's surface forms an interface between the air, in which electromagnetic waves travel at the speed of light, and the Earth, which is electrically conductive and therefore attenuates the propagating waves. A source at or near the Earth's surface generates waves in the air and in the ground. Since we are interested in the response of the ground, the air wave is considered to be a problem. For a source in the Earth, the air wave is a refraction. The choice of the source time function is affected by considerations of both the air wave and the noise.

2.1 The Current Dipole Source

Two source types are in use for CSEM: magnetic and electric. Magnetic sources are actually insulated loops carrying a current and are known as magnetic dipole sources. Electric sources are usually straight lengths of insulated wire connected to the earth through low-resistance electrodes at each end. This is known as a current dipole source. Magnetic sources are often used in exploration for electrically conductive mineral deposits. Airborne CSEM surveys use a magnetic loop source and various magnetic receivers in the search for shallow mineral deposits. They have also been successful in mapping shallow (less than 150 m deep) oil sands, notably near Fort McMurray, Alberta (Smith, 2010). For deeper hydrocarbon targets, electric sources have been found to be superior. A magnetic loop source has self-inductance, which makes it difficult to reverse the current flow rapidly without huge voltages. Normally a steady current is applied for a sufficiently long time for turn-on currents to dissipate, and then the current is switched off in a controlled manner. The rapid reduction of the transmitter current causes a rapid change in the magnetic dipole field, which induces an electromotive force in nearby conductors, according to Faraday's law of induction. When these have dissipated, the process can be repeated with the same polarity or opposite polarity (Nekut and Spies, 1989).

We established in Chapter 1 that the displacement of salt water by hydrocarbons in a porous rock increases the resistivity of the rock. Therefore, electromagnetic exploration for hydrocarbon reservoirs is the search for resistive formations in an electrically conductive background. By the 1980s it had been established that resistive layers are best resolved by measuring the electric fields from electric sources, sometimes known as galvanic sources, or grounded-wire antennas (Chave and Cox, 1982; Kaufman and Keller, 1983; Edwards and Chave, 1986; Nekut and Spies, 1989; Strack et al., 1989). One of the really attractive properties of electric current dipoles is that the direction of current flow can be reversed very rapidly – in a microsecond. In practice this means the source time function of a current dipole source can be designed to have specific desirable properties. We consider land and marine operations with current dipole sources.

Figure 2.1 shows a current dipole source. A current $I(t)$ flows from the transmitter through an insulated wire to the positive electrode, through the ground to the negative electrode, and through an insulated wire back to the transmitter. The transmitter contains switches to reverse the direction of the current flow. The strength of the source is the dipole moment $I(t)\Delta x_s$, defined as the product of the current $I(t)$ and the length of the dipole Δx_s – the distance between the electrodes. The dipole moment has SI units of A m. In the marine case the electrodes and insulated wire are immersed in the water, which has high conductivity – typically about 3 S m^{-1} – and it is easy to generate currents of 1000 A or more. On land, the resistance of the ground is often high, making it difficult to generate high currents, since, in most countries, safety restrictions prevent the use of voltages higher than 1000 volts. To generate sufficient current at the source, considerable effort often has to be devoted to reducing the contact resistance at the source electrodes, including

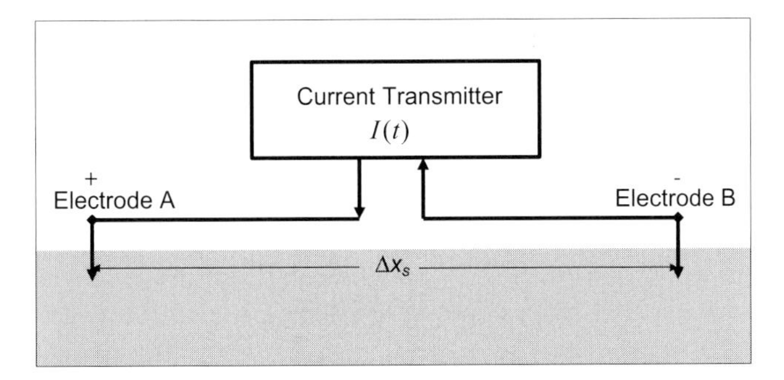

Figure 2.1 Current dipole with dipole moment $I(t)\Delta x_s$.

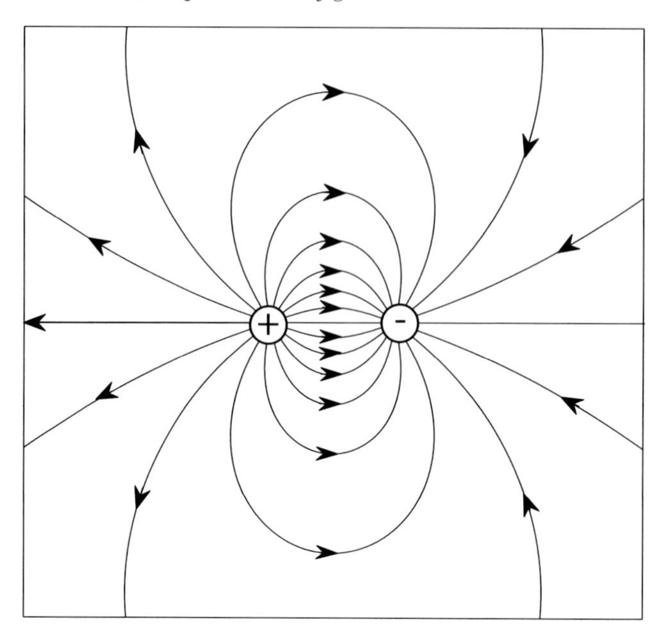

Figure 2.2 Electric field of a physical dipole in a whole space.

the use of metal electrodes that penetrate several metres into the ground to reach the water table, and using electrodes in parallel at each end of the dipole.

The field of a physical dipole source in a whole space is illustrated in Figure 2.2. The distance between the poles of a physical dipole is finite. With a finite current, the dipole moment is also finite. We also use the concept of a point dipole source, in which the distance between the poles tends to zero as the dipole moment is kept constant. A point dipole still has direction.

The electric field has symmetry about the dipole axis. In Figure 2.2 the plane shown is any plane through the axis of the dipole. For example, if the dipole is on the x-axis, the figure could represent the horizontal plane $z = 0$, or the vertical plane $y = 0$, or any other plane that contains the x-axis. The strength of the field at any point is proportional to the *line density* – the number of field lines per unit area, where the area is perpendicular to the field. At distances r large compared with the dipole length ($r \gg \Delta x_s$), the physical dipole field is almost identical with the field of a point dipole with the same dipole moment and the field strength is proportional to $1/r^3$.

2.2 Receivers

Present-day controlled-source electromagnetic methods owe a great debt to the development of the magnetotelluric method, which was pioneered in France, where

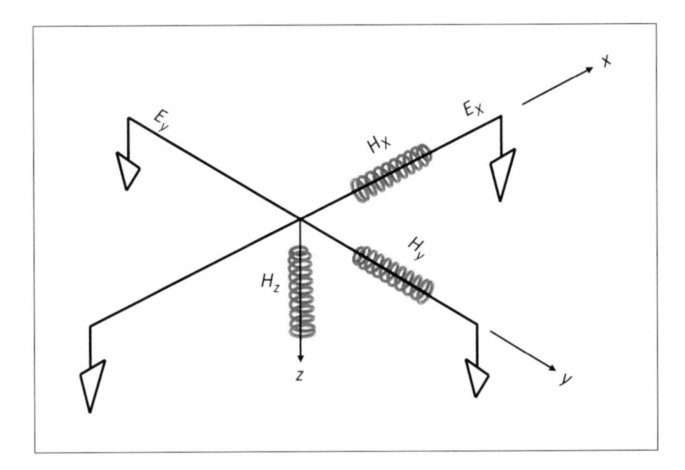

Figure 2.3 Magnetotelluric receiver, which measures three orthogonal components of the magnetic field, H_x, H_y and H_z, with magnetometers, and the two horizontal components of the electric field E_x and E_y, as the voltage gradients between orthogonal pairs of electrodes.

the principles were recognised in the 1920s by Conrad Schlumberger, and where the first practical applications were made a few years before the Second World War (Cagniard, 1953). The magnetotelluric method uses variations in the magnetic field incident on the Earth. For distances on the Earth's surface of the order of tens of kilometres, the incident magnetic field can be regarded as a plane wave propagating vertically from above – in the z-direction. The incident magnetic variations induce electric fields in conductors in the Earth. If the incident magnetic field has a single frequency and is polarised in a particular direction, for example the x-direction, it can be shown (for example, Cagniard 1953) that the resistivity of the Earth is proportional to the square of the magnetic field in the x-direction divided by the square of the electric field in the y-direction. Magnetotelluric measurements therefore consist of orthogonal components of both the electric and magnetic fields. A diagram of a magnetotelluric receiver on the Earth's surface is shown in Figure 2.3.

Typically, all three components of the magnetic field are measured as a function of time; that is, $H_x(t)$, $H_y(t)$ and $H_z(t)$. Usually only the two horizontal components of the electric field $E_x(t)$ and $E_y(t)$ are measured because the vertical component at the surface of the ground $E_z(t)$ is zero, as shown in Chapter 4.

The resistivity derived from magnetotellurics is associated with a frequency-dependent depth δ, known as the *skin depth*:

$$\delta = \sqrt{\frac{2\rho}{\omega\mu_0}}, \tag{2.1}$$

where ρ is the resistivity, ω is the angular frequency and $\mu_0 = 4\pi \times 10^{-7}$ H m^{-1} is the magnetic permeability of free space. The skin depth is the distance at which the amplitude of a plane electromagnetic wave propagating in a conducting medium is reduced to $1/e$ of its original value. For water of resistivity 0.3 ohm-m and a frequency of 0.25 Hz, the skin depth is about 550 m. In water depths exceeding 1100 m, magnetotelluric signals at frequencies above 0.25 Hz are so attenuated that they are negligible. Magnetotellurics in these water depths does not work at these frequencies. This is an important reason for the development of the marine CSEM method.

Deep water marine CSEM uses a current dipole source towed near the sea floor and receiver nodes on the sea floor that measure up to three components of the magnetic field and up to three components of the electric field. Each receiver node is similar to a magnetotelluric receiver, but with the addition of a vertical electric field component. Since the nodes are autonomous and do not transmit data to the vessel, they digitise and record the signals measured on all six channels and also record the time signal from a clock, which is an essential component of the equipment of the node. The electric field electrodes are silver–silver chloride, designed to prevent them becoming polarised in the water. The distance between the electrodes of a given pair is about 8 m. Ellingsrud et al. (2002) emphasise that it is the two horizontal components of the electric field that are important for CSEM. Since the receivers are recording all the time, they act, in principle, as autonomous magnetotelluric stations when there is no detectable signal from the CSEM source. Because of the attenuation in the water, the usable magnetotelluric signal is at low frequencies where the attenuation is insignificant.

Edwards and Chave (1986) and Edwards (1988) provided a theory for an inline dipole–dipole source–receiver configuration for mapping the conductivity of the sea floor using the response of the earth to a step in current at the source. They called this the *transient response*.

2.3 Source–Receiver Configurations

Cox et al. (1981) describe a deep water active source electromagnetic experiment near the East Pacific Rise using a grounded 800 m current dipole source and three receiver units, each with two perpendicular horizontal 9 m electric dipole receivers, dropped on the sea floor. They established that deep water CSEM was feasible. Bannister (1984) developed simplified formulae for the fields produced by the four principal source dipoles for the subsurface-to-subsurface propagation case: horizontal electric dipole (HED); vertical electric dipole (VED); horizontal magnetic dipole (HMD); and vertical magnetic dipole (VMD). The formulae are applicable

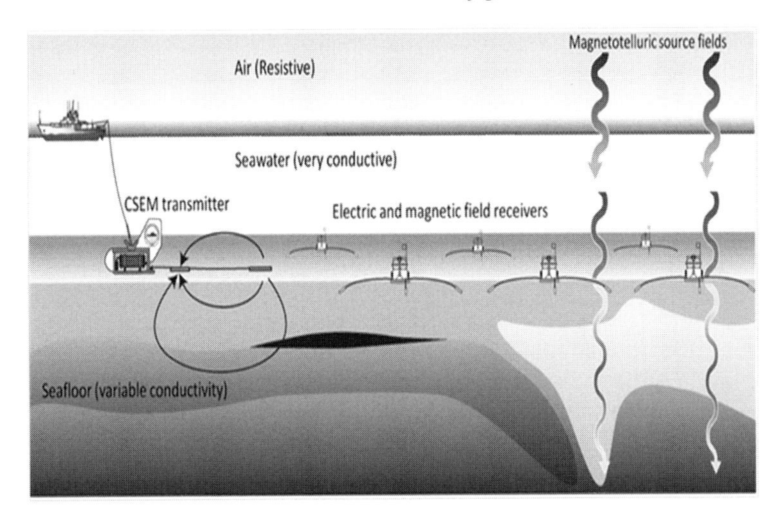

Figure 2.4 Towed neutrally buoyant horizontal current dipole source with ocean-bottom receiver nodes (redrawn from Constable and Srnka, 2007).

particularly for the air–sea and sea–seabed propagation cases. Cox et al. (1986) used a towed 600 m horizontal electric dipole source on the sea floor and 600 m horizontal seafloor electric receivers in a dipole–dipole configuration to determine the resistivity of the oceanic crust and upper mantle in the North Pacific. Constable and Srnka (2007) give an excellent brief history of the academic development of marine CSEM, leading to the particular configuration they describe, widely used today in hydrocarbon exploration, especially in deep water, and shown in Figure 2.4. In parallel with the deep water marine CSEM development, land CSEM was also developing. Spies and Frischknecht (1987) describe a range of source–receiver geometries, most of which apply to mineral exploration on land.

The commercialisation of CSEM for hydrocarbon exploration was largely due to exploration moving into deep water, where the method developed by the Scripps Institution of Oceanography had been applied. The impetus for the commercial development of the deep water application came from scientists in Statoil, whose first publication on this topic, Ellingsrud et al. (2002), was the result of a test over an oil field in deep water offshore Angola, carried out in collaboration with scientists from the Scripps Institution of Oceanography and the University of Southampton. In 2002 the company Electromagnetic Geoservices ASA (EMGS) started, using the configuration shown in Figure 2.4. In this configuration the source is a towed neutrally buoyant horizontal electric current dipole, as described by Sinha et al. (1990), and the receiver nodes are lowered from the vessel into the sea, where they descend under gravity to the sea floor. They are essentially marine versions of magnetotelluric receivers, as described in Section 2.2. The source is conventionally

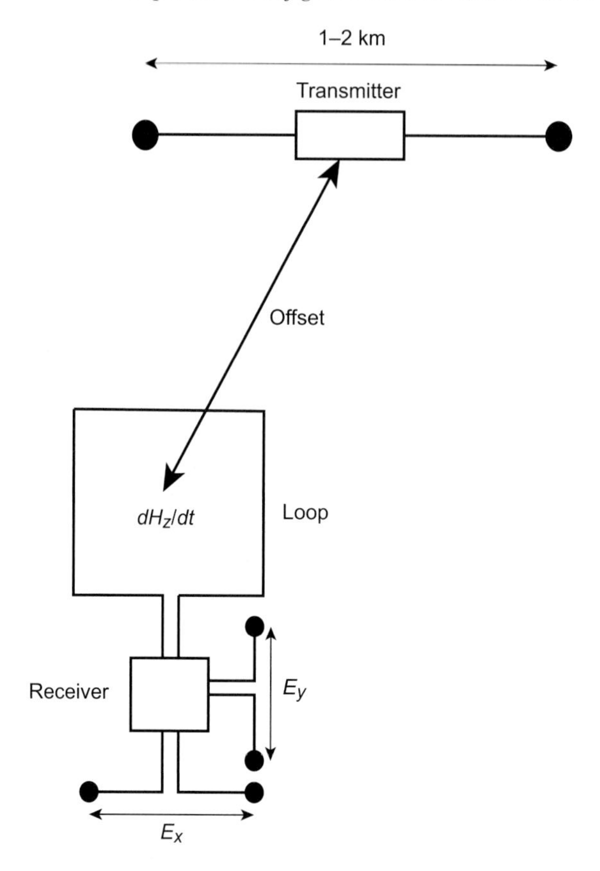

Figure 2.5 Schematic plan of LOTEM setup showing 1–2 km current dipole transmitter, two orthogonal electric field receiver dipoles and a horizontal receiver loop that measures the rate of change of the vertical component of the magnetic field (not to scale).

towed directly behind the vessel and close to the sea floor, and the receivers are deployed in a variety of patterns, but now commonly on a rectangular grid.

For land surveys, Strack et al. (1989) used a large current dipole source, 1–2 km in length. Each receiver station consisted of two orthogonal horizontal electric field dipole receivers $E_x(t)$ and $E_y(t)$ and a large horizontal loop with, typically, 24 turns and area 2500 m^2 to measure the rate of change of the vertical component of the magnetic field dH_z/dt. The setup for a source and one receiver is shown in Figure 2.5. The distance from the source to a receiver is known as the offset, as shown in Figure 2.5. When the offset is greater than or equal to the exploration depth, Strack et al. (1989) define it as long. The method is called *long offset transient electro-magnetics*, or *LOTEM*.

Following the ideas of Strack et al. (1989) for the single receiver, and combining this with the configuration used in 2D seismic reflection, a time-lapse experiment

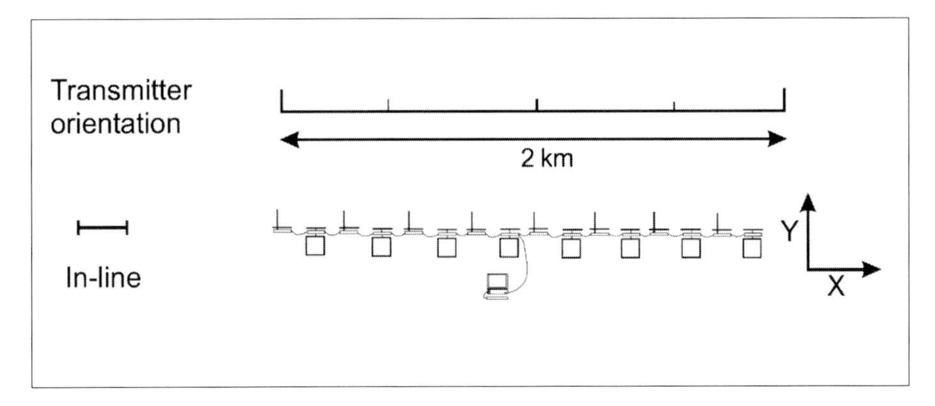

Figure 2.6 Field layout for 1994 and 1996 electromagnetic surveys over a gas storage reservoir in northern France. The 2 km receiver spread consisted of 16 two-channel stations: eight stations at 250 m intervals measured $E_x(t)$ and $E_y(t)$, and eight alternate stations measured and $E_x(t)$ and dH_z/dt, all connected to a computer recording system (redrawn from Wright et al., 2002).

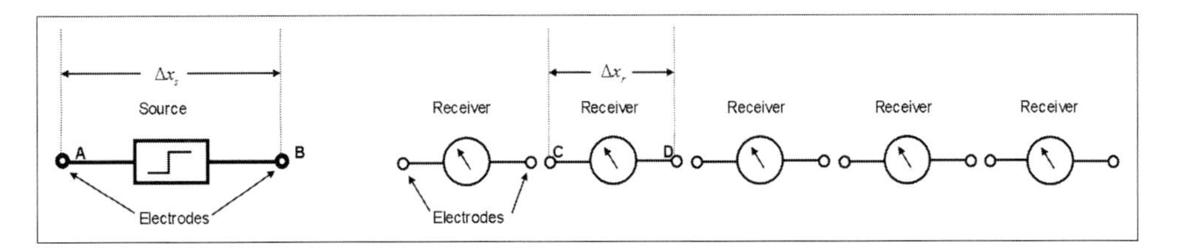

Figure 2.7 Schematic plan view of multi-channel dipole–dipole setup. The separation of the source dipole electrodes is Δx_s; the electric field receivers are in-line with Δx_r separation of electrodes for each channel (redrawn from Ziolkowski, 2007).

was performed in 1994 and 1996 by the University of Edinburgh, University of Cologne, Deutsche Montan Technologie and Compagnie Générale de Géophysique over a gas storage reservoir in France (Wright et al., 2002). The field setup used a line of receivers measuring $E_x(t)$, $E_y(t)$ and dH_z/dt, as shown in Figure 2.6. Two orientations of the dipole source were used: in-line (x) and cross-line (y). Wright et al. (2002) found that the in-line source and in-line electric field measurements produced the most significant data. The other receiver components and the cross-line source did not contribute further information. As a result, Ziolkowski et al. (2007) used only an in-line current dipole source and in-line dipole electric receivers for a 2004 land survey over a different underground gas storage reservoir in France.

The in-line horizontal current dipole source and in-line electric field dipole receivers, with layout shown in Figure 2.7, became established as a viable configuration for 2D land operations. The figure shows all receivers the same length, but this does not have to be the case. If they are the same length, however, roll-along acquisition for common mid-point gathers can be applied, following well-established seismic data acquisition practice.

This same *dipole–dipole* configuration had already been established theoretically for marine operations by Edwards and Chave (1986), Edwards (1988) and Chave et al. (1987). It was used in a two-channel 'yo-yo' system, developed by the University of Toronto, to detect gas hydrates on the Cascadia Margin west of Vancouver Island (Schwalenberg et al., 2005). Commercial operations for the marine dipole–dipole system with over 20 inline electric field receivers were established by MTEM Limited in 2007, who were taken over later the same year by Petroleum Geo-Services (PGS). The marine MTEM system used one vessel for the source and another for the receiver cable, with both source and receiver deployed on the sea floor. This system was used for several surveys, including a time-lapse survey over the Harding field in the North Sea (Ziolkowski et al., 2010) in 2007 and 2008. It was also used in a 2007 survey for ENI of Italy, conducted in the Mediterranean Sea off the coast of Tunisia. This survey showed a resistive target that was drilled immediately; drilling confirmed the presence of hydrocarbons (D'Arienzo et al., 2010). A similar configuration is used by Geomar.

Compared with marine towed-streamer seismic reflection data, the two-vessel sea-floor CSEM operation was very expensive per data point. It had the advantage, however, that there was essentially no limit to the time available to collect the data and the data quality was good. Nevertheless, it was clearly an expensive way to collect data. Before acquiring MTEM Limited in 2007, PGS had been planning a towed multi-channel dipole–dipole system, which has now been developed and is acquiring data commercially. The source is typically 10 m below the sea surface and the multi-channel receiver streamer can be towed at any depth down to about 200 m, but is normally towed at about 100 m. The source and receiver can be towed at 4–5 knots and can be operated simultaneously with a marine seismic reflection system on the same vessel. The seismic and CSEM systems do not interfere with each other. The rate of data acquisition is obviously much greater than with the two-ship sea-floor systems and the cost of data acquisition is acceptable. The setup is shown in Figure 2.8.

More recently a VED source and VED receiver have been developed for offshore hydrocarbon exploration (Holten et al., 2009) by the Norwegian company Petromarker. An advantage of this configuration is that the source–receiver offset can be small compared with the depth to the target (Barsukov and Fainberg, 2017). A disadvantage is that the vertical component of the electric field is small compared with the horizontal component, as pointed out by Holten et al. (2009). It is necessary, therefore, to constrain the receiver dipole to be vertical, which requires a

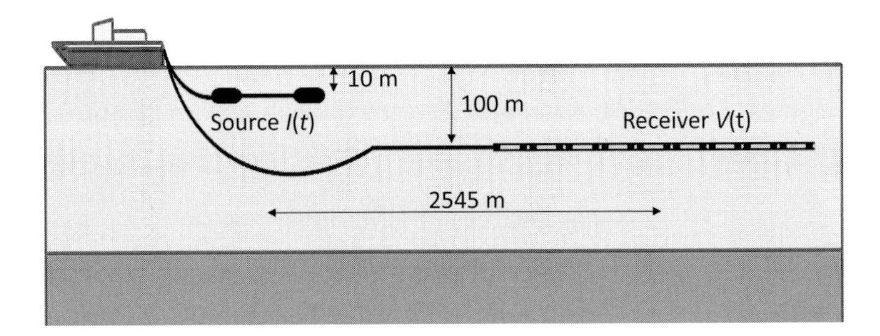

Figure 2.8 Towed multi-channel dipole–dipole system (redrawn from Mattsson et al., 2012).

structure on the sea floor. This system operates in shallower water (typically 100 m) than the deep water horizontal dipole–dipole system shown in Figure 2.4 and is used for looking vertically below the source, which is attached to the stationary vessel.

2.4 The Air Wave Problem

Electromagnetic waves travel through the air at close to the speed of light with negligible attenuation, because air is a very poor conductor of electricity. If a horizontal dipole–dipole system is used on the surface of the Earth, for instance in a land CSEM survey, the signal from the source travelling through the air and coupled to the ground arrives at the receivers almost instantaneously, while the signal through the ground travels more slowly and is attenuated. If the source signal is a single frequency, say $A\sin(\omega t)$, the received signal contains two components, the air wave $B\sin(\omega t)$ (in which the travel time through the air is negligible), and a wave through the ground $C\sin(\omega t+\phi)$, where ϕ is an unknown phase delay. The received signal is thus

$$E(t) = B\sin(\omega t) + C\sin(\omega t+\phi). \tag{2.2}$$

The air wave problem is that this is one equation with three unknowns: B, C and ϕ. In addition, B can often be much larger than C, so the desired signal from the ground $C\sin(\omega t+\phi)$ can be drowned in the air wave. This problem occurs at all frequencies.

In the marine case the source is in the water. An upward-travelling wave from the source is refracted at the surface and travels at the surface at close to the speed of light with negligible attenuation. It is coupled to the water and therefore a downgoing wave arises from this propagation. The upward and downward propagation in the water is attenuated, as described above. A receiver in the water will receive this attenuated air wave in addition to other arrivals. In shallow water at long offsets, the air wave can be much larger than the direct wave or the waves that have travelled through the sea floor.

Much ingenuity in CSEM systems is devoted to dealing with this problem.

2.5 Sources of EM Noise

An electric dipole receiver consists of two electrodes a distance Δx_r apart, connected by *telluric cables* to a device for measuring the potential difference between them. The measurement is a time-varying voltage $V_T(t)$. The electric field is the gradient of the voltage and is usually estimated by dividing the measured voltage by Δx_r, the distance between the electrodes. The noise at the receiver is what would be measured if there were no source signal. So the air wave is not noise, it is part of the response to the action of the source.

There are five primary sources of noise in CSEM: electrode noise; magnetotelluric noise; cultural noise; motionally induced electromagnetic induction noise; and electronic system noise. The total noise voltage at the receiver may therefore be expressed as

$$N_T(t) = V_E(t) + \Delta x_r[E_M(t) + E_C(t) + E_I(t)] + V_S(t), \tag{2.3}$$

where $V_E(t)$ is the noise voltage of the two electrodes, $E_M(t)$ is the magnetotelluric electric field, $E_C(t)$ is the cultural noise field, $E_I(t)$ is the motionally induced induction noise and $V_S(t)$ is the electronic system noise. The electrode noise and the system noise are voltages generated locally, whereas the other terms are fields. The impact of the local voltage terms relative to the electric field terms depends on the receiver length Δx_r. In modern systems the electronic system noise is usually much smaller than the other components.

Electrode noise is caused by differences in electric potential between the electrode and the contact medium, causing electrochemical reactions and consequent variations in the potential of the electrode. Electrodes are designed to minimise this noise. When the electrodes are stationary, the noise is often small, but when they are moving, for instance in water, the noise increases.

Magnetotelluric signals are natural variations in the Earth's magnetic field caused by outside influences, particularly the interaction of the solar wind with the ionosphere, and with signals from thunderstorms, known as sferics, which are guided between the Earth and the ionosphere and can travel thousands of kilometres. The magnetic variations induce variations in the electric field in the conducting Earth through Faraday's law of induction, and create variations in the potential difference between any two points in the Earth, or on its surface. The magnetotelluric method uses magnetotelluric signals to determine conductivity

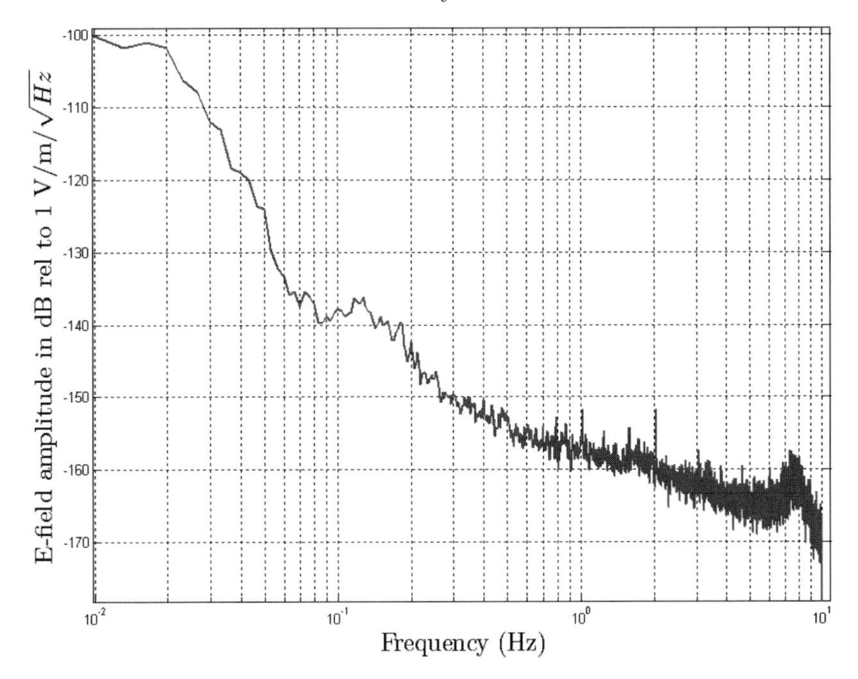

Figure 2.9 Spectrum of magnetotelluric noise measured on the floor of the North Sea (redrawn from Ziolkowski et al., 2011).

variations in the Earth. For the purposes of CSEM, however, these signals are noise, in much the same way as earthquake vibrations are noise for exploration seismology. The spectrum of magnetotelluric noise recorded on the floor of the North Sea in 100 m of water is shown in Figure 2.9. It increases dramatically with decreasing frequency below about 1 Hz.

Cultural noise is man-made noise, typically power-line noise and noise from electric railways, which often run at different frequencies from the national grid. Electric machinery is grounded, causing currents to flow in the ground, with associated variations in electric potential. As electricity usage varies with time and place, the cultural noise at any place changes with time. The spectrum of a short time window of cultural noise in the United Kingdom is shown in Figure 2.10.

The noise term $E_I(t)$ is caused by motion of the telluric cable in the Earth's magnetic field according to Faraday's law of electromagnetic induction. For stationary receivers, this term is zero. This term is important for marine systems towed in the sea, particularly if wave motion is significant (Djanni et al., 2016).

Finally, $V_S(t)$ is the noise of the recording system, which is always present. System noise is usually very small and is usually reduced with each upgrade to the system.

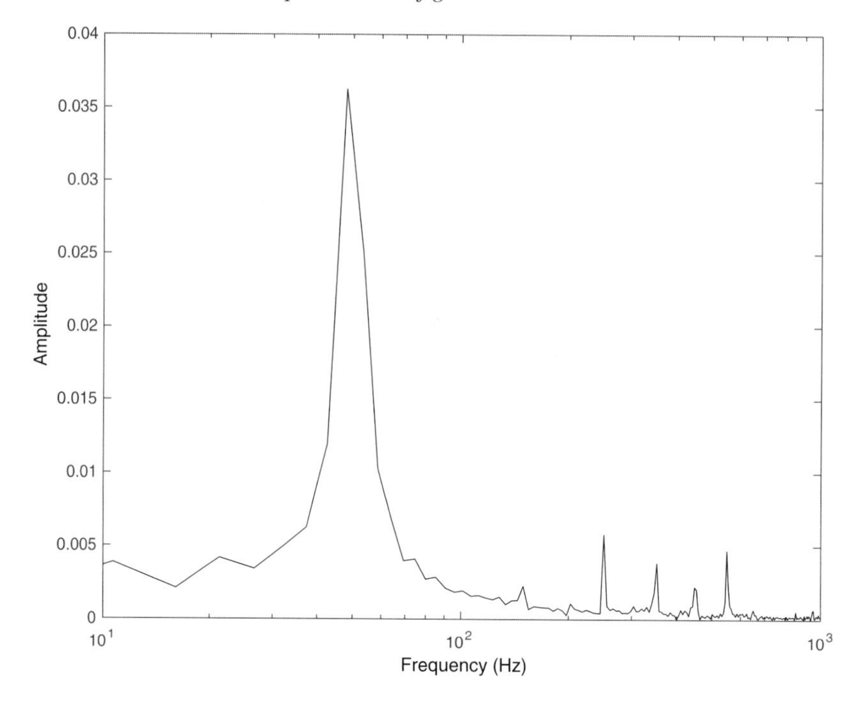

Figure 2.10 Amplitude spectrum of typical cultural noise in the United Kingdom, showing fundamental frequency of 50 Hz and odd harmonics at 150, 250, 350, 450 and 550 Hz.

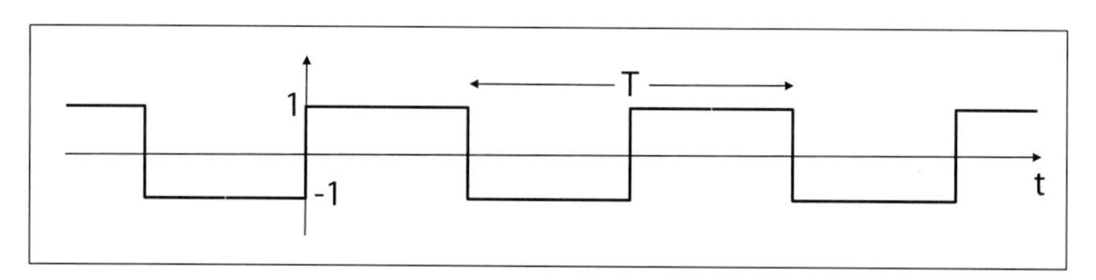

Figure 2.11 Square wave with amplitude 1 and period T.

2.6 Source Time Functions

There are two kinds of source time function in use in CSEM: continuous and transient. Continuous source time functions are usually periodic – for example, a square wave, as shown in Figure 2.11.

When deep water CSEM began in the first decade of this millennium, using the configuration shown in Figure 2.4, the square wave was the source time function of choice (Ellingsrud et al., 2002; Constable and Srnka, 2007), with period $T = 4$ s and fundamental frequency $f = 1/T = 0.25$ Hz. The square wave shown in Figure 2.11 may be represented as an infinite series of sine waves:

$$x_{sq}(t) = \frac{4}{\pi}(\sin(2\pi ft) + \frac{1}{3}\sin(6\pi ft) + \frac{1}{5}\sin(10\pi ft) + \frac{1}{7}\sin(14\pi ft) + \ldots). \quad (2.4)$$

It consists of one sine wave at the fundamental frequency f and a further sine wave at every odd harmonic frequency $3f$, $5f$, $7f$ and so on, with amplitudes diminishing as $1/3$, $1/5$, $1/7$ and so on. The air wave problem did not arise for these surveys because the water was deep, the source was near the sea floor, the receivers were on the sea floor, and both the upgoing wave from the source and the downgoing wave from the sea surface were attenuated by the deep water. The water depth in the survey of Ellingsrud et al. (2002) was about 1200 m, so the influence of the air wave was negligible. Furthermore, the effect of magnetotelluric noise was also negligible for the same fundamental reason: the noise at all frequencies was attenuated by the water column, the attenuation increasing with each harmonic. The magnetotelluric noise field is attenuated by travelling through the water column only once. The air wave attenuation is greater because the upgoing wave from the source, which gives rise to the air wave, is attenuated by the water column, so the air wave is already small before it starts its journey down to the receivers. Early deep water surveys using this configuration used only the fundamental frequency for interpretation of the data (Ellingsrud et al., 2002; Srnka et al., 2006). Later a richer spectrum was required.

On land the problem is tougher. There is no escape from the air wave or magnetotelluric noise, plus there is also cultural noise. The traditional way to avoid the air wave is to use a switch-off source time function and to record the response (Nekut and Spies, 1989; Streich, 2016). The switch-off source is, mathematically, a scaled version of $H(-t)$, where $H(t)$ is the Heaviside function. $H(-t)$ and $H(t)$ are illustrated in Figure 2.12.

Before the switch, the electric field at a receiver has an initial value $E(0)$, say, caused by the source current which has been on forever (in principle). At $t = 0$ when the source current is switched off, the magnitude of the electric field

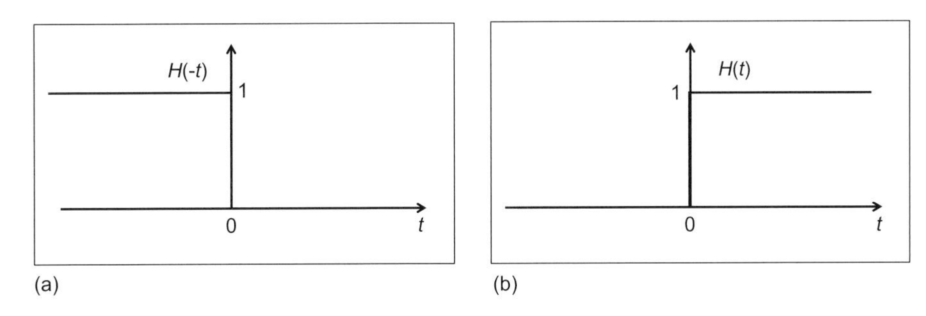

Figure 2.12 (a) Switch-off source time function $H(-t)$; (b) switch-on source time function $H(t)$.

immediately jumps to a lower value and, because of attenuation in the Earth, this is followed by a decay to zero, with the magnitude of the jump and the exact shape of the decay curve depending on the electrical conductivity structure beneath the source and receiver.

For switch-on the source time function is a scaled version of the Heaviside function $H(t)$ and the reverse happens. The electric field at the receiver due to the source is initially zero. When the source current is switched on there is an immediate jump in the field followed by further increases in magnitude, tending to the final value $E(0)$. The switch-off response E_- and switch-on response E_+ are related as

$$E_+(t) = -E_-(t) + E(0). \tag{2.5}$$

That is, the switch-on response is minus the switch-off response plus a constant which is known from the initial value of the switch-off response or from the late times of the switch-on response. The constant may be hard to determine if the data are noisy. A solution to this is to use a long-period square wave as the source time function, as shown in Figure 2.11. If the half period $T/2$ is longer than the time it takes for the switch-off or switch-on response to reach a steady state, the response at each polarity reversal is doubled, relative to switch-on or switch-off, and the response at each polarity change is identical, apart from a change in polarity. It follows that the signal-to-noise ratio can be increased by reversing the polarity of alternate responses and summing, or stacking. This approach was used by Strack (1992), Wright et al. (2002) and Ziolkowski et al. (2007).

Duncan et al. (1980) introduced periodic pseudo-random binary sequences (PRBSs) for the acquisition of land CSEM data. An example of a portion of a periodic PRBS is shown in Figure 2.13.

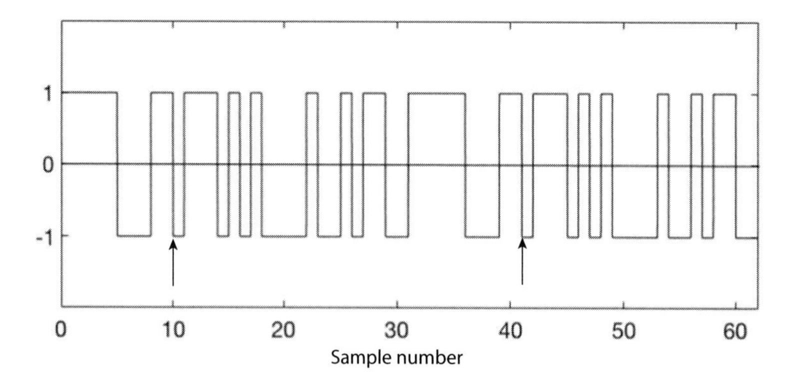

Figure 2.13 An example of a periodic PRBS; the arrows show the length of one period. In this case the period is $2^5 - 1 = 31$ samples.

A PRBS is a programmed periodic sequence that switches in a pseudo-random manner between one value and another; in this case, the values are $+1$ and -1. Each period of the sequence contains $N = 2^n - 1$ samples, where the integer n is known as the *order* of the sequence. If the temporal sampling interval is Δt the amplitude spectrum of the PRBS is flat in the frequency bandwidth

$$\frac{1}{N\Delta t} \leq f \leq \frac{1}{2\Delta t}. \tag{2.6}$$

Thus the available frequency bandwidth is determined by the two parameters N and Δt. The reason for using such a sequence is in the power of *cross-correlation*, discussed in Chapter 3. The principle is similar to the use of swept-frequency signals in Vibroseis for seismic exploration. In CSEM cross-correlation of the received signal with the input PRBS is equivalent to using a source time function that is the autocorrelation of the PRBS; that is, a periodic series of impulses with period $N\Delta t$. In summary, the principle is to put as much energy as possible into the Earth over a long time period and then to concentrate all that energy into a single time sample by data processing. This enhances the signal-to-noise ratio of the data, as is discussed in Chapter 6.

Ziolkowski et al. (2011) used a single period of a PRBS. This is very similar to the Vibroseis technique in seismic exploration. The source time function is then one period of the PRBS. The signal measured at the receiver begins when the PRBS starts, and finishes a certain time after the PRBS finishes, the extra time being equal to the time for the response to switch off or switch on to reach a steady state.

In node-based marine CSEM the practice has been to use a continuous signal. As mentioned above, the traditional square wave has acknowledged shortcomings for the purpose of CSEM, and operators desired to have a source time function with more frequencies at the same amplitude, but not necessarily the full spectrum that is provided by a PRBS. Candidate signal designs have been published by Lu and Srnka (2009), Mittet and Schaug-Pettersen (2008) and Myer et al. (2011).

3

Fourier Analysis and Linear Filters

This chapter discusses Fourier transforms and linear filters, which are central to the subsequent analysis. It establishes the notation and the sign convention, and introduces concepts that are needed in later chapters. These include: resolution and its Fourier counterpart, bandwidth; causality; impulse function, linear filters and impulse response; Earth as a linear filter; Parseval's theorem and the energy in a signal; convolution and the convolution theorem; cross-correlation, autocorrelation and time reverse; time derivative and the derivative theorem. The theory for source control and signature deconvolution of digital electromagnetic data is presented here, including sampling theory, the discrete Fourier transform, frequency-domain and time-domain deconvolution and the Wiener filter. The chapter concludes by introducing the Laplace transform, which is essential for the analysis in Chapter 4. Principal sources for this material are Bracewell (1965), Rabiner and Gold (1975), Robinson (1967) and Wiener (1949).

3.1 Temporal and Spatial Fourier Transformation

We may describe a wavefield as a function $a(x, y, z, t)$ that varies with both position (x, y, z) and time (t). The temporal Fourier transform of the space–time domain function $a(x, y, z, t)$ transforms the function to the space–frequency domain. We define the temporal Fourier transformation as

$$\hat{a}(x, y, z, \omega) = \int_{-\infty}^{\infty} a(x, y, z, t) e^{i\omega t} dt, \tag{3.1}$$

with inverse

$$a(x.y, z, t) = \frac{1}{2\pi} \int_{-\infty}^{\infty} \hat{a}(x, y, z, \omega) e^{-i\omega t} d\omega, \tag{3.2}$$

where ω is the *angular frequency*. It can be written as $\omega = 2\pi f$, with natural frequency f in hertz. Equations 3.1 and 3.2 are a *Fourier transform pair*. The hat ˆ indicates the change of domain from time to frequency, which is accompanied by a change of dimension, and therefore of units. We have chosen the sign of the exponential to be positive in the forward transform and it is therefore negative in the inverse transform.

Geophysical measurements of wavefields are generally made at the Earth's surface, and sometimes in boreholes. Surface measurements are made as a function of time (t) and horizontal coordinates (x, y), with z positive downwards in our convention. The double spatial Fourier transform of the space–frequency domain function $\hat{a}(x, y, z, \omega)$ is defined as

$$\tilde{a}(k_x, k_y, z, \omega) = \int_{-\infty}^{\infty} \int_{-\infty}^{\infty} \hat{a}(x, y, z, \omega) e^{-i(k_x x + k_y y)} dx dy, \tag{3.3}$$

with inverse

$$\hat{a}(x, y, z, \omega) = \frac{1}{4\pi^2} \int_{-\infty}^{\infty} \int_{-\infty}^{\infty} \tilde{a}(k_x, k_y, z, \omega) e^{i(k_x x + k_y y)} dk_x dk_y, \tag{3.4}$$

where k_x and k_x are the *horizontal wavenumbers* and the tilde ˜ indicates the further change of domain. Here we have chosen the negative sign for the exponential for transformation from space to wavenumber and therefore the positive sign for the inverse transformation.

The forward temporal and spatial Fourier transforms in equations 3.1 and 3.3 can be combined to give the forward triple Fourier transform:

$$\tilde{a}(k_x, k_y, z, \omega) = \int_{-\infty}^{\infty} \int_{-\infty}^{\infty} \int_{-\infty}^{\infty} a(x, y, z, t) e^{i(\omega t - k_x x - k_y y)} dx dy dt. \tag{3.5}$$

Similarly, the inverse transforms in equations 3.2 and 3.4 can be combined to give the inverse triple Fourier transform:

$$a(x, y, z, t) = \frac{1}{8\pi^3} \int_{-\infty}^{\infty} \int_{-\infty}^{\infty} \int_{-\infty}^{\infty} \tilde{a}(k_x, k_y, z, \omega) e^{-i(\omega t - k_x x - k_y y)} dk_x dk_y d\omega. \tag{3.6}$$

Very often, wavefields, diffusive fields and potential fields are measured on a horizontal plane, say $z = h_1$. The measured field would then be $p(x, y, z = h_1, t)$. We might want to manipulate the measurements to see what the field would look like at some other plane – for example, at $z = h_2$. The transformations of equations 3.5 and 3.6 facilitate the manipulation.

3.2 Example of a Plane Wave

Consider the function $a(x, y, z, t) = f(t - x/c)$. This is a plane wave travelling in the positive x-direction with velocity c. The wavefront is perpendicular to the x-axis and independent of the y and z coordinates. The temporal Fourier transform of this function is

$$\hat{a}(x, y, z, \omega) = \int_{-\infty}^{\infty} a(x, y, z, t) e^{i\omega t} dt = \int_{-\infty}^{\infty} f(t - x/c) e^{i\omega t} dt. \tag{3.7}$$

Substituting $\tau = t - x/c$, with $d\tau = dt$, gives

$$\begin{aligned} \hat{a}(x, y, z, \omega) &= \int_{-\infty}^{\infty} f(\tau) e^{i\omega(\tau + x/c)} d\tau \\ &= e^{i\omega x/c} \int_{-\infty}^{\infty} f(\tau) e^{i\omega\tau} d\tau \\ &= \hat{f}(\omega) e^{i\omega x/c}, \end{aligned} \tag{3.8}$$

where $\hat{f}(\omega)$ is the Fourier transform of $f(t)$. There is a positive phase delay of $\omega x/c$. This is a direct result of the choice of sign in the temporal Fourier transform. The quantity ω/c has dimensions of 1/length and is a wavenumber. We choose to have a positive wavenumber for waves travelling in the positive x-direction and in this case, therefore, the wavenumber $k_x = \omega/c$.

3.3 Resolution and Bandwidth

There is an intimate relationship between resolution and bandwidth. The greater the required resolution in the time domain, the greater the bandwidth that must be available in the frequency domain. Equally, the greater the required resolution in the space domain, the greater must be the bandwidth in the wavenumber domain. In fact, if a function exists only over a finite interval in one domain, or is said to have finite support, it must have infinite support in the transformed domain. This is fundamental, and perhaps not always as well understood in geophysics as it should be. Berkhout (1984) recognised that the subject deserved a book to itself.

In this section, only the principle is illustrated, not all the consequences. Following Bracewell (1965), we use the rectangle function $\Pi(t)$, shown in Figure 3.1(a), to illustrate the principle. The rectangle function is defined as

$$\Pi(t) = \begin{cases} 0, & |t| > \frac{1}{2} \\ \frac{1}{2}, & |t| = \frac{1}{2}. \\ 1, & |t| < \frac{1}{2} \end{cases} \tag{3.9}$$

It is zero everywhere, except in the finite time interval $-0.5 \le t \le 0.5$. Its Fourier transform, part of which is shown in Figure 3.1(b), is

$$\begin{aligned} \int_{-\frac{1}{2}}^{\frac{1}{2}} e^{i\omega t} dt &= \left. \frac{e^{i\omega t}}{i\omega} \right|_{-\frac{1}{2}}^{\frac{1}{2}} = \frac{e^{i\omega/2} - e^{-i\omega/2}}{i\omega} \\ &= \frac{\sin(\omega/2)}{\omega/2} = \frac{\sin(\pi f)}{\pi f}, \end{aligned} \tag{3.10}$$

and has infinite bandwidth. The function $\sin(\pi f)/\pi f$ is known as the *normalised sinc function* $\mathrm{sinc}(f)$ and the Fourier transform of the rectangle function is therefore

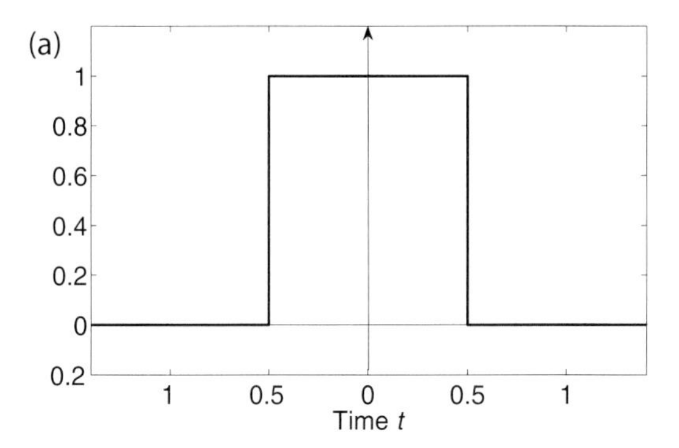

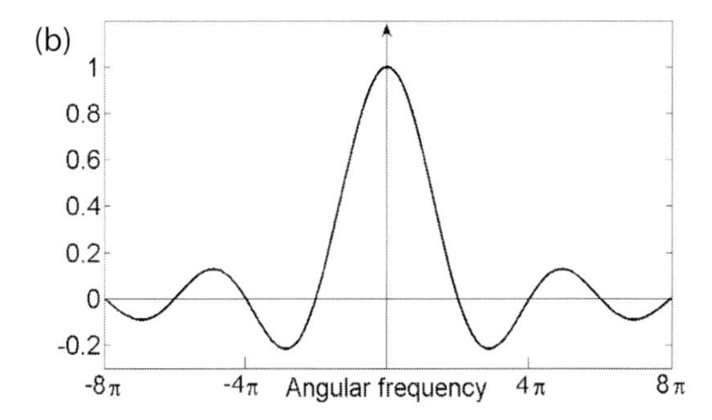

Figure 3.1 (a) Rectangle function $\Pi(t)$; (b) the central part of its Fourier transform $\mathrm{sinc}(\omega/2\pi)$.

$\text{sinc}(\omega/2\pi)$. This is an example of the principle that a function that has finite support in one domain must have infinite support in the transform domain.

3.4 Similarity Theorem

To see the relationship between resolution and bandwidth, it is better to see how a scale factor on the time axis affects the Fourier transform. If $a(t)$ has a Fourier transform $\hat{a}(\omega)$, then $a(\alpha t)$ has a Fourier transform $|\alpha|^{-1}\,\hat{a}(\omega/\alpha)$, which can be seen as follows. The Fourier transform of $a(\alpha t)$ is

$$\begin{aligned}\int_{-\infty}^{\infty} a(\alpha t)e^{i\omega t}dt &= \frac{1}{|\alpha|}\int_{-\infty}^{\infty} a(\alpha t)e^{i\alpha t(\omega/\alpha)}d(\alpha t) \\ &= \frac{1}{|\alpha|}\hat{a}(\omega/\alpha).\end{aligned} \tag{3.11}$$

This theorem is well known in its application to waveforms and spectra, where compression of the time axis, for example from seconds to microseconds, corresponds to expansion of the frequency axis from hertz to megahertz. Note, however, that as one member of the Fourier transform contracts horizontally, it also grows vertically so as to keep the area beneath it constant.

3.5 Impulse Function (δ)

The notation $\delta(t)$ for an impulse was introduced by Paul Dirac, but the concept had been in use by mathematicians for many decades before. An impulse is an infinitely strong pulse of unit area that can be defined as the two conditions

$$\begin{aligned}\delta(t) &= 0, t \neq 0 \\ \int_{-\infty}^{\infty} \delta(t)dt &= 1.\end{aligned} \tag{3.12}$$

It may be considered to be the limiting case of the scaled rectangular pulse $\tau^{-1}\Pi(t/\tau)$. This pulse is illustrated in Figure 3.2. It has a width of τ and a height of τ^{-1}. Therefore its area equals 1, whatever the value of τ. It behaves as the impulse function as its width τ tends to zero:

$$\begin{aligned}\lim_{\tau\to 0} \tau^{-1}\Pi(t/\tau) &= 0, t \neq 0 \\ \lim_{\tau\to 0}\int_{-\infty}^{\infty} \tau^{-1}\Pi(t/\tau)dt &= 1.\end{aligned} \tag{3.13}$$

So the scaled rectangle function has the properties required for an impulse, provided τ is small enough.

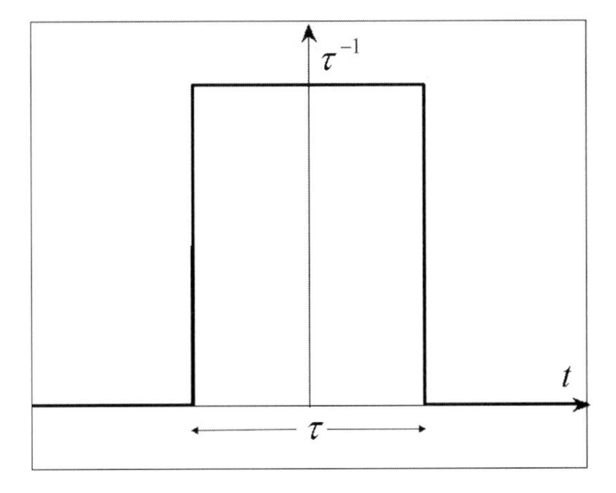

Figure 3.2 The scaled rectangle function $\tau^{-1}\Pi(t/\tau)$ (redrawn from Bracewell, 1965).

Since the scaled rectangle function is symmetric, it follows that

$$\delta(t) = \delta(-t). \tag{3.14}$$

3.6 The Sifting Property

Now consider the integral

$$\int_{-\infty}^{\infty} \delta(t) g(t) dt, \tag{3.15}$$

where $g(t)$ is a function of time. Figure 3.3 illustrates $g(t)$ and the integrand $\tau^{-1}\Pi(t/\tau)g(t)$. The area of the integrand is τ^{-1} times the shaded area. The shaded area has a width τ and an average height of $g(0)$. Thus the area under the integrand approaches $g(0)$ as τ tends to zero, and it follows that

$$\int_{-\infty}^{\infty} \delta(t) g(t) dt = g(0). \tag{3.16}$$

Since the characteristics of the pulse shape do not enter the right-hand side of equation 3.16, it is clear that $\delta(t)$ can be a pulse of any suitable shape; it does not have to be a rectangular pulse. Any function that in the limit satisfies the criteria for the impulse function can be used. The impulse $\delta(t)$, therefore, simply stands for an impulse of unit area whose duration is much smaller than any time interval of interest. It is normally represented as a spike of height equal to unity, as shown in Figure 3.4 (Bracewell, 1965).

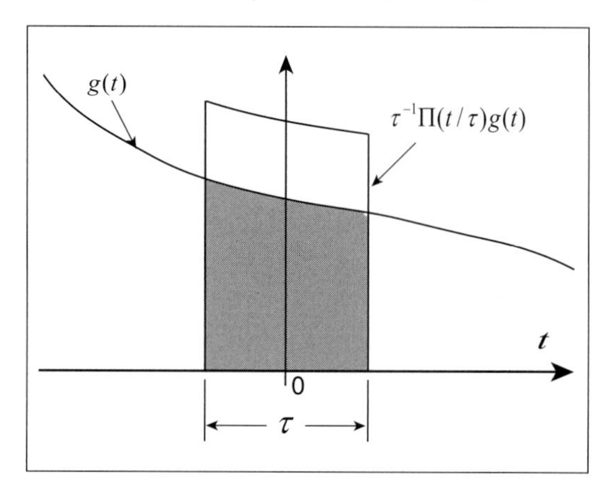

Figure 3.3 The sifting property. The area of the shape $\tau^{-1}\Pi(t/\tau)g(t)$ is approximately $g(0)$ (redrawn from Bracewell, 1965, Figure 5.2.).

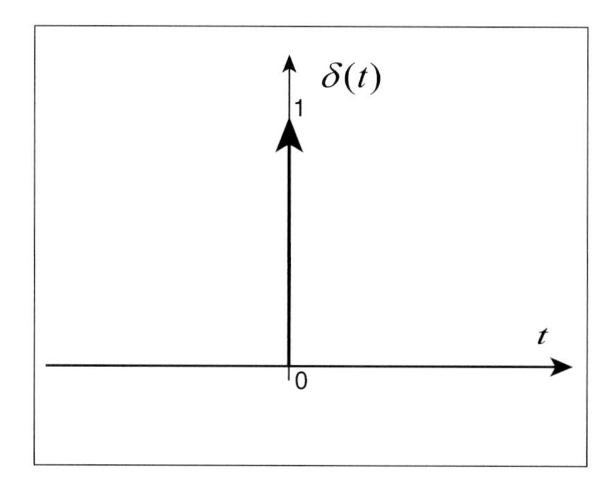

Figure 3.4 Graphical representation of the unit area impulse $\delta(t)$ as a spike of unit height (redrawn from Bracewell, 1965, Figure 5.3.).

Now consider the integral

$$\int_{-\infty}^{\infty} \delta(t-a)g(t)dt. \tag{3.17}$$

This is illustrated in Figure 3.5, using the scaled rectangular function, as before. It is clear that $\delta(t-a)$ is a pulse at time $t = a$. Therefore, in the case where the width of the scaled rectangular pulse tends to zero, it follows that

$$\int_{-\infty}^{\infty} \delta(t-a)g(t)dt = g(a). \tag{3.18}$$

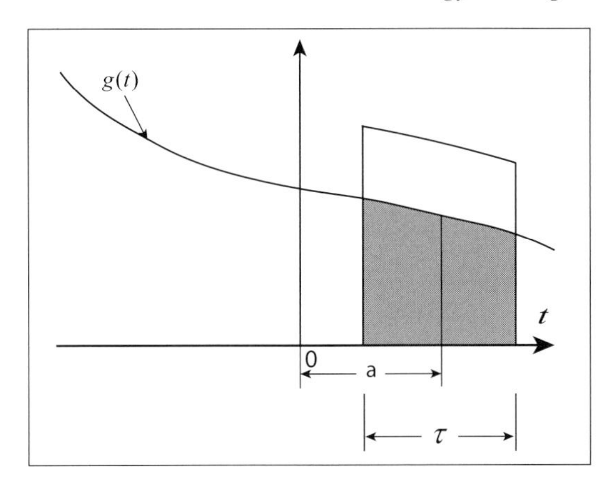

Figure 3.5 The sifting property. The shaded area is approximately $\tau g(a)$.

These simple results are used extensively in later sections. As an example, consider the Fourier transform of the delta function $\delta(t)$. This can be found by substituting $e^{i\omega t}$ for $g(t)$ in equation 3.16, giving

$$\int_{-\infty}^{\infty} \delta(t) e^{i\omega t} dt = 1, \tag{3.19}$$

with inverse transform

$$\delta(t) = \frac{1}{2\pi} \int_{-\infty}^{\infty} e^{-i\omega t} d\omega, \tag{3.20}$$

in which t and ω are interchangeable, thus

$$\delta(\omega) = \frac{1}{2\pi} \int_{-\infty}^{\infty} e^{-i\omega t} dt. \tag{3.21}$$

Equation 3.19 means that the impulse function, or delta function $\delta(t)$, is composed of an infinite number of cosines, with frequencies from $-\infty$ to ∞, each with amplitude equal to 1. It follows that an impulse, or delta function, contains an infinite amount of energy concentrated in an infinitesimally short time period. It is physically impossible to achieve, but is nevertheless a very useful concept.

We demonstrate this energy argument in the next section, using Rayleigh's energy theorem, which is a special case of the generalisation of Parseval's theorem.

3.7 Parseval's Theorem and the Energy in a Signal

The generalised form of Parseval's theorem is

$$\int_{-\infty}^{\infty} f(t) g^*(t) dt = \frac{1}{2\pi} \int_{-\infty}^{\infty} \hat{f}(\omega) \hat{g}^*(\omega) d\omega, \tag{3.22}$$

where the asterisk superscript * denotes the complex conjugate. For real signals, $g^*(t) = g(t)$ and

$$\hat{g}^*(\omega) = \int_{-\infty}^{\infty} g(t)e^{-i\omega t}dt = \hat{g}(-\omega). \tag{3.23}$$

To prove the general case, we begin with the inverse temporal Fourier transform,

$$f(t) = \frac{1}{2\pi}\int_{-\infty}^{\infty} \hat{f}(\omega)e^{-i\omega t}d\omega, \tag{3.24}$$

and write the left-hand side of equation 3.22 as

$$\int_{-\infty}^{\infty} f(t)g^*(t)dt = \int_{-\infty}^{\infty}\left(\frac{1}{2\pi}\int_{-\infty}^{\infty}\hat{f}(\omega)e^{-i\omega t}d\omega\right)\left(\frac{1}{2\pi}\int_{-\infty}^{\infty}\hat{g}^*(\omega')e^{i\omega' t}d\omega'\right)dt. \tag{3.25}$$

Rearranging the order of integration, we obtain

$$\int_{-\infty}^{\infty} f(t)g^*(t)dt = \frac{1}{2\pi}\int_{-\infty}^{\infty}\int_{-\infty}^{\infty}\hat{f}(\omega)\hat{g}^*(\omega')\left(\frac{1}{2\pi}\int_{-\infty}^{\infty}e^{-it(\omega-\omega')}dt\right)d\omega d\omega'. \tag{3.26}$$

Following equation 3.21, we recognise the integral in parentheses as the inverse Fourier transform of the delta function

$$\delta(\omega-\omega') = \frac{1}{2\pi}\int_{-\infty}^{\infty} e^{-it(\omega-\omega')}dt. \tag{3.27}$$

Using this in equation 3.26, we obtain

$$\int_{-\infty}^{\infty} f(t)g^*(t)dt = \frac{1}{2\pi}\int_{-\infty}^{\infty}\hat{f}(\omega)\left(\int_{-\infty}^{\infty}\hat{g}^*(\omega')\delta(\omega-\omega')d\omega'\right)d\omega. \tag{3.28}$$

In equation 3.28 the integral in parentheses is simply

$$\hat{g}^*(\omega) = \int_{-\infty}^{\infty}\hat{g}^*(\omega')\delta(\omega-\omega')d\omega'. \tag{3.29}$$

Therefore, equation 3.28 may be written as

$$\int_{-\infty}^{\infty} f(t)g^*(t)dt = \frac{1}{2\pi}\int_{-\infty}^{\infty}\hat{f}(\omega)\hat{g}^*(\omega)d\omega. \tag{3.30}$$

This is the generalised form of Parseval's theorem we wished to prove. For $g(t) = f(t)$ it becomes

$$\int_{-\infty}^{\infty} f(t)f^*(t) = \frac{1}{2\pi}\int_{-\infty}^{\infty}\hat{f}(\omega)\hat{f}^*(\omega)d\omega, \tag{3.31}$$

or

$$\int_{-\infty}^{\infty}|f(t)|^2dt = \frac{1}{2\pi}\int_{-\infty}^{\infty}|\hat{f}(\omega)|^2d\omega. \tag{3.32}$$

This is known as Parseval's theorem or as Rayleigh's theorem, or simply as the energy theorem. Each integral represents the amount of energy in the signal, the integral on the left being taken over all time, the integral on the right taken over all frequencies. It can obviously be applied to other coordinates, such as space and wavenumber.

To apply Parseval's theorem to a delta function, we note from equation 3.19 that the Fourier transform of a delta function is 1. So from the right-hand side of 3.32 we see that the energy of a delta function is

$$\frac{1}{2\pi}\int_{-\infty}^{\infty} d\omega, \tag{3.33}$$

which is infinite. Therefore, an impulse or delta function contains infinite energy concentrated in an infinitesimally short time. It is physically impossible to create a perfect impulse.

3.8 Convolution and the Convolution Theorem

The convolution of two functions $s(t)$ and $h(t)$ is

$$x(t) = \int_{-\infty}^{\infty} s(a)h(t-a)da, \tag{3.34}$$

which is often written as $x(t) = s(t) * h(t)$. The order of convolution is irrelevant: the result of convolving $s(t)$ with $h(t)$ is the same as convolving $h(t)$ with $s(t)$. This can be seen by substituting $\tau = t - a$ in equation 3.34 with $d\tau = -da$, to give

$$x(t) = \int_{\infty}^{-\infty} s(t-\tau)h(\tau)(-d\tau) = \int_{-\infty}^{\infty} s(t-\tau)h(\tau)d\tau. \tag{3.35}$$

If one of the functions is an impulse, the result of convolution is the other function:

$$s(t) = \int_{-\infty}^{\infty} s(a)\delta(t-a)da. \tag{3.36}$$

The convolution theorem states that convolution in one Fourier domain is equivalent to multiplication in the other Fourier domain. Thus, the result of convolving two time signals is equivalent, in the frequency domain, to multiplying their Fourier transforms. Equally, convolution of two (complex) signals in the frequency domain is equivalent, in the time domain, to multiplication of their Fourier transforms. This result applies of course to all Fourier-transformable functions, including functions of space and wavenumber. The theorem may be stated mathematically as

$$FT[h(t) * s(t)] = \hat{h}(\omega)\hat{s}(\omega), \tag{3.37}$$

in which $FT[x]$ means 'Fourier transform of x'.

The theorem may be proved as follows. First, the Fourier transform of the convolution of $h(t)$ and $s(t)$ is written out in full:

$$FT[h(t) * s(t)] = \int_{-\infty}^{\infty} \left[\int_{-\infty}^{\infty} s(a)h(t-a)da \right] e^{i\omega t} dt. \tag{3.38}$$

Then, the integrand is multiplied by $1 = e^{-i\omega a}e^{i\omega a}$, and the order of integration is changed:

$$FT[h(t) * s(t)] = \int_{-\infty}^{\infty} s(a) \left[\int_{-\infty}^{\infty} h(t-a)e^{i\omega(t-a)} dt \right] e^{i\omega a} da, \tag{3.39}$$

in which the Fourier transform of $h(t)$ may be recognised in the square brackets. Thus:

$$FT[h(t) * s(t)] = \int_{-\infty}^{\infty} s(a)\hat{h}(\omega)e^{i\omega a} da. \tag{3.40}$$

$\hat{h}(\omega)$ may be taken outside the integral, as it is independent of the integration variable a, to yield

$$\begin{aligned} FT[h(t) * s(t)] &= \hat{h}(\omega) \int_{-\infty}^{\infty} s(a)e^{i\omega a} da \\ &= \hat{h}(\omega)\hat{s}(\omega). \end{aligned} \tag{3.41}$$

This is the result to be proved.

To prove that convolution in the frequency domain is equivalent to multiplication in the time domain, one begins with $FT[\hat{h}(\omega)\hat{s}(\omega)]$ and then follows the same steps as above, except that the exponent has a negative sign throughout.

3.9 Linear Filters and Impulse Response

A filter is a system that has an input and an output. Figure 3.6 illustrates a physical linear time-invariant system, with one input and one output, whose response to the impulsive input $\delta(t)$ is $h(t)$, which is known as the *impulse response* of the system. Since the system is physical, it cannot respond before there is an input; that is, it is causal. The causality of the system may be expressed simply as

$$h(t) = 0, t < 0. \tag{3.42}$$

Since the system is time invariant, the response is independent of the time origin. If the input is delayed until a time $t = a$, the output is also delayed until the same time. An impulse at time $t = a$ is $\delta(t-a)$; the delayed impulse response is $h(t-a)$,

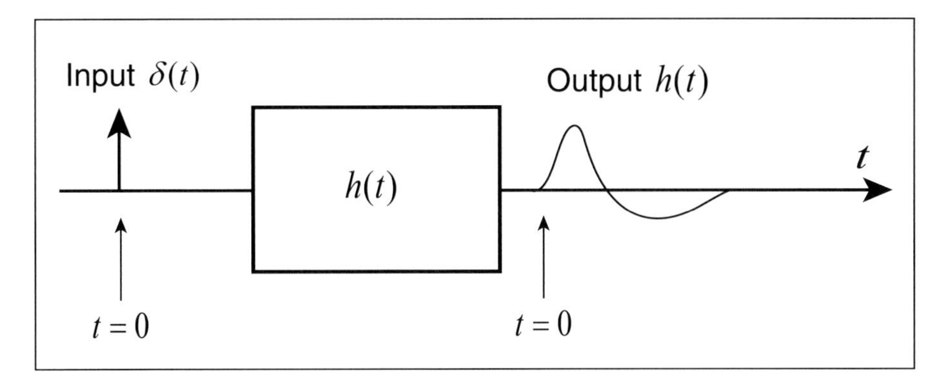

Figure 3.6 Response of physical linear time-invariant system $h(t)$ to an impulse at $t = 0$.

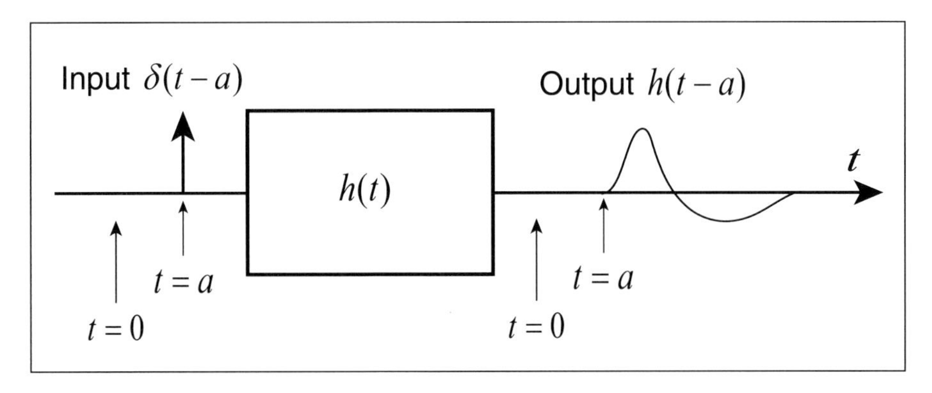

Figure 3.7 Response of physical linear time-invariant system $h(t)$ to an impulse at $t = a$.

as illustrated in Figure 3.7. It is shown using the sifting property, given in equation 3.18, in the convolution integral

$$h(t-a) = \int_{-\infty}^{\infty} \delta(\tau - a)h(t-\tau)d\tau. \tag{3.43}$$

Consider an input $s(t)$ to the causal linear filter with impulse response $h(t)$ shown in Figure 3.8. The input can be divided into a series of parallel strips, as shown in Figure 3.9. Each strip is very narrow and has finite area. Using the sifting property again, the value $s(a)$ can be expressed as

$$s(a) = \int_{-\infty}^{\infty} \delta(t-a)s(t)dt. \tag{3.44}$$

The response to an impulse of unit area at $t = a$ is $h(t - a)$, as illustrated in Figure 3.7. Since the system is linear, the response to a narrow rectangular strip

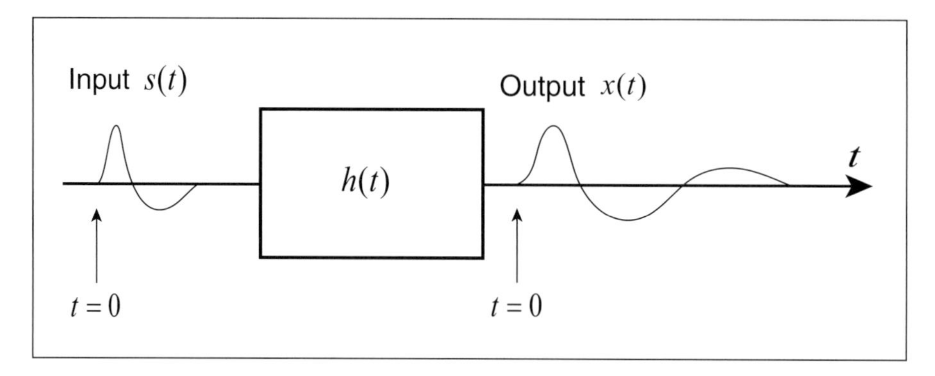

Figure 3.8 Response of filter $h(t)$ to input $s(t)$.

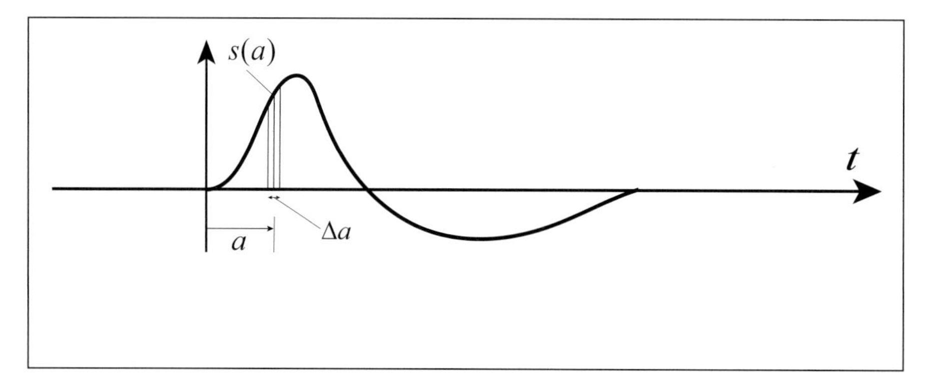

Figure 3.9 A strip of $s(t)$ at time $t = a$.

of area $s(a)\Delta a$ at time $t = a$ is $s(a)h(t - a)\Delta a$. The total response is obtained by integrating the responses to all the strips:

$$\begin{aligned} x(t) &= \int_{-\infty}^{\infty} \left(\int_{-\infty}^{\infty} \delta(\tau - a)s(\tau)d\tau \right) h(t - a)da \\ &= \int_{-\infty}^{\infty} s(a)h(t - a)da. \end{aligned} \tag{3.45}$$

We recognise equation 3.45 as a *convolution*: the output $x(t)$ is the convolution of the input $s(t)$ with the impulse response of the filter $h(t)$. Using the convolution theorem, the Fourier transform of this result is

$$\hat{x}(\omega) = \hat{s}(\omega)\hat{h}(\omega). \tag{3.46}$$

The Fourier transform of the impulse response $h(t)$ is $\hat{h}(\omega)$, which is known as the *transfer function* of the system. If the input signal is $A\cos(\omega_0 t)$, the output has the same frequency, but has a different amplitude, say B, and phase, say ϕ, and may be written as $B\cos(\omega_0 t - \phi)$. The input and output for angular frequency ω_0 are related by the complex *transfer factor*

$$\hat{h}(\omega_0) = \frac{B}{A} e^{i\phi}. \tag{3.47}$$

The transfer factor is the constant of proportionality of the linear filter that relates a harmonic input signal to the corresponding output of the filter. Each frequency component $\hat{s}(\omega)$ of the input signal $s(t)$ is multiplied by the corresponding transfer factor $\hat{h}(\omega)$ of the system $h(t)$.

If a system is causal, the output $x(t)$ at time t is the response to all the inputs up to time t. Since $h(t)$ is causal, $h(t-a)$ is zero for values of a greater than t. Therefore, the upper limit of the integral in equation 3.45 is t. Similarly, if the input signal $s(t)$ is zero before $t = 0$, the lower limit of the integral is 0. If $s(t)$ and $h(t)$ are both causal, we have

$$x(t) = \int_0^t s(a)h(t-a)da = \int_0^t h(a)s(t-a)da. \tag{3.48}$$

The results of this section are summarised in Figure 3.10. Since the output $x(t)$ is independent of the order of the sequence of linear filters, it is impossible to deduce the order of the sequence from the output alone.

3.10 Earth as a Linear Filter

Hooke's law is linear and Ohm's law is linear. We therefore expect the Earth to behave as a linear time-invariant system in seismic and electromagnetic

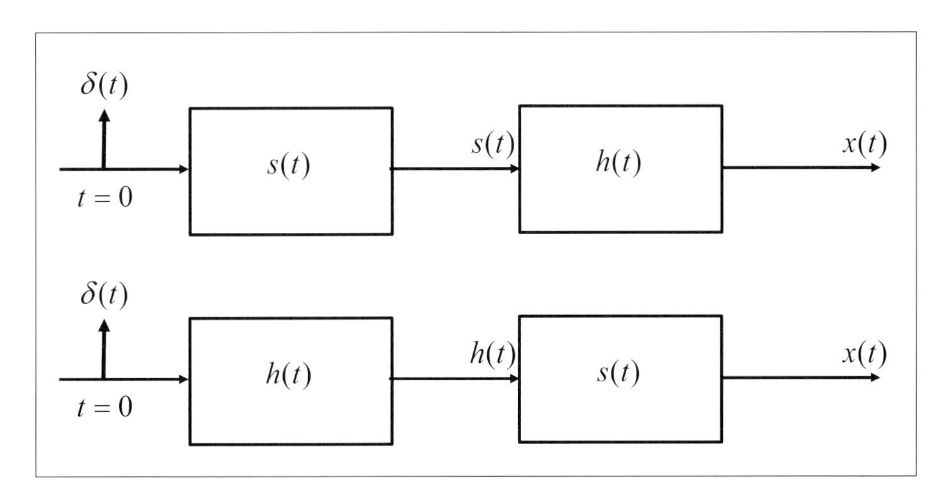

Figure 3.10 The impulse response of two cascaded linear filters: the order of convolution is irrelevant.

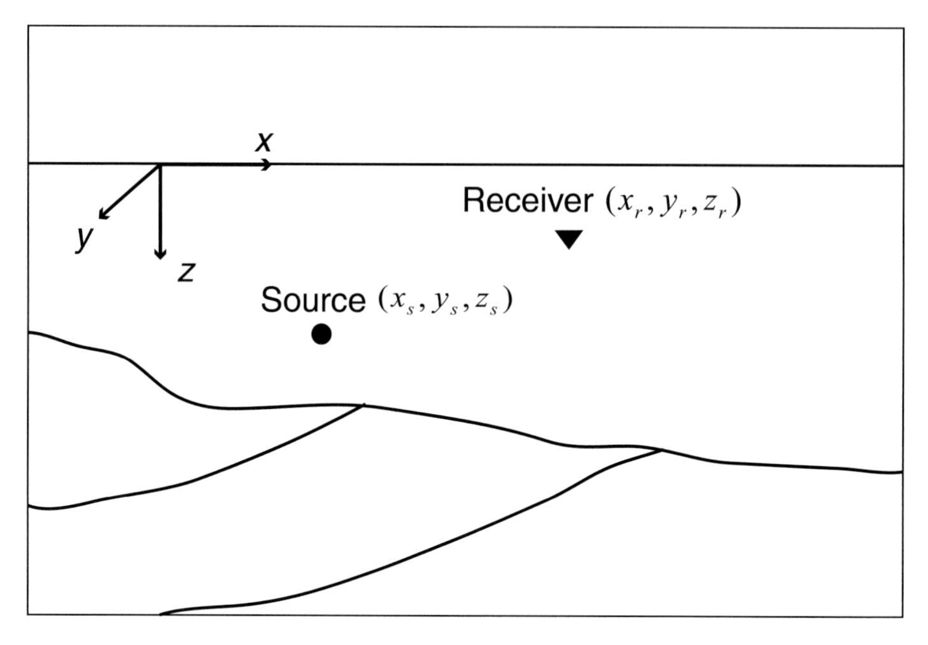

Figure 3.11 The impulse response depends on the positions of both source and receiver.

investigations. The concept of impulse response is then very useful in the analysis of seismic and electromagnetic data. Since Earth is a complicated body, the response obtained depends on the positions of the source and receiver. This is illustrated in Figure 3.11. These positions can be included in the response explicitly as

$$G(x_r, y_r, z_r; x_s, y_s, z_s; t) = G(\mathbf{r}_r, \mathbf{r}_s, t), \tag{3.49}$$

where $\mathbf{r}_r = x_r\mathbf{i} + y_r\mathbf{j} + z_r\mathbf{k}$ is the position vector of the receiver and $\mathbf{r}_s = x_s\mathbf{i} + y_s\mathbf{j} + z_s\mathbf{k}$ is the position vector of the source. This function G is known as the Green's function. It is the response seen at the receiver point to an impulse $\delta(t)$ at the source point. It is an impulse response. In principle there is an infinite number of such possible responses in the Earth. The sampling problems in space and time to obtain an adequate picture of the Earth have engrossed geophysicists for many years, and are likely to do so for some time.

In practice the concept is further complicated when vector, rather than scalar, field quantities are considered. In seismology the particle velocity at a receiver is a vector; in electromagnetics the electric field and magnetic field are both vectors. Electromagnetic sources are also normally regarded as vectors and a current dipole source, for example, has a particular direction as well as a position. Correct specification of the Green's function is, in general, complicated. In what follows much of this complication is omitted, but it should be clear from the context what is implied.

3.11 Cross-Correlation, Autocorrelation and Time-Reverse

Correlation is related to convolution. The correlation of signal $s(t)$ with $h(t)$ may be defined as

$$\begin{aligned}\phi_{sh}(\tau) &= \lim_{T\to\infty}\frac{1}{2T}\int_{-T}^{T} s(t+\tau)h^*(t)dt,\\ &= \lim_{T\to\infty}\frac{1}{2T}\int_{-T}^{T} s(t)h^*(t-\tau)dt.\end{aligned} \tag{3.50}$$

For real-valued signals the superscript asterisk may be omitted. The two forms of the cross-correlation are equivalent, for the result is the same whether the first signal is shifted to the left along the time axis by an amount τ (that is, a time advance of the first signal to obtain $s(t+\tau)$), or the second signal is shifted to the right by τ (that is, a delay of the second signal to obtain $h^*(t-\tau)$).

It is clear that the order of the two signals being cross-correlated is important. Writing $\phi_{hs}(\tau)$ in the first of the two equivalent forms,

$$\phi_{hs}(\tau) = \lim_{T\to\infty}\frac{1}{2T}\int_{-T}^{T} h(t+\tau)s^*(t)dt, \tag{3.51}$$

it is easily recognised that

$$\phi_{hs}(\tau) = \phi^*_{sh}(-\tau), \tag{3.52}$$

which says that the cross-correlation of one signal with a second signal for time shift τ is equal to the complex conjugate of the cross-correlation of the second signal with the first for a time shift $-\tau$.

Autocorrelation is a special case of cross-correlation, when $s(t) = h(t)$:

$$\begin{aligned}\phi_{ss}(\tau) &= \lim_{T\to\infty}\frac{1}{2T}\int_{-T}^{T} s(t+\tau)s^*(t)dt,\\ &= \lim_{T\to\infty}\frac{1}{2T}\int_{-T}^{T} s(t)s^*(t-\tau)dt,\end{aligned} \tag{3.53}$$

and therefore

$$\phi_{ss}(\tau) = \phi^*_{ss}(-\tau). \tag{3.54}$$

This states that for a complex-valued signal the autocorrelation coefficient for time shift τ is equal to the complex conjugate of the autocorrelation coefficient for time shift $-\tau$. Thus the autocorrelation function exhibits conjugate symmetry: its real part is an even function of τ, while its imaginary part is an odd function of τ. If the signal $s(t)$ is real, $s(t) = s^*(t)$, and

$$\phi_{ss}(\tau) = \phi_{ss}(-\tau). \tag{3.55}$$

That is, the autocorrelation is real and symmetric. In practice, the time signals we deal with are almost always real. We allow them to be complex in this analysis not only because it is more general, but also because it is then often easier to see the relationships between the time domain and the frequency domain.

It is convenient to define the time-reverse of a signal. If the signal is $s(t)$, its *time-reverse* is $s^*(-t)$, where the asterisk indicates the complex conjugate. That is, the time-reverse of a signal is obtained by flipping the signal about the time origin, thus reversing the time axis, and taking the complex conjugate in case the signal is complex. The formula for cross-correlation may be written as

$$\phi_{sh}(\tau) = \lim_{T\to\infty} \frac{1}{2T} \int_{-T}^{T} s(\tau - t)h^*(-t)dt, \tag{3.56}$$

and it is apparent that cross-correlation of $s(t)$ with $h(t)$ is the same as the convolution of $s(t)$ with the time-reverse of $h(t)$. This is the fundamental relationship between the operations of convolution and cross-correlation.

3.12 Derivative Theorem

If $h(t)$ has the Fourier transform $\hat{h}(\omega)$, then $dh(t)/dt = h'(t)$ has the Fourier transform $-i\omega\hat{h}(\omega)$. To prove this, we begin with the inverse temporal Fourier transform

$$h(t) = \frac{1}{2\pi} \int_{-\infty}^{\infty} \hat{h}(\omega)e^{-i\omega t}d\omega. \tag{3.57}$$

We differentiate this with respect to t to obtain

$$h'(t) = \frac{1}{2\pi} \int_{-\infty}^{\infty} -i\omega\hat{h}(\omega)e^{-i\omega t}d\omega, \tag{3.58}$$

and then take the temporal Fourier transform to yield

$$\begin{aligned} \int_{-\infty}^{\infty} h'(t)e^{i\omega' t}dt &= -\frac{i}{2\pi} \int_{-\infty}^{\infty} \omega\hat{h}(\omega) \left[\int_{-\infty}^{\infty} e^{i(\omega'-\omega)t}dt \right] d\omega, \\ &= -\frac{i}{2\pi} \int_{-\infty}^{\infty} \omega\hat{h}(\omega) \left[2\pi\delta(\omega' - \omega) \right] d\omega \\ &= -i\omega'\hat{h}(\omega'). \end{aligned} \tag{3.59}$$

This is the result to be proved. On the right-hand side of equation 3.59 we have used the Fourier transform of the delta function from equation 3.27 for the factor in square brackets. The second step is simply the sifting property.

The theorem is very useful. One of the uses is the separation of variables of partial differential equations such as the wave equation and the diffusion equation.

3.13 Wavefield Transformation

In Section 3.1 the wavenumber was defined as $k = \omega/c = 2\pi f/c$, where f is frequency and c is propagation velocity. The wavelength is then $\lambda = 2\pi c/\omega = 2\pi/k$. Sometimes wavenumber is defined as the reciprocal of wavelength, and is the number of cycles per unit distance. In some ways this is a more intuitive definition. Nevertheless, we use the one we have already used because it makes the formulae more compact and easier to read.

Consider now the Fourier transform of the wavefield $p(x, y, z, t)$, defined in space and time. Following the example of equation 3.5, the transformation from time and two spatial coordinates to frequency and the corresponding wavenumbers, the Fourier transform of $p(x, y, z, t)$ is now defined as

$$\breve{p}(k_x, k_y, k_z, \omega) = \int_{-\infty}^{\infty}\int_{-\infty}^{\infty}\int_{-\infty}^{\infty}\int_{-\infty}^{\infty} p(x, y, z, t)e^{i[\omega t - k_x x - k_y y - k_z z]}dxdydzdt. \quad (3.60)$$

The sign of the exponent is positive for time t, and negative for the spatial coordinates. This is consistent with our desire for waves travelling in the positive x, y and z directions to have positive wavenumbers. The inverse Fourier transform of equation 3.60 is

$$p(x, y, z, t) = \frac{1}{16\pi^4}\int_{-\infty}^{\infty}\int_{-\infty}^{\infty}\int_{-\infty}^{\infty}\int_{-\infty}^{\infty} \breve{p}(k_x, k_y, k_z, \omega)e^{-i[\omega t - k_x x - k_y y - k_z z]}dk_x dk_y dk_z d\omega. \quad (3.61)$$

From the derivative theorem we have established that $\partial/\partial t$ transforms to $-i\omega$, which we may write as

$$\frac{\partial}{\partial t} \rightarrow -i\omega, \quad (3.62)$$

and it follows that

$$\frac{\partial^2}{\partial t^2} \rightarrow -\omega^2. \quad (3.63)$$

The transforms of the spatial derivatives may be written as

$$\frac{\partial}{\partial x} \rightarrow ik_x, \qquad \frac{\partial^2}{\partial x^2} \rightarrow -k_x^2, \quad (3.64)$$

$$\frac{\partial}{\partial y} \rightarrow ik_y, \qquad \frac{\partial^2}{\partial y^2} \rightarrow -k_y^2, \quad (3.65)$$

$$\frac{\partial}{\partial z} \rightarrow ik_z, \qquad \frac{\partial^2}{\partial z^2} \rightarrow -k_z^2. \quad (3.66)$$

We can now take the Fourier transform of a partial differential equation, for example the wave equation for a pressure field $p(x, y, z, t)$,

$$\nabla^2 p - \frac{1}{c^2}\frac{\partial^2 p}{\partial t^2} = 0. \tag{3.67}$$

We first write this as

$$\frac{\partial^2 p}{\partial x^2} + \frac{\partial^2 p}{\partial y^2} + \frac{\partial^2 p}{\partial z^2} - \frac{1}{c^2}\frac{\partial^2 p}{\partial t^2} = 0, \tag{3.68}$$

and then employ relations 3.60, 3.63, 3.64, 3.65 and 3.66 to give

$$\left(-k_x^2 - k_y^2 - k_z^2 + \frac{\omega^2}{c^2}\right)\breve{p}(k_x, k_y, k_z, \omega) = 0. \tag{3.69}$$

This equation is satisfied if

$$k_x^2 + k_y^2 + k_z^2 = k^2 = \frac{\omega^2}{c^2}. \tag{3.70}$$

This transformation is needed for the analysis of the whole-space response to sources in Chapter 4.

3.14 Sampling and Aliasing

The Fourier theory has so far used continuous signals and integrals. In transformation of data in computers the integrals are replaced by summations. The continuous signal $a(t)$ becomes the discrete signal or time series a_k, in which k is an integer, and the sampling has taken place at regular intervals Δt. (We have used k as a wavenumber, for instance in equation 3.70. Now we are using k as an index. It should be obvious from the context whether k is a wavenumber or an index.) Sampling, or analogue-to-digital conversion, may be expressed as

$$a_k = a(k\Delta t) = \int_{-\infty}^{\infty} a(t)\delta(t - k\Delta t)dt. \tag{3.71}$$

Thus the discrete signal corresponds to the continuous signal at times $k\Delta t$. In equation 3.71 the suffix k is used to indicate the sample number. Of course, any suffix can be used and, for example, a_t may be used to indicate a discrete time series.

It is necessary to ensure adequate sampling. If a sinusoidal signal is undersampled, as shown in Figure 3.12, the sampled signal appears to have a lower frequency than the original. This is known as *aliasing*.

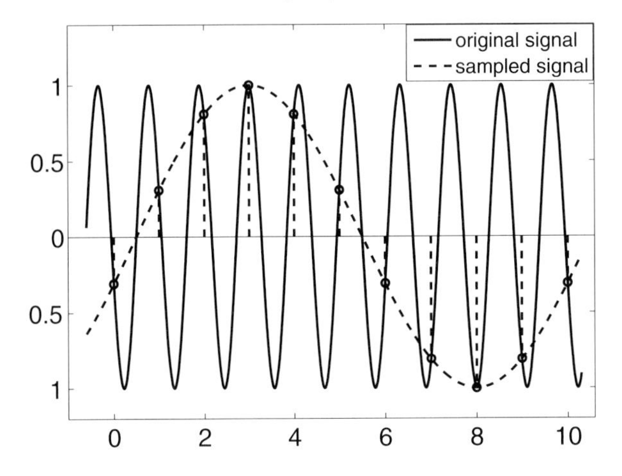

Figure 3.12 If a sinusoidal signal is under-sampled at regular intervals it appears to have a lower frequency.

3.15 Sampling Theorem

The rule for correctly sampling a continuous signal is now derived. We begin with the Fourier transform pair

$$\hat{a}(\omega) = \int_{-\infty}^{\infty} a(t) e^{i\omega t} dt, \tag{3.72}$$

$$a(t) = \frac{1}{2\pi} \int_{-\infty}^{\infty} \hat{a}(\omega) e^{-i\omega t} d\omega. \tag{3.73}$$

The signal $a(t)$ is now sampled at regular intervals $t = k\Delta t$, $k = \cdots, -2, -1, 0, 1, 2, \cdots$. To find the Fourier transform of the discrete function, the integral is replaced by a summation, using the trapezoidal rule

$$\hat{a}(\omega) = \Delta t \sum_{k=-\infty}^{\infty} a(k\Delta t) e^{i\omega k\Delta t}. \tag{3.74}$$

Now we evaluate the integral in equation 3.73 at the discrete times $k\Delta t$:

$$a(k\Delta t) = \frac{1}{2\pi} \int_{-\infty}^{\infty} \hat{a}(\omega) e^{-i\omega k\Delta t} d\omega. \tag{3.75}$$

This integral may be replaced by an infinite sum of pieces of the integral, each of which has width $2\pi/\Delta t$:

$$a(k\Delta t) = \frac{1}{2\pi} \sum_{m=-\infty}^{\infty} \int_{\frac{(2m-1)\pi}{\Delta t}}^{\frac{(2m+1)\pi}{\Delta t}} \hat{a}(\omega) e^{-i\omega k\Delta t} d\omega. \tag{3.76}$$

Substituting $\omega = \nu + 2\pi m/\Delta t$ yields

$$a(k\Delta t) = \frac{1}{2\pi} \sum_{m=-\infty}^{\infty} \int_{\frac{-\pi}{\Delta t}}^{\frac{\pi}{\Delta t}} \hat{a}\left(\nu + \frac{2\pi}{\Delta t} m\right) e^{-i\nu k\Delta t} e^{-2\pi imk} d\nu. \tag{3.77}$$

Changing the order of the integration and summation and noting that $e^{-2\pi imk} = 1$, since m and k are both integers, gives

$$a(k\Delta t) = \frac{1}{2\pi} \int_{\frac{-\pi}{\Delta t}}^{\frac{\pi}{\Delta t}} \sum_{m=-\infty}^{\infty} \hat{a}\left(\nu + \frac{2\pi}{\Delta t} m\right) e^{-i\nu k\Delta t} d\nu. \tag{3.78}$$

Since ν is simply the integration variable in equation 3.78 it can be replaced by any parameter; we choose ω. The Fourier transform of the discrete time series

$$a_k = a(t), \text{when } t = k\Delta t, \tag{3.79}$$

is thus

$$a_k = \frac{1}{2\pi} \int_{\frac{-\pi}{\Delta t}}^{\frac{\pi}{\Delta t}} \hat{a}_D(\omega) e^{-i\omega k\Delta t} d\omega, \tag{3.80}$$

provided

$$\hat{a}_D(\omega) = \sum_{m=-\infty}^{\infty} \hat{a}(\omega + \frac{2\pi}{\Delta t} m). \tag{3.81}$$

Equation 3.81 shows that the periodic frequency response of the sequence $a(k\Delta t)$ is a sum of an infinite number of components of the frequency response of the analogue signal $a(t)$. Now, where the analogue frequency response is bandlimited to the range $|\omega| \le \pi/\Delta t$, that is, $\hat{a}(\omega) = 0, |\omega| > \pi/\Delta t$, equation 3.80 shows that, in the frequency range $|\omega| \le \pi/\Delta t$,

$$\hat{a}_D(\omega) = \hat{a}(\omega). \tag{3.82}$$

The discrete Fourier transform pair for an infinite time series is thus

$$\hat{a}_D(\omega) = \Delta t \sum_{k=-\infty}^{\infty} a_k e^{i\omega k\Delta t}, \tag{3.83}$$

$$a_k = \frac{1}{2\pi} \int_{\frac{-\pi}{\Delta t}}^{\frac{\pi}{\Delta t}} \hat{a}_D(\omega) e^{-i\omega k\Delta t} d\omega. \tag{3.84}$$

The angular frequency

$$\omega_N = \frac{\pi}{\Delta t} \tag{3.85}$$

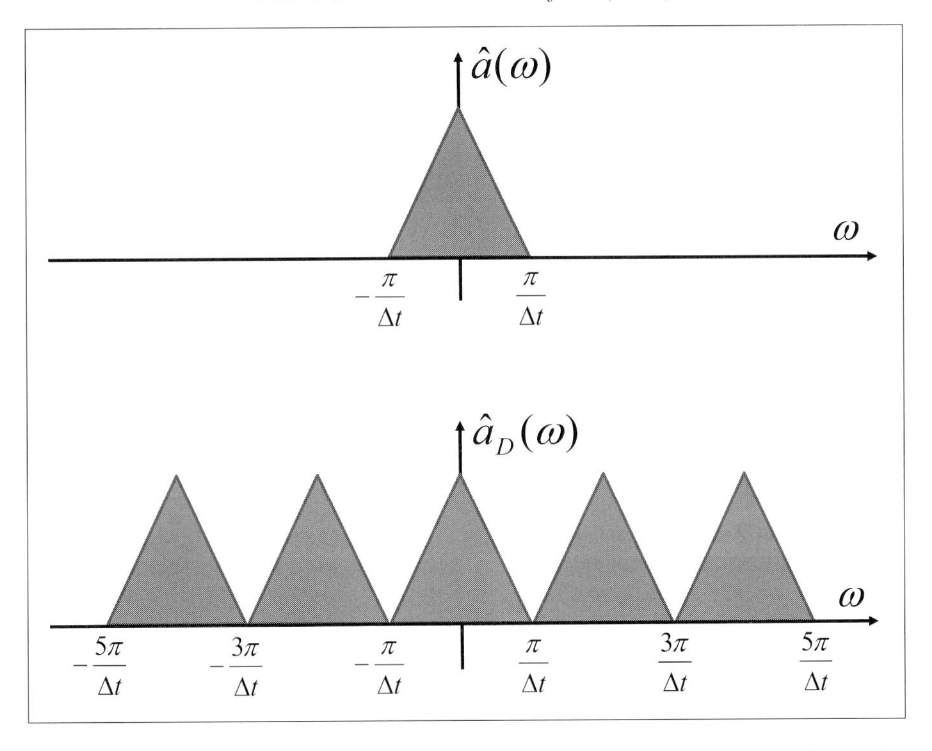

Figure 3.13 Relationship between the bandlimited Fourier transform $\hat{a}(\omega)$ of an analogue (continuous) signal $a(t)$ and the Fourier transform $\hat{a}_D(\omega)$ of the corresponding properly sampled digital signal $a(k\Delta t)$ (redrawn from Rabiner and Gold, 1975, Figure 2.13).

is known as the angular Nyquist frequency. Clearly, if there is no information in the continuous time signal $a(t)$ at angular frequencies above ω_N, the maximum sampling interval is $\Delta t = \pi/\omega_N$. This is the sampling theorem.

The relationship between the discrete Fourier transform $\hat{a}_D(\omega)$ and the continuous bandlimited Fourier transform $\hat{a}(\omega)$ for this case is shown in Figure 3.13. To guarantee that there will be no aliasing, as illustrated in Figure 3.12, it is necessary to ensure there is no energy in angular frequencies above the angular Nyquist frequency ω_N. This is normally achieved by applying an adequate analogue low-pass filter to $a(t)$ before digitisation at the regular sample interval Δt.

3.16 Discrete Fourier Transform (DFT)

In practice, the number of samples N in a time series a_k, sampled at regular intervals $t = k\Delta t$, is always finite. It is then convenient to consider the series as a single period, of duration $N\Delta t$, of a periodic sequence $a_{[k]}$, in which the square brackets around the index k denote a periodic sequence. Now, we showed in Section 3.15 that regular sampling of a time signal led to a periodic frequency response. Regular sampling in one domain leads to periodicity in the transform domain. Equally, periodicity in one domain leads to regular sampling in the transform domain. The angular frequencies ω of which $a_{[k]}$ is composed must therefore be regularly sampled and must be the fundamental angular frequency $2\pi/(N\Delta t)$ and its harmonics, because these are the only frequencies that are integrally related to $N\Delta t$. That is, $\omega = 2\pi n/(N\Delta t)$ for $-\infty < n < \infty$. In this case we may express $a_{[k]}$ as

$$a_{[k]} = \frac{\Delta\omega}{2\pi} \sum_{n=-\infty}^{\infty} \hat{a}_{[n]} e^{-\frac{2\pi ink}{N}}. \tag{3.86}$$

In this equation the multiplication by $\Delta\omega$ on the right-hand side preserves the dimensions, since an integral corresponds to a summation (compare equations 3.86 and 3.80). Because angular frequency has become a discrete parameter, we can write it as $\omega = n\Delta\omega$ with $\Delta\omega = 2\pi/(N\Delta t)$, as the fundamental frequency.

Equation 3.86 contains much redundant information because the exponential is periodic:

$$e^{-\frac{2\pi ink}{N}} = e^{-\frac{2\pi ik(n\pm mN)}{N}}, \qquad 0 < m < \infty. \tag{3.87}$$

Therefore equation 3.86 may be written simply as

$$a_{[k]} = \frac{1}{N\Delta t} \sum_{n=0}^{N-1} \hat{a}_{[n]} e^{-\frac{2\pi ink}{N}}. \tag{3.88}$$

This states that there are only N frequency components for a periodic time series of period N time samples. To obtain the inverse transform, that is $\hat{a}_{[n]}$ in terms of $a_{[k]}$, we begin by multiplying both sides of equation 3.88 by $e^{\frac{2\pi ikm}{N}}$ and summing over k to give

$$\sum_{k=0}^{N-1} a_{[k]} e^{\frac{2\pi ikm}{N}} = \frac{1}{N\Delta t} \sum_{k=0}^{N-1} \sum_{n=0}^{N-1} \hat{a}_{[n]} e^{-\frac{2\pi ik(n-m)}{N}}. \tag{3.89}$$

Interchanging the order of summation on the right-hand side of 3.89 and using the relation

$$\sum_{k=0}^{N-1} e^{-\frac{2\pi ik(n-m)}{N}} = \begin{cases} N & \text{if} \quad n = m \\ 0 & \text{otherwise,} \end{cases} \tag{3.90}$$

gives

$$\sum_{k=0}^{N-1} a_{[k]} e^{\frac{2\pi ikm}{N}} = \frac{1}{\Delta t} \sum_{n=0}^{N-1} \hat{a}_{[n]} \delta_{n-m}, \tag{3.91}$$

where

$$\delta_{n-m} = \begin{cases} 1 & \text{if} \quad n = m \\ 0 & \text{otherwise.} \end{cases} \tag{3.92}$$

Reversing the left-hand side and right-hand side of equation 3.91 and substituting the index n for m gives

$$\hat{a}_{[n]} = \Delta t \sum_{k=0}^{N-1} a_{[k]} e^{\frac{2\pi ikn}{N}}. \tag{3.93}$$

Equation 3.93 is called the discrete Fourier transform (DFT) and equation 3.88 is called the inverse discrete Fourier transform (IDFT). Both sequences $a_{[k]}$ and $\hat{a}_{[n]}$ are periodic, with period N samples. It is also clear that $\hat{a}_{[n]}$ may be obtained from one period of $a_{[k]}$. Rabiner and Gold (1975: 52) state that

> The DFT coefficients of a finite duration sequence constitute a unique representation of that sequence because the inverse DFT relations may be used to reconstruct the desired sequence exactly from the DFT coefficients. Thus even though the DFT, IDFT relations are derived on the basis of periodic sequences, they are even more important in their ability to represent finite duration sequences.

These relations will mostly be applied to finite duration sequences, and it is therefore convenient to drop the square brackets to form the DFT pair

$$\hat{a}_n = \Delta t \sum_{k=0}^{N-1} a_k e^{\frac{2\pi ikn}{N}}, \tag{3.94}$$

$$a_k = \frac{1}{N \Delta t} \sum_{n=0}^{N-1} \hat{a}_n e^{-\frac{2\pi ink}{N}}. \tag{3.95}$$

We note that a_k and $\hat{a}_n$ have different units. If a_k has units of volts (V), say, then $\hat{a}_n$ has units volt-seconds (V s). It is common practice to drop the factor Δt, which means it is combined with a_k. When this is done, the sampled time function and its discrete Fourier transform have the same units (for example (V s) instead of (V)), but the domains are different and are related by the following more compact form of the DFT pair:

$$\hat{a}_n = \sum_{k=0}^{N-1} \grave{a}_k e^{\frac{2\pi ikm}{N}}, \tag{3.96}$$

$$\grave{a}_k = \frac{1}{N}\sum_{n=0}^{N-1} \hat{a}_n e^{-\frac{2\pi ink}{N}}, \tag{3.97}$$

where $\grave{a}_k = \Delta t a_k$. The sign of the exponential is positive for the forward transform and negative for the inverse transform. Of course the sign of the exponential may be negative for the forward transform, provided it is positive for the inverse transform.

3.17 Filtering of Sampled Signals: Discrete Convolution

Consider the convolution

$$x(t) = \int_{-\infty}^{\infty} s(\tau)h(t-\tau)d\tau = s(t) * h(t). \tag{3.98}$$

Let the discrete, or sampled, versions of the continuous signals $s(t)$, $h(t)$ and $x(t)$ be s_k, h_k and x_k, respectively, at time $t = k\Delta t$. The convolution of equation 3.98 becomes

$$x_k = \Delta t \sum_{l=-\infty}^{\infty} s_l h_{k-l}, \tag{3.99}$$

where $\tau = l\Delta t$ to discretise τ in the integral in the same way as t to ensure the summation runs over sampled values of both signals. The sampling interval factor Δt ensures the discrete convolution has the same units as the continuous version. The discrete time functions are defined as in equation 3.79: $x_k = x(k\Delta t)$, $s_l = s(l\Delta t)$ and $h_{k-l} = h((k-l)\Delta t)$.

Let s_k and h_k be causal signals of finite length:

$$\begin{aligned} s_k = \cdots, &s_0, s_1, s_2, \cdots, s_n; \\ &\uparrow \\ &t = 0 \end{aligned} \tag{3.100}$$

$$\begin{aligned} h_k = \cdots, &h_0, h_1, h_2, \cdots, h_m. \\ &\uparrow \\ &t = 0 \end{aligned} \tag{3.101}$$

Since both s_k and h_k are causal, there are no contributions to the summation for negative values of the sampling index, so equation 3.99 may be written as

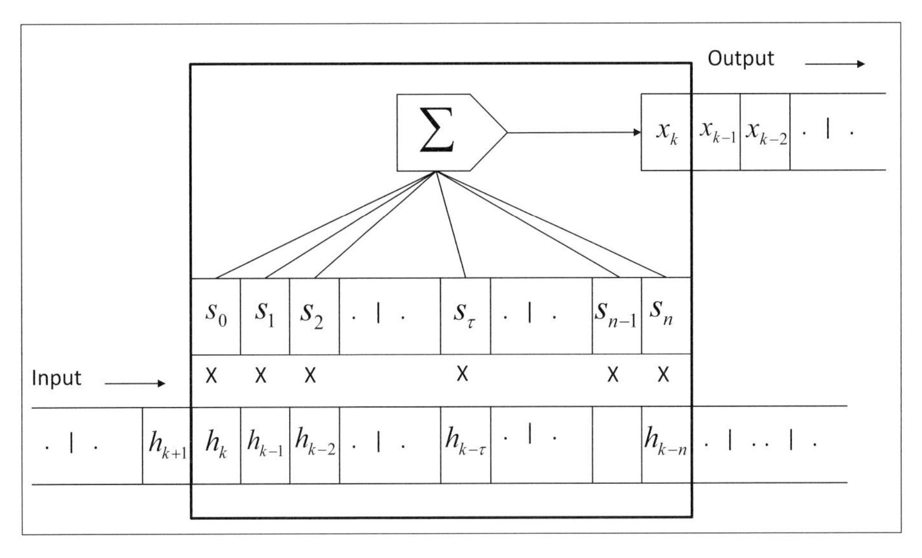

Figure 3.14 Convolution of s_k and h_k: situation at time $k\Delta t$.

$$\begin{aligned} x_k &= \Delta t \sum_{l=\max(0,k-m)}^{\min(k,n)} s_l h_{k-l}, \quad k = 0, 1, 2, \ldots, n+m, \\ &= \Delta t \sum_{l=\max(0,k-n)}^{\min(k,m)} h_l s_{k-l}, \quad k = 0, 1, 2, \ldots, n+m, \\ &= s_k * h_k. \end{aligned} \tag{3.102}$$

The first form of the discrete convolution of equation 3.102 is shown pictorially in Figure 3.14. It can be seen from these two forms of the discrete convolution that the order of convolution does not matter. It can also be seen that the result of convolving the two signals of length $n + 1$ and $m + 1$ is to produce a signal of length $n + m + 1$. Thus, in general (when neither n nor m is equal to zero), the result of the convolution is longer than either of the two convolved signals. This has consequences in the frequency domain, which are discussed in Section 3.18.

3.18 Frequency Domain Deconvolution

Convolution in the time domain transforms to multiplication in the frequency domain. The discrete convolution described in equation 3.102 can be performed by an alternative process involving three steps: (1) Fourier transformation of the two time series s_k and h_k to the frequency domain to yield two Fourier transforms $\hat{s}_n$ and $\hat{h}_n$; (2) multiplication of the Fourier transforms, frequency by frequency, to yield the Fourier transform of the convolution $\hat{x}_n$

$$\hat{x}_n = \hat{s}_n \hat{h}_n, \quad n = 0, 1, 2, \ldots, N; \tag{3.103}$$

and (3) inverse Fourier transformation of $\hat{x}_n$ to time to yield x_k. To perform this operation the series s_k and h_k must be the same length, which must be at least as long as the series resulting from convolution. At minimum, m zeros must be added to s_k and n zeros to h_k. All series then have the same length $N = n + m + 1$. This procedure provides an exact computation of the discrete convolution. The addition of zeros in the time domain results in exact interpolation in the frequency domain.

Deconvolution can be formulated as the inverse process, consisting of three steps: (1) Fourier transformation of the $n+m+1$-length time series x_k and, say, the $n+m+1$-length time series s_k (after the addition of m zeros) to the frequency domain to yield the Fourier transforms $\hat{x}_n$ and $\hat{s}_n$; (2) frequency-by-frequency division of $\hat{x}_n$ by $\hat{s}_n$ to yield the Fourier transform $\hat{h}_n$ and (3) inverse Fourier transformation of $\hat{h}_n$ to time to yield the series h_k (plus n zeros). Step 2 is written as

$$\hat{h}_n = \frac{\hat{x}_n}{\hat{s}_n}, \quad n = 0, 1, 2, \ldots, N = n + m + 1. \tag{3.104}$$

If $|\hat{s}_n|$, the amplitude of $\hat{s}_n$, tends to zero, the amplitude of $\hat{h}_n$ tends to infinity. So this step in the deconvolution can be a problem.

3.19 The Wiener Filter

The Wiener filter has been used extensively in geophysical data processing, following the pioneering work of Robinson (1954) and Robinson and Treitel (1967). The notation of Robinson and Treitel (1967) is adopted, with the modification that discrete time has index k, to be consistent with our preceding notation. Figure 3.15, showing the general filter design model, is redrawn from figure 1 of Robinson and Treitel (1967). There are three signals in the model: the input signal x_k, the desired output signal z_k and the actual output signal y_k. These three signals may all be transient, time-limited signals, also known as wavelets. Or they may be continuous stationary time series. The Wiener approach is to consider energies for transient signals and powers for stationary series. The analysis for these two kinds of signal is similar.

The basic principle in Wiener filtering is the least-squares criterion: one minimises the energy or power (depending on the problem) existing in the difference between the desired filter output z_k and the actual filter output

$$y_k = \sum_{\tau=0}^{n} f_\tau x_{k-\tau}. \tag{3.105}$$

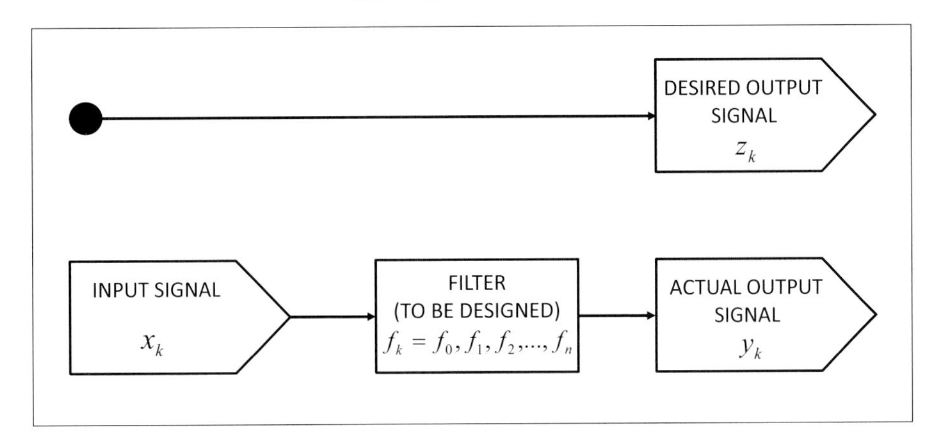

Figure 3.15 The general filter design model (redrawn from figure 1 of Robinson and Treitel, 1967).

The error at sample k is $z_k - y_k$, and the sum of the squares of the errors is

$$I = \sum_k (z_k - y_k)^2 = \sum_k \left(z_k - \sum_{\tau=0}^{n} f_\tau x_{k-\tau} \right)^2 . \tag{3.106}$$

The quantity I is minimised by setting

$$\frac{\partial I}{\partial f_j} = 0, \quad j = 0, 1, 2, \ldots, n. \tag{3.107}$$

The partial derivative of I with respect to f_j is

$$\begin{aligned}
\frac{\partial I}{\partial f_j} &= \sum_k \left[2 \left(z_k - \sum_{\tau=0}^{n} f_\tau x_{k-\tau} \right) \frac{\partial}{\partial f_j} \left(z_k - \sum_{\tau=0}^{n} f_\tau x_{k-\tau} \right) \right], \\
&= 2 \sum_k \left[\left(z_k - \sum_{\tau=0}^{n} f_\tau x_{k-\tau} \right) (-x_{k-j}) \right], \\
&= 2 \sum_k \left[-z_k x_{k-j} + \sum_{\tau=0}^{n} f_\tau x_{k-\tau} x_{k-j} \right], \\
&= 2 \left[-\sum_k z_k x_{k-j} + \sum_{\tau=0}^{n} f_\tau \sum_k x_{k-\tau} x_{k-j} \right], \\
&= 2 \left[-\phi_{zx}(j) + \sum_{\tau=0}^{n} f_\tau \phi_{xx}(j - \tau) \right].
\end{aligned} \tag{3.108}$$

Setting the partial derivatives to zero gives the equations

$$\sum_{\tau=0}^{n} f_\tau \phi_{xx}(j-\tau) = \phi_{zx}(j), \quad j = 0, 1, 2, \dots, n, \tag{3.109}$$

where $\phi_{xx}(\tau)$ is the autocorrelation of the input signal x_k and $\phi_{zx}(\tau)$ is the cross-correlation of the desired output signal z_k with the input signal x_k. These are known as the 'normal equations'.

Levinson (1946) developed a fast solution of these equations exploiting the symmetry of the autocorrelation function $\phi_{xx}(\tau)$ for real x_k. He also showed that a convenient expression for the error energy can be found when the normal equations are divided by the total energy of the desired output, which is the zero-lag coefficient of the autocorrelation of the desired output

$$\phi_{zz}(0) = \sum_k z_k^2. \tag{3.110}$$

Using the property that the autocorrelation of real signals is symmetric, $\phi_{xx}(\tau) = \phi_{xx}(-\tau)$, equations 3.109 may be written as

$$\begin{bmatrix} a_0 & a_1 & \dots & a_n \\ a_1 & a_0 & \dots & a_{n-1} \\ \dots & \dots & \dots & \dots \\ a_n & a_{n-1} & \dots & a_0 \end{bmatrix} \begin{bmatrix} f_0 \\ f_1 \\ \cdot \\ f_n \end{bmatrix} = \begin{bmatrix} b_0 \\ b_1 \\ \cdot \\ b_n \end{bmatrix}, \tag{3.111}$$

with

$$a_\tau = \frac{\phi_{xx}(\tau)}{\phi_{zz}(0)}, \tag{3.112}$$

$$b_\tau = \frac{\phi_{zx}(\tau)}{\phi_{zz}(0)}. \tag{3.113}$$

The error I is minimum, by definition. The normalised mean square error can be expressed as

$$\epsilon = \frac{I}{\phi_{zz}(0)} = 1 - \sum_{\tau=0}^{n} f_\tau b_\tau. \tag{3.114}$$

This is a sum of squares, so it cannot be negative, and it cannot be greater than 1 because its value can always be reduced to 1 by letting all the filter coefficients f_τ be zero. Therefore

$$0 \le \epsilon \le 1. \tag{3.115}$$

The filter quality q may be defined as the complementary quantity,

$$q = 1 - \epsilon = \sum_{\tau=0}^{n} f_\tau b_\tau, \quad \text{with} \quad 0 \le q \le 1. \tag{3.116}$$

The smaller the error, the higher the value of q. If $q = 1$, the error is zero and the filter is perfect.

3.20 Time Domain Deconvolution

The Wiener filter approach can be applied to the deconvolution problem. Let the series x_k be the convolution of s_k and h_k, as given in equation 3.102, and let x_k and s_k be known. The problem is to find h_k. For this problem the input is s_k and the desired output is x_k. The normal equations are then

$$\begin{bmatrix} A_0 & A_1 & \dots & A_n \\ A_1 & A_0 & \dots & A_{n-1} \\ \dots & \dots & \dots & \dots \\ A_n & A_{n-1} & \dots & A_0 \end{bmatrix} \begin{bmatrix} h_{e,0} \\ h_{e,1} \\ \cdot \\ h_{e,n} \end{bmatrix} = \begin{bmatrix} B_0 \\ B_1 \\ \cdot \\ B_n \end{bmatrix}, \tag{3.117}$$

in which the Wiener filter $h_{e,k}$ is the estimate of the filter h_k, and A_k and B_k are the normalised autocorrelation of s_k, and normalised cross-correlation of x_k with s_k, respectively:

$$A_\tau = \frac{\phi_{ss}(\tau)}{\phi_{xx}(0)}, \tag{3.118}$$

$$B_\tau = \frac{\phi_{xs}(\tau)}{\phi_{xx}(0)}. \tag{3.119}$$

3.21 Laplace Transform

In the preceding discussion, time, space, frequency and wavenumber have been taken to be real variables. For the consideration of causal electrical transient phenomena, the subject of this book, it is necessary to let frequency be complex. Consider the integral

$$\int_{-\infty}^{\infty} a(x, y, z, t) e^{-st} dt. \tag{3.120}$$

This is known as the two-sided time-Laplace transform of $a(x, y, z, t)$ and is the same as the temporal Fourier transform for the substitution $s = -i\omega$, where s is complex.

When the real part of s is zero, it becomes identical with the Fourier transform. If $a(x, y, z, t) = 0$ for $t < 0$, only half the integral is required.

The one-sided time-Laplace transform is defined as

$$\int_{0^+}^{\infty} a(x, y, z, t)e^{-st}dt, \tag{3.121}$$

in which the lower limit of the integral is 0^+. Normally it is written as 0, with the assumption that

$$\int_{0}^{\infty} a(x, y, z, t)e^{-st}dt \qquad \text{means} \qquad \lim_{h\downarrow 0}\int_{h}^{\infty} a(x, y, z, t)e^{-st}dt. \tag{3.122}$$

Because s is complex, the inverse transform is a contour integration in the complex plane

$$a(x, y, z, t) = \frac{1}{2\pi i}\int_{c-i\infty}^{c+i\infty} \hat{a}(x, y, z, s)e^{st}ds, \tag{3.123}$$

where c is a positive constant and

$$\hat{a}(x, y, z, s) = \int_{-\infty}^{\infty} a(x, y, z, t)e^{-st}dt. \tag{3.124}$$

For wave propagation problems with no losses, for instance in seismology, it is useful to decompose the time signals into frequency components, or sines and cosines, using Fourier transforms. For electrical propagation in conducting media the fundamental components are decaying exponentials, which are obtained via Laplace transforms. Under the diffusive field approximation the instant $t = 0$ represents a singularity in the Earth response and time-reversal symmetry does not apply. The one-sided Laplace transform is better suited to deal with such functions than the two-sided Fourier transform.

It is very common to combine the time-Laplace transform with spatial Fourier transforms. For example, the time-Laplace and two-dimensional spatial Fourier transform of the wavefield $a(x, y, z, t)$ is

$$\tilde{a}(k_x, k_y, z, s) = \int_{-\infty}^{\infty}\int_{-\infty}^{\infty}\int_{-\infty}^{\infty} a(x, y, z, t)e^{-st-i(k_x x+k_y y)}dxdydt. \tag{3.125}$$

This Laplace–wavenumber domain is used extensively in Chapter 4. The inverse two-dimensional spatial Fourier transform of $\tilde{a}(k_x, k_y, z, s)$ is

$$\hat{a}(x, y, z, s) = \int_{-\infty}^{\infty}\int_{-\infty}^{\infty} \tilde{a}(k_x, k_y, z, s)e^{i(k_x x+k_y y)}dk_x dk_y. \tag{3.126}$$

4

Electromagnetic Fields in a Horizontally Layered VTI Medium

Beginning with Maxwell's equations, this chapter develops equations for the electric and magnetic fields in free space and in conducting media. In free space the fields are waves propagating at the speed of light; in conducting media there are losses and for time variations that are large compared with the 'charge relaxation time' the propagation is diffusive. The Earth's surface forms a boundary between these two regimes. In conducting media the losses attenuate the propagating wave and this effect is characterised by the *skin depth*, which is well known in one-dimensional propagation. It can be used in three-dimensions as *skin range* in an isotropic whole space and in a vertical transverse isotropic (VTI) whole space as skin depth and skin radial distance. In the time domain these notions are called diffusion distances, and they are introduced in this chapter for a VTI whole space. Because the Earth is considered a linear time-invariant system in the time window of the measurements, the generated fields are linearly related to the action of the source. For this reason, the impulsive electric current source is introduced and the Earth's impulse response is derived. The Earth's impulse response is called the Green's function. In controlled-source electromagnetics (CSEM) the source is usually an electric current dipole. The response to a point electric current dipole in a conducting whole space is derived, followed by the response of a layered medium. This is an important model for the analysis of marine CSEM data. The response at the surface of a conducting layered medium is an important model for the analysis of land multi-transient electromagnetic land data. The principal sources for the material in this chapter are Nabighian and Corbett (1994), Slob et al. (2010) and Hunziker et al. (2015).

In Section 4.1 we give solutions for the whole space electric and magnetic fields in terms of the impulse responses, or Green's functions. This is performed in the three-dimensional wavenumber Laplace transformed domain. Because we investigate a conductive VTI space, separation of the fields into two independent modes follows naturally from this solution. We finish this section with the separate mode field equations and Green's functions in the horizontal wavenumber–frequency domain. We then give closed form expressions in the space–frequency and space–time domains for the whole space in Section 4.2, for the half-space in Section 4.3, and explain how to obtain numerical solutions for a layered half-space in the marine CSEM setting in Section 4.4 and in the land CSEM setting in Section 4.5. Detailed derivations are given in Appendices A and B.

4.1 Basic Equations

We begin with Maxwell's equations in the space–time domain in free space:

$$\nabla \times \mathbf{E} + \mu_0 \frac{\partial \mathbf{H}}{\partial t} = 0, \tag{4.1}$$

$$\nabla \times \mathbf{H} - \varepsilon_0 \frac{\partial \mathbf{E}}{\partial t} = 0, \tag{4.2}$$

in which $\mathbf{E}(x, y, z, t) = E_x(x, y, z, t)\mathbf{i} + E_y(x, y, z, t)\mathbf{j} + E_z(x, y, z, t)\mathbf{k}$ is the vector electric field intensity [V m^{-1}] and the vector $\mathbf{H}(x, y, z, t)$ is the magnetic field intensity [A m^{-1}]. In the International System of units (SI system) the values of two electromagnetic parameters in free space are defined: $\mu_0 = 4\pi \times 10^{-7}$ [H m^{-1} or s Ω m^{-1}] is the magnetic permeability, and $c_0 = 299792458$ [m s^{-1}] is the electromagnetic wave propagation velocity. The free space electric permittivity is denoted ε_0 [F m^{-1} or s (Ω m)$^{-1}$] and is defined in terms of the magnetic permeability and the propagation velocity as $\varepsilon_0 = 1/(\mu_0 c_0^2) \approx 8.85 \times 10^{-12}$ [s (Ω m)$^{-1}$].

An important observation from equations 4.1 and 4.2 can be made. By multiplying equation 4.1 with $\mathbf{E}\cdot$ and equation 4.2 with $\mathbf{H}\cdot$, where $\cdot$ represents a scalar product of two vectors, we obtain $\mathbf{E} \cdot \mathbf{H} = 0$ in both equations, because $\mathbf{F} \cdot (\nabla \times \mathbf{F}) = 0$ for any vector function $\mathbf{F}(x, y, z, t)$. Hence, the electric field is perpendicular to the magnetic field. This demonstrates that the electromagnetic field is not a full vector field, because the electromagnetic field vector directions are dependent.

4.1.1 Maxwell's Equations

In the material world the equations are given by

$$\nabla \times \mathbf{E} + \frac{\partial \mathbf{B}}{\partial t} = -\mathbf{J}^m, \qquad \text{(Faraday's law)} \tag{4.3}$$

$$-\nabla \times \mathbf{H} + \frac{\partial \mathbf{D}}{\partial t} + \mathbf{J} = -\mathbf{J}^e, \qquad \text{(Ampère's law)} \tag{4.4}$$

in which the vector $\mathbf{B}(x, y, z, t)$ is the magnetic induction [Wb m^{-2}, T or V s m^{-2}] the vector $\mathbf{D}(x, y, z, t)$ is the electric flux density [C m^{-2}], the vector $\mathbf{J}(x, y, z, t)$ is the volume density of the induced electric current [A m^{-2}], $\mathbf{J}^m(x, y, z, t)$ is the volume density of the external magnetic current, which describes a fictitious magnetic current source [V m^{-2}] and $\mathbf{J}^e(x, y, z, t)$ is the volume density of the external electric current, which describes an electric current source [A m^{-2}]. The equations are rightfully named Maxwell's equations, but it was Oliver Heaviside who condensed the 21 equations of Maxwell into the present four vector equations by eliminating the electric and magnetic field potentials that Maxwell had used (Nahin, 2002). Equation 4.3 is *Faraday's law*, which states that the time rate of change of a magnetic field induces an electric field, whose direction is orthogonal to its own space rate of change and to the magnetic field and whose magnitude is proportional to the time rate of change of the magnetic field. The resulting induced current flows in such a way as to produce a magnetic flux that tends to oppose the change. This is known as *Lenz's law*. Equation 4.4 is *Ampère's law*, but after it was repaired by Maxwell. The original formulation may be written as

$$\nabla \times \mathbf{H} = \mathbf{J}^{tot}, \tag{4.5}$$

which states that a current flowing in a conductor generates a magnetic field which curls around the current, as illustrated in Figure 4.1. The current I is related to the current density $\mathbf{J}$ as $I = A\mathbf{n} \cdot \mathbf{J}$, where A is the cross-sectional area of the conductor and $\mathbf{n}$ is the unit normal vector of the cross-section. Maxwell discovered

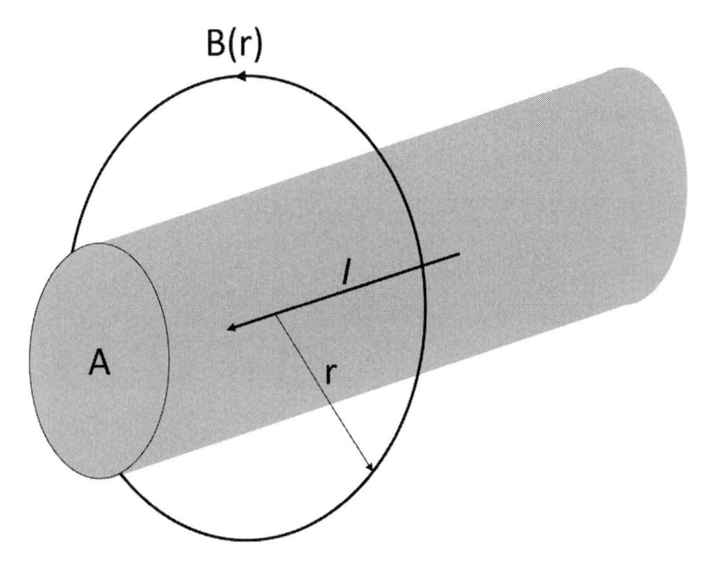

Figure 4.1 The magnetic field lines **B** around a long, straight conductor carrying a current I are concentric circles. Notice the right-hand rule.

that Ampère's law was not correct for non-steady currents. The term that Maxwell added corrected this flaw and also predicted that a changing electric field induces a magnetic field. By doing this, Maxwell was able to demonstrate that light is an electromagnetic wave.

From Maxwell's equations, two auxiliary equations can be found. By taking the divergence of equation 4.3 we find

$$\frac{\partial \nabla \cdot \mathbf{B}}{\partial t} = -\nabla \cdot \mathbf{J}^m = 0. \qquad \text{(Gauss's law for magnetic fields)} \tag{4.6}$$

Equation (4.6) is Gauss's law for magnetic fields. This equation shows that the magnetic induction is divergence-free. The magnetic current source is fictitious because no magnetic charges have been observed, which is equivalent to saying that no magnetic monopoles have been observed. The fourth Maxwell equation is given by

$$\nabla \cdot \mathbf{D} = \varrho_f, \qquad \text{(Gauss's law for electrostatics)} \tag{4.7}$$

with the historic choice of ϱ_f being the volume density of free charge [C m^{-3}]. Equation 4.7 is Gauss's law for electrostatics. The electric field diverges away from a positive charge. Electric field lines originate on positive charges and terminate on negative ones, as shown in Figure 2.2.

4.1.2 Constitutive Relations and the Diffusive Field Approximation

Maxwell's equations are two seemingly uncoupled differential equations of the vector functions **E**, **H**, **B**, **D** and **J**. These equations are coupled through three *constitutive relations*. In a linear, instantaneously reacting, anisotropic and heterogeneous medium these are given by

$$\mathbf{D}(x, y, z, t) = \boldsymbol{\varepsilon}(x, y, z) \cdot \mathbf{E}(x, y, z, t), \tag{4.8}$$

$$\mathbf{B}(x, y, z, t) = \boldsymbol{\mu}(x, y, z) \cdot \mathbf{H}(x, y, z, t), \tag{4.9}$$

$$\mathbf{J}(x, y, z, t) = \boldsymbol{\sigma}(x, y, z) \cdot \mathbf{E}(x, y, z, t). \tag{4.10}$$

Maxwell called **D** the *displacement current*. The quantities ε, μ and σ are the *electric permittivity* the *magnetic permeability*, and the *electrical conductivity*, respectively. As noted in Section 1.3, resistivity, the reciprocal of conductivity, is a tensor. So conductivity is a tensor. In general, all three quantities are tensors and are functions of position and are also pressure- and temperature-dependent. This book is primarily concerned with finding the conductivity as a function of position $\sigma(x, y, z)$.

All media are now assumed to be independent of temperature and pressure. The magnetic permeability μ is assumed to be equal to that of free space; that is, $\mu =$

μ_0. The conductivity is assumed to have a value σ in the horizontal direction and a possibly different value σ_v in the vertical direction, and to depend only on depth z. We write the conductivity tensor as

$$\boldsymbol{\sigma}(x, y, z) = \begin{pmatrix} \sigma(z) & 0 & 0 \\ 0 & \sigma(z) & 0 \\ 0 & 0 & \sigma_v(z) \end{pmatrix}. \tag{4.11}$$

The influence of permittivity is negligible in CSEM applications, so $\varepsilon = 0$ F/m. To show this, we begin by taking the divergence of equation 4.4 and use equations 4.7, 4.8 and 4.10 to find

$$\frac{\partial \varrho_f}{\partial_t} = -\nabla \cdot (\mathbf{J} + \mathbf{J}^e), \tag{4.12}$$

which is known as the equation of continuity of electric current or the charge conservation law. We now replace the source by assuming an initial charge distribution ϱ_0 released at $t = 0$ s. We use $\nabla \cdot \mathbf{J} = \sigma \nabla \cdot \mathbf{D}/\varepsilon = \sigma \varrho_f/\varepsilon$ in equation 4.12 and find an equation for the volume density of free charge,

$$\frac{\partial \varrho_f}{\partial_t} = -\frac{\sigma}{\varepsilon}\varrho_f. \tag{4.13}$$

This is a first-order differential equation for which the solution is

$$\varrho_f(t) = \varrho_0 \exp(-\sigma t/\varepsilon), \tag{4.14}$$

given an initial charge distribution ϱ_0 at $t = 0$ s. From this result it is clear that the initial free charge vanishes exponentially with time in a conductive medium. We observe that the consequence of equation 4.12 is the following theorem: *within a region of non-vanishing conductivity there can be no permanent distribution of free charge*.

The ratio ε/σ, whose inverse is present in the argument of the exponent, is known as the charge relaxation time and is a measure of the time it takes the medium to return to its equilibrium state after it has been disturbed by an electromagnetic wave. Because $\varepsilon \lesssim 10^{-9}$ F/m and $\sigma \gtrsim 10^{-4}$ S/m for most earth materials under investigation, the relaxation time is less than 10 μs. This means that when the alternating source currents have periods much larger than the relaxation time, the influence of the electric permittivity is negligible and we can safely take $\varepsilon = 0$ F/m. For transient sources the first sample is usually taken well after 100 μs and we can still take $\varepsilon = 0$ F/m. In air, conductivity is zero or close to zero and in this case we can assume that the field in air behaves as a static field. This is because taking the permittivity zero means that the wave propagation velocity is infinite and hence all waves arrive everywhere at time zero. The arrival time error made is well below the measurement sampling time, and we may therefore take the electric permittivity of air to be zero. This is used in later sections when the air is used as an upper half-space in the models. It has the advantage that the half-space response can be evaluated analytically and the fields in the conductive half-space can be given explicitly as a combination of ordinary and special functions in both the space–frequency and space–time domains. When we discuss the source and receiver located in a homogeneous half-space, we will come back to this and show that taking $\varepsilon = 0$ F/m in the air removes the TM-mode airwave, but this contribution is so small it can never be measured. From here onwards the electric permittivity is neglected in Maxwell's equations.

4.1.3 Electromagnetic Field Equations in a VTI Whole Space

In this section we derive field equations for the electric field **E** and the magnetic field **H** generated by a point source of electric current in a whole space. The point sources are modelled with an arbitrary time function and are located at $\mathbf{r}_s = (0, 0, z_s)$. The space is anisotropic but homogeneous, and conductivity has two constant values, σ and σ_v. From equations 4.3 and 4.9 we can write the equations in components as

$$\frac{\partial E_z}{\partial y} - \frac{\partial E_y}{\partial z} + \mu_0 \frac{\partial H_x}{\partial t} = -I_x^m \delta(x, y, z - z_s), \tag{4.15}$$

$$\frac{\partial E_x}{\partial z} - \frac{\partial E_z}{\partial x} + \mu_0 \frac{\partial H_y}{\partial t} = -I_y^m \delta(x, y, z - z_s), \tag{4.16}$$

$$\frac{\partial E_y}{\partial x} - \frac{\partial E_x}{\partial y} + \mu_0 \frac{\partial H_z}{\partial t} = -I_z^m \delta(x, y, z - z_s). \tag{4.17}$$

In these equations, $\mathbf{I}^m(t)$ [Vm] is the magnetic dipole moment. It is the time-dependent part of the magnetic source current, such that $\mathbf{J}^m(\mathbf{r}, t) = \mathbf{I}^m(t)\delta(x, y, z - z_s)$. From equations 4.4, 4.8 and 4.10 we have

$$\frac{\partial H_z}{\partial y} - \frac{\partial H_y}{\partial z} - \sigma E_x = I_x^e \delta(x, y, z - z_s), \tag{4.18}$$

$$\frac{\partial H_x}{\partial z} - \frac{\partial H_z}{\partial x} - \sigma E_y = I_y^e \delta(x, y, z - z_s), \tag{4.19}$$

$$\frac{\partial H_y}{\partial x} - \frac{\partial H_x}{\partial y} - \sigma_v E_z = I_z^e \delta(x, y, z - z_s). \tag{4.20}$$

In these equations, $\mathbf{I}^e(t)$ [Am] is the electric moment. It is the time-dependent part of the electric source current, such that $\mathbf{J}^e(\mathbf{r}, t) = \mathbf{I}^e(t)\delta(x, y, z - z_s)$. We observe that the electric and magnetic fields are perpendicular to each other in a VTI medium, for the same reason as in the free-space situation. In a whole space, shift invariance can be exploited in all dimensions. We take $z_s = 0$ and perform a time-Laplace and a three-dimensional spatial Fourier transformation, as defined in Chapter 3. This gives

$$\mathrm{i}k_y\breve{E}_z - \mathrm{i}k_z\breve{E}_y + \zeta\breve{H}_x = -\hat{I}_x^m, \tag{4.21}$$

$$\mathrm{i}k_z\breve{E}_x - \mathrm{i}k_x\breve{E}_z + \zeta\breve{H}_y = -\hat{I}_y^m, \tag{4.22}$$

$$\mathrm{i}k_x\breve{E}_y - \mathrm{i}k_y\breve{E}_x + \zeta\breve{H}_z = -\hat{I}_z^m, \tag{4.23}$$

in which $\zeta = s\mu_0$, and

$$\mathrm{i}k_y\breve{H}_z - \mathrm{i}k_z\breve{H}_y - \sigma\breve{E}_x = \hat{I}_x^e, \tag{4.24}$$

$$\mathrm{i}k_z\breve{H}_x - \mathrm{i}k_x\breve{H}_z - \sigma\breve{E}_y = \hat{I}_y^e, \tag{4.25}$$

$$\mathrm{i}k_x\breve{H}_y - \mathrm{i}k_y\breve{H}_x - \sigma_v\breve{E}_z = \hat{I}_z^e. \tag{4.26}$$

In equations 4.21–4.26, the field quantities depend on the wavenumber vector and frequency, as indicated with the diacritical ˘ symbol, whereas the sources depend only on frequency, indicated with the diacritical ˆ symbol. With the aid of equations 4.21–4.23 we eliminate the magnetic field from equations 4.24–4.26 to obtain the following algebraic matrix equation

$$\begin{pmatrix} k_y^2 + k_z^2 + \gamma^2 & -k_xk_y & -k_xk_z \\ -k_xk_y & k_x^2 + k_z^2 + \gamma^2 & -k_yk_z \\ -k_xk_z & -k_yk_z & k_x^2 + k_y^2 + \gamma_v^2 \end{pmatrix} \begin{pmatrix} \breve{E}_x \\ \breve{E}_y \\ \breve{E}_z \end{pmatrix} = \begin{pmatrix} -\zeta\hat{I}_x^e - \mathrm{i}k_y\hat{I}_z^m + \mathrm{i}k_z\hat{I}_y^m \\ -\zeta\hat{I}_y^e - \mathrm{i}k_z\hat{I}_x^m + \mathrm{i}k_x\hat{I}_z^m \\ -\zeta\hat{I}_z^e - \mathrm{i}k_x\hat{I}_y^m + \mathrm{i}k_y\hat{I}_x^m \end{pmatrix}, \tag{4.27}$$

in which $\gamma = \sqrt{\zeta\sigma}$, with $\Re\{\gamma\} \geq 0$, meaning that γ has a non-negative real part, and $\gamma_v = \sqrt{\zeta\sigma_v}$, with $\Re\{\gamma_v\} \geq 0$, are the wavenumbers with the horizontal and vertical conductivities, respectively. The electric field is found in terms of the sources by inverting the matrix on the left-hand side of equation 4.27. It is given by

$$\begin{pmatrix} \breve{E}_x \\ \breve{E}_y \\ \breve{E}_z \end{pmatrix} = \begin{pmatrix} \frac{k_x^2(k_z^2+\Gamma^2)+\gamma_v^2(k_z^2+k_x^2+\lambda^2k_y^2+\gamma^2)}{\gamma_v^2(k_z^2+\Gamma^2)(k_z^2+\Gamma_v^2)} & \frac{k_xk_y(k_z^2+\kappa^2+\gamma_v^2)}{\gamma_v^2(k_z^2+\Gamma^2)(k_z^2+\Gamma_v^2)} & \frac{k_xk_z}{\gamma_v^2(k_z^2+\Gamma_v^2)} \\ \frac{k_xk_y(k_z^2+\kappa^2+\gamma_v^2)}{\gamma_v^2(k_z^2+\Gamma^2)(k_z^2+\Gamma_v^2)} & \frac{k_y^2(k_z^2+\Gamma^2)+\gamma_v^2(k_z^2+\lambda^2k_x^2+k_y^2+\gamma^2)}{\gamma_v^2(k_z^2+\Gamma^2)(k_z^2+\Gamma_v^2)} & \frac{k_yk_z}{\gamma_v^2(k_z^2+\Gamma_v^2)} \\ \frac{k_xk_z}{\gamma_v^2(k_z^2+\Gamma_v^2)} & \frac{k_yk_z}{\gamma_v^2(k_z^2+\Gamma_v^2)} & \frac{k_z^2+\gamma^2}{\gamma_v^2(k_z^2+\Gamma_v^2)} \end{pmatrix} \begin{pmatrix} -\zeta\hat{I}_x^e - \mathrm{i}k_y\hat{I}_z^m + \mathrm{i}k_z\hat{I}_y^m \\ -\zeta\hat{I}_y^e - \mathrm{i}k_z\hat{I}_x^m + \mathrm{i}k_x\hat{I}_z^m \\ -\zeta\hat{I}_z^e - \mathrm{i}k_x\hat{I}_y^m + \mathrm{i}k_y\hat{I}_x^m \end{pmatrix}, \tag{4.28}$$

in which the coefficient of anisotropy is given by $\lambda = \sqrt{\sigma/\sigma_v}$, the horizontal wavenumber is given by $\kappa = \sqrt{k_x^2 + k_y^2}$ and the vertical wavenumbers are given by $\Gamma = \sqrt{\kappa^2 + \gamma^2}$ with $\Re\{\Gamma\} \geq 0$ and $\Gamma_v = \sqrt{\lambda^2\kappa^2 + \gamma^2}$ with $\Re\{\Gamma_v\} \geq 0$. By looking at the denominators in the elements of the matrix expression on the right-hand side of equation 4.28, we can see that the vertical electric field component depends only on Γ_v and the same applies to all electric field components generated by a vertical electric dipole source. The horizontal electric field generated by horizontal electric current dipole sources contains the two vertical wavenumbers.

Similarly, with the aid of equations 4.24–4.26 we eliminate the electric field from equations 4.21–4.23 to obtain the following algebraic matrix equation:

$$\begin{pmatrix} \lambda^2 k_y^2 + k_z^2 + \gamma^2 & -\lambda^2 k_x k_y & -k_x k_z \\ -\lambda^2 k_x k_y & \lambda^2 k_x^2 + k_z^2 + \gamma^2 & -k_y k_z \\ -k_x k_z & -k_y k_z & k_x^2 + k_y^2 + \gamma^2 \end{pmatrix} \begin{pmatrix} \breve{H}_x \\ \breve{H}_y \\ \breve{H}_z \end{pmatrix} = \begin{pmatrix} -\sigma \hat{I}_x^m + \mathrm{i}k_y\lambda^2\hat{I}_z^e - \mathrm{i}k_z\hat{I}_y^e \\ -\sigma \hat{I}_y^m + \mathrm{i}k_z\hat{I}_x^e - \mathrm{i}k_x\lambda^2\hat{I}_z^e \\ -\sigma \hat{I}_z^m + \mathrm{i}k_x\hat{I}_y^e - \mathrm{i}k_y\hat{I}_x^e \end{pmatrix}. \tag{4.29}$$

This equation can be solved in a similar way by inverting the matrix on the left-hand side of the equation, which results in

$$\begin{pmatrix} \breve{H}_x \\ \breve{H}_y \\ \breve{H}_z \end{pmatrix} = \begin{pmatrix} \frac{k_x^2(k_z^2+\Gamma_v^2)+\gamma^2(k_z^2+\lambda^2k_x^2+k_y^2+\gamma^2)}{\gamma^2(k_z^2+\Gamma^2)(k_z^2+\Gamma_v^2)} & \frac{k_xk_y(k_z^2+\lambda^2(\kappa^2+\gamma^2))}{\gamma^2(k_z^2+\Gamma^2)(k_z^2+\Gamma_v^2)} & \frac{k_xk_z}{\gamma^2(k_z^2+\Gamma^2)} \\ \frac{k_xk_y(k_z^2+\lambda^2(\kappa^2+\gamma^2))}{\gamma^2(k_z^2+\Gamma^2)(k_z^2+\Gamma_v^2)} & \frac{k_y^2(k_z^2+\Gamma_v^2)+\gamma^2(k_z^2+k_x^2+\lambda^2k_y^2+\gamma^2)}{\gamma^2(k_z^2+\Gamma^2)(k_z^2+\Gamma_v^2)} & \frac{k_yk_z}{\gamma^2(k_z^2+\Gamma^2)} \\ \frac{k_xk_z}{\gamma^2(k_z^2+\Gamma^2)} & \frac{k_yk_z}{\gamma^2(k_z^2+\Gamma^2)} & \frac{k_z^2+\gamma^2}{\gamma^2(k_z^2+\Gamma^2)} \end{pmatrix} \begin{pmatrix} -\sigma \hat{I}_x^m + \mathrm{i}k_y\lambda^2\hat{I}_z^e - \mathrm{i}k_z\hat{I}_y^e \\ -\sigma \hat{I}_y^m + \mathrm{i}k_z\hat{I}_x^e - \mathrm{i}k_x\lambda^2\hat{I}_z^e \\ -\sigma \hat{I}_z^m + \mathrm{i}k_x\hat{I}_y^e - \mathrm{i}k_y\hat{I}_x^e \end{pmatrix}. \tag{4.30}$$

By looking at the denominators in the elements of the matrix expression on the right-hand side of equation 4.30, we can see that the vertical magnetic field component depends only on Γ and the same applies to all magnetic field components generated by a vertical magnetic dipole source. The horizontal magnetic field generated by horizontal magnetic current dipole sources contains the two vertical wavenumbers. Comparing the expressions in equations 4.28 and 4.30, we see that the vertical electric field does not depend on the vertical magnetic field. Horizontal electric and magnetic field components can be decomposed such that each is associated with either the vertical electric field or the vertical magnetic field. To do this we need to separate the common fractional term in the elements in the top-left 2×2 submatrices of equations 4.28 and 4.30, given by $[(k_z^2+\Gamma^2)(k_z^2+\Gamma_v^2)]^{-1}$, into two separate terms.

One term is then proportional to $(k_z^2 + \Gamma_v^2)^{-1}$, associated with the vertical electric field, and the other term is proportional to $(k_z^2 + \Gamma^2)^{-1}$, associated with the vertical magnetic field. This separation is achieved by applying partial fraction decomposition. Before we do that we first observe that the only singular points in these expressions are $k_z = \pm i\Gamma$ and $k_z = \pm i\Gamma_v$. The partial fraction decomposition leads to

$$\frac{1}{(k_z^2 + \Gamma^2)(k_z^2 + \Gamma_v^2)} = \frac{\gamma_v^2}{(\gamma_v^2 - \gamma^2)\kappa^2} \left(\breve{G}_V - \breve{G}_H \right), \tag{4.31}$$

with the two scalar Green's functions given by

$$\breve{G}_V = \frac{1}{(k_z^2 + \Gamma_v^2)}, \tag{4.32}$$

$$\breve{G}_H = \frac{1}{(k_z^2 + \Gamma^2)}, \tag{4.33}$$

in which $\breve{G}_V$ is associated with the vertical electric field and we use the subscript V. The function $\breve{G}_H$ is associated with the field for which the vertical electric field is zero and we use the subscript H. This field is associated with the vertical magnetic field. These two independent parts of the field are called modes. The field associated with the vertical electric field is called the *transverse magnetic* mode, or TM mode, because the vertical magnetic field is zero in this mode and *transverse* is related to depth. The other mode has no vertical electric field and is known as the *transverse electric* mode, or TE mode. As can be seen in the right-hand side of equation 4.31, an apparent singularity is introduced at $\kappa = 0$. It has no contribution in the space–frequency or space–time domain to the total fields, but it introduces non-physical events in each mode with opposite sign. We deal with these when we look at the results for the modes separately. Substituting the result of equation 4.31 into equation 4.28 allows the electric field to be expressed in terms of electric field impulse responses generated by electric and magnetic dipole sources, or Green's functions. Each Green's function can be separated into matrices acting on the scalar Green's functions $\breve{G}_H$ and $\breve{G}_V$.

Combining the results for the electric and magnetic sources and, after partial fraction decomposition, we obtain general expressions for the electric and magnetic fields in the form

$$\breve{\mathbf{E}} = (\breve{\mathbf{G}}_H^{ee} + \breve{\mathbf{G}}_V^{ee}) \cdot \hat{\mathbf{I}}^e + (\breve{\mathbf{G}}_H^{em} + \breve{\mathbf{G}}_V^{em}) \cdot \hat{\mathbf{I}}^m, \tag{4.34}$$

$$\breve{\mathbf{H}} = (\breve{\mathbf{G}}_H^{me} + \breve{\mathbf{G}}_V^{me}) \cdot \hat{\mathbf{I}}^e + (\breve{\mathbf{G}}_H^{mm} + \breve{\mathbf{G}}_V^{mm}) \cdot \hat{\mathbf{I}}^m. \tag{4.35}$$

In the Green's functions the first superscript refers to the field and the second superscript refers to the source type; in both cases ^{e} refers to electric and ^{m} to magnetic. The Green's function matrix of mode $_M$ for a receiver of type ^{r} and source of type ^{s} has the general structure given by

$$\breve{\mathbf{G}}_M^{rs} = \begin{pmatrix} \breve{G}^{rs}_{xx;M} & \breve{G}^{rs}_{xy;M} & \breve{G}^{rs}_{xz;M} \\ \breve{G}^{rs}_{yx;M} & \breve{G}^{rs}_{yy;M} & \breve{G}^{rs}_{yz;M} \\ \breve{G}^{rs}_{zx;M} & \breve{G}^{rs}_{zy;M} & \breve{G}^{rs}_{zz;M} \end{pmatrix}, \tag{4.36}$$

where $M = H$ for the TE mode and $M = V$ for the TM mode, r can be e for electric or m for magnetic field, s can be e for electric or m for magnetic source. In each element of the Green's function matrix, the first subscript refers to the vector component of the field and the second subscript refers to the vector direction of the source. As an example, $\breve{G}^{em}_{xy;V}$ represents the TM mode part of the x-component of the electric field generated by a magnetic dipole source pointing in the y-direction. For each field and source, the Green's function matrices are now written as TE mode and TM mode functions as

$$\breve{\mathbf{G}}_H^{ee} = \frac{\zeta}{\kappa^2} \begin{pmatrix} -k_y^2 & k_x k_y & 0 \\ k_x k_y & -k_x^2 & 0 \\ 0 & 0 & 0 \end{pmatrix} \breve{G}_H. \tag{4.37}$$

The TM mode Green's function is obtained as

$$\breve{\mathbf{G}}_V^{ee} = -\left[\begin{pmatrix} k_x^2 & k_x k_y & k_x k_z \\ k_x k_y & k_y^2 & k_y k_z \\ k_x k_z & k_y k_z & \gamma^2 + k_z^2 \end{pmatrix} + \frac{\gamma_v^2}{\kappa^2} \begin{pmatrix} k_x^2 & k_x k_y & 0 \\ k_x k_y & k_y^2 & 0 \\ 0 & 0 & 0 \end{pmatrix} \right] \frac{\breve{G}_V}{\sigma_v}. \tag{4.38}$$

If the whole space were isotropic we would have $\breve{G}_V = \breve{G}_H$ and the TE mode matrix added to the second matrix in the TM mode would result in a unit 2×2 submatrix. We deal with this situation in the next paragraph.

The electric field generated by a magnetic dipole source is given by

$$\breve{\mathbf{G}}_H^{em} = \begin{pmatrix} \frac{k_x k_y \mathrm{i} k_z}{\kappa^2} & \frac{k_y^2 \mathrm{i} k_z}{\kappa^2} & -\mathrm{i} k_y \\ -\frac{k_x^2 \mathrm{i} k_z}{\kappa^2} & -\frac{k_x k_y \mathrm{i} k_z}{\kappa^2} & \mathrm{i} k_x \\ 0 & 0 & 0 \end{pmatrix} \breve{G}_H, \tag{4.39}$$

where the third row is zero because the vertical electric field is purely TM mode. The TM mode Green's function is found as

$$\breve{\mathbf{G}}_V^{em} = \begin{pmatrix} -\frac{k_x k_y \mathrm{i}k_z}{\kappa^2} & \frac{k_x^2 \mathrm{i}k_z}{\kappa^2} & 0 \\ -\frac{k_y^2 \mathrm{i}k_z}{\kappa^2} & \frac{k_x k_y \mathrm{i}k_z}{\kappa^2} & 0 \\ \lambda^2 \mathrm{i}k_y & -\lambda^2 \mathrm{i}k_x & 0 \end{pmatrix} \breve{G}_V, \tag{4.40}$$

where the last column is zero because a vertical magnetic current source does not generate TM mode fields. By reciprocity we have $\breve{\mathbf{G}}_M^{me} = \left(\breve{\mathbf{G}}_M^{em}\right)^t$, where the superscript ^{t} denotes matrix transposition. The magnetic field generated by a magnetic dipole source is found as

$$\breve{\mathbf{G}}_H^{mm} = -\left[\begin{pmatrix} k_x^2 & k_x k_y & k_x k_z \\ k_x k_y & k_y^2 & k_y k_z \\ k_x k_z & k_y k_z & \gamma^2 + k_z^2 \end{pmatrix} + \frac{\gamma^2}{\kappa^2}\begin{pmatrix} k_x^2 & k_x k_y & 0 \\ k_x k_y & k_y^2 & 0 \\ 0 & 0 & 0 \end{pmatrix}\right] \frac{\breve{G}_H}{\zeta}. \tag{4.41}$$

The TM mode Green's function is obtained as

$$\breve{\mathbf{G}}_V^{mm} = \frac{\sigma}{\kappa^2}\begin{pmatrix} -k_y^2 & k_x k_y & 0 \\ k_x k_y & -k_x^2 & 0 \\ 0 & 0 & 0 \end{pmatrix} \breve{G}_V, \tag{4.42}$$

where the last column is zero because a vertical magnetic current source does not generate TM mode fields and the third row is zero because the vertical magnetic field is purely TE mode. With these two sets of electric and magnetic fields for the modes we have determined the whole electromagnetic field.

We have seen that the TM mode is characterised by the vertical component of the electric field and the TE mode is characterised by the vertical component of the magnetic field. Equations 4.26 and 4.23 show that we can associate $\mathrm{i}k_x\breve{H}_y - \mathrm{i}k_y\breve{H}_x$ and $-\mathrm{i}k_x\breve{E}_y + \mathrm{i}k_y\breve{E}_x$ with the TM mode and TE mode, respectively. The horizontal components of the electromagnetic field are uniquely determined by their horizontal curl and divergence. This means we can also associate $\mathrm{i}k_x\breve{E}_x + \mathrm{i}k_y\breve{E}_y$ and $\mathrm{i}k_x\breve{H}_x + \mathrm{i}k_y\breve{H}_y$ with the TM mode and TE mode, respectively. We multiply equation 4.21 by k_y and equation 4.22 by k_x and subtract the results. We then use equation 4.26 to rewrite the vertical component of the electric field in terms of the horizontal components of the magnetic fields. This gives the first TM mode equation. We multiply equation 4.24 by $-k_x$ and equation 4.25 by $-k_y$ and add the results. This leads to the second TM mode equation. We then transform the resulting equations to depth by a one-dimensional inverse spatial Fourier transformation and use an arbitrary source depth location z_s. The TM mode equations are then given by

$$\frac{\partial \tilde{E}_V}{\partial z} + \sigma^{-1}\Gamma_v^2 \tilde{H}_V = (\hat{I}_V^m + \sigma_v^{-1}\kappa^2 \hat{I}_z^e)\delta(z - z_s), \tag{4.43}$$

$$\frac{\partial \tilde{H}_V}{\partial z} + \sigma \tilde{E}_V = \hat{I}_V^e \delta(z - z_s), \tag{4.44}$$

in which the TM mode 'fields' and 'sources' are given by

$$\tilde{H}_V = -\mathrm{i}k_y\tilde{H}_x + \mathrm{i}k_x\tilde{H}_y, \qquad \hat{I}^m_V = \mathrm{i}k_y\hat{I}^m_x - \mathrm{i}k_x\hat{I}^m_y, \tag{4.45}$$

$$\tilde{E}_V = \mathrm{i}k_x\tilde{E}_x + \mathrm{i}k_y\tilde{E}_y, \qquad \hat{I}^e_V = -\mathrm{i}k_x\hat{I}^e_x - \mathrm{i}k_y\hat{I}^e_y. \tag{4.46}$$

We combine the same equations and apply the same procedure to obtain TE mode equations, which are given by

$$\frac{\partial\tilde{E}_H}{\partial z} + \zeta\tilde{H}_H = \hat{I}^m_H\delta(z - z_s), \tag{4.47}$$

$$\frac{\partial\tilde{H}_H}{\partial z} + \zeta^{-1}\Gamma^2\tilde{E}_H = (\hat{I}^e_H + \zeta^{-1}\kappa^2\hat{I}^m_z)\delta(z - z_s), \tag{4.48}$$

in which the TE mode 'fields' and 'sources' are given by

$$\tilde{E}_H = \mathrm{i}k_y\tilde{E}_x - \mathrm{i}k_x\tilde{E}_y, \qquad \hat{I}^e_H = -\mathrm{i}k_y\hat{I}^e_x + \mathrm{i}k_x\hat{I}^e_y, \tag{4.49}$$

$$\tilde{H}_H = \mathrm{i}k_x\tilde{H}_x + \mathrm{i}k_y\tilde{H}_y, \qquad \hat{I}^m_H = -\mathrm{i}k_x\hat{I}^m_x - \mathrm{i}k_y\hat{I}^m_y. \tag{4.50}$$

Equations 4.47 and 4.48 do not depend on σ_v, but are otherwise very similar to equations 4.44 and 4.43, and each set forms an independent set of two equations. This implies that when we solve the TM mode equations the solution for the TE mode can be obtained by substitution. The necessary substitutions are given in Table 4.1. Note that the reverse substitution cannot be performed because we did not start with a magnetic VTI medium, otherwise this would also have been possible. From this we conclude that we need only solve a scalar problem. The next observation is that $\tilde{E}_V$ can be found from $\tilde{H}_V$ using equation 4.44. The final conclusion is that we need to solve for only one scalar field and then the whole electromagnetic field is known and can be written down by substitutions and taking derivatives.

From here onwards we will write equations and find solutions for $\tilde{E}_V$ and $\tilde{H}_V$ for electric and magnetic sources. Then we give the final expressions for the electric field generated by an electric current source. Full derivations and expressions for the electromagnetic field in a VTI whole space are given in Appendix A. Equations 4.43 and 4.44 are supplemented by boundary conditions when the conductivity values

Table 4.1 *Substitutions from TM mode to TE mode.*

Mode	Field	Source	Medium
TM	$\tilde{E}_V$ $\tilde{H}_V$	$\hat{I}^e_V$ $\hat{I}^e_z$ $\hat{I}^m_V$	σ σ_v ζ
TE	$\tilde{H}_H$ $\tilde{E}_H$	$\hat{I}^m_H$ $\hat{I}^m_z$ $\hat{I}^e_H$	ζ ζ σ

change abruptly across a depth level z_n, in which case $\tilde{H}_V$ and $\tilde{E}_V$ are continuous across such a source-free jump discontinuity.

Finally, the Cartesian components of the electric and magnetic fields are recovered as follows:

$$\tilde{E}_x = -\frac{ik_y}{\kappa^2}\tilde{E}_H - \frac{ik_x}{\kappa^2}\tilde{E}_V, \qquad \tilde{H}_x = -\frac{ik_x}{\kappa^2}\tilde{H}_H + \frac{ik_y}{\kappa^2}\tilde{H}_V, \tag{4.51}$$

$$\tilde{E}_y = \frac{ik_x}{\kappa^2}\tilde{E}_H - \frac{ik_y}{\kappa^2}\tilde{E}_V, \qquad \tilde{H}_y = -\frac{ik_y}{\kappa^2}\tilde{H}_H - \frac{ik_x}{\kappa^2}\tilde{H}_V, \tag{4.52}$$

$$\tilde{E}_z = -\sigma_v^{-1}\left(\hat{I}_z^e\delta(z-z_s) - \tilde{H}_V\right), \quad \tilde{H}_z = -\zeta^{-1}\left(\hat{I}_z^m\delta(z-z_s) - \tilde{E}_H\right). \tag{4.53}$$

In the next section the solution for a point source in a VTI homogeneous whole space is given in the wavenumber–frequency, space–frequency and space–time domains. This is followed by repeating the analysis for two homogeneous half-spaces: the marine CSEM setting and the land CSEM setting.

4.2 The Electromagnetic Field for a Source in a VTI Whole Space

The electromagnetic field in a whole space is given in terms of the mode Green's functions by equations 4.37–4.42. In the remainder of this chapter we focus on the electric field generated by an electric current dipole. We begin by giving the TM mode and TE mode scalar Green's functions as the result of performing the inverse spatial Fourier transformation of equations 4.32 and 4.33 to depth,

$$\tilde{G}_V = \frac{\exp(-\Gamma_v|z-z_s|)}{2\Gamma_v}, \tag{4.54}$$

$$\tilde{G}_H = \frac{\exp(-\Gamma|z-z_s|)}{2\Gamma}, \tag{4.55}$$

in which the TM–mode Green's function satisfies

$$\left(\frac{\partial^2}{\partial z^2} - \Gamma_v^2\right)\tilde{G}_V = -\delta(z-z_s). \tag{4.56}$$

This is easily verified by evaluating the vertical derivatives

$$\frac{\partial \tilde{G}_V}{\partial z} = -\text{sign}(z-z_s)\Gamma_v\tilde{G}_V, \tag{4.57}$$

$$\frac{\partial^2 \tilde{G}_V}{\partial z^2} = \Gamma_v^2\tilde{G}_V - \delta(z-z_s), \tag{4.58}$$

in which $\text{sign}(z-z_s)$ is the sign-function given by

$$\text{sign}(z-z_s) = \frac{z-z_s}{|z-z_s|} = H(z-z_s) - H(z_s-z), \tag{4.59}$$

and $[\mathrm{sign}(z - z_s)]^2 = 1$. The derivative of the sign-function is

$$\frac{\partial \mathrm{sign}(z - z_s)}{\partial z} = 2\delta(z - z_s). \tag{4.60}$$

The Green's matrix functions that describe the modes in the electric field of the Earth's response to the action of an impulsive electric dipole source are obtained from equations 4.37 and 4.38 and given by

$$\tilde{\mathbf{G}}_H^{ee} = \zeta \begin{pmatrix} -\frac{k_y^2}{\kappa^2} & \frac{k_x k_y}{\kappa^2} & 0 \\ \frac{k_x k_y}{\kappa^2} & -\frac{k_x^2}{\kappa^2} & 0 \\ 0 & 0 & 0 \end{pmatrix} \tilde{G}_H \tag{4.61}$$

and

$$\tilde{\mathbf{G}}_V^{ee} = -\left[\frac{1}{\sigma_v} \begin{pmatrix} k_x^2 & k_x k_y & \frac{-\mathrm{i}k_x \partial}{\partial z} \\ k_x k_y & k_y^2 & \frac{-\mathrm{i}k_y \partial}{\partial z} \\ \frac{-\mathrm{i}k_x \partial}{\partial z} & \frac{-\mathrm{i}k_y \partial}{\partial z} & \gamma^2 - \frac{\partial^2}{\partial z^2} \end{pmatrix} + \zeta \begin{pmatrix} \frac{k_x^2}{\kappa^2} & \frac{k_x k_y}{\kappa^2} & 0 \\ \frac{k_x k_y}{\kappa^2} & \frac{k_y^2}{\kappa^2} & 0 \\ 0 & 0 & 0 \end{pmatrix} \right] \tilde{G}_V. \tag{4.62}$$

The first matrix expression of the TM mode Green's function of equation 4.62 contains regular functions of κ and can be transformed back to the space domain without difficulty. The horizontal components of the mode Green's functions generated by horizontal electric current sources present angular singularities as a function of k_x and k_y, as can be seen in the matrix of equation 4.61 and in the second matrix in equation 4.62. This singularity arises because of the decomposition, which is a mathematical construct. If the whole space were isotropic we would have $\tilde{G}_V = \tilde{G}_H$ and the TE mode matrix added to the second matrix in the TM mode would result in a unit 2×2 submatrix. We deal with this situation in the next paragraph.

The notion of skin depth comes from a normal-incidence plane wave in the frequency domain. For $\kappa = 0$ the TE and TM mode scalar Green's functions are equal and given by

$$\tilde{G}_{V,H}(0, h^-, -\mathrm{i}\omega) = \frac{\exp(-\sqrt{-\mathrm{i}\omega\sigma\mu_0}|h^-|)}{2\sqrt{-\mathrm{i}\omega\sigma\mu_0}}, \tag{4.63}$$

where $h^- = z - z_s$ denotes the vertical distance to the source depth location. The square root should be taken such that the exponent has a negative real part, hence

$$\tilde{G}_{V,H}(0, h^-, -\mathrm{i}\omega) = \frac{\exp[-(1-\mathrm{i})\sqrt{\omega\sigma\mu_0/2}|h^-|]}{(1-\mathrm{i})\sqrt{2\omega\sigma\mu_0}}. \tag{4.64}$$

This is written as

$$\tilde{G}_{V,H}(0, h^-, -\mathrm{i}\omega) = \frac{1}{2}(1+\mathrm{i})\delta \exp(-(1-\mathrm{i})|h^-|/\delta), \tag{4.65}$$

where $\delta = \sqrt{2/(\omega\sigma\mu_0)}$ is known as the skin depth. This value can be computed as $\delta \approx 503.3/\sqrt{f\sigma}$ and depends only on the value of the horizontal conductivity. For a single frequency we can make an analogous argument related to propagating waves and introduce the diffusive field velocity, v, such that we write

$$\tilde{G}_{V,H}(0, h^-, -\mathrm{i}\omega) = \frac{1}{2}(1+\mathrm{i})\delta \exp(-|h^-|/\delta + \mathrm{i}\omega|h^-|/v), \tag{4.66}$$

and the velocity can be found as $v = \delta\omega = 2\pi\delta f \approx 3162.3\sqrt{f/\sigma}$. The wavelength is therefore given by $2\pi\delta$. From this result it can be seen that the attenuation has the same strength as the phase. When the distance from the source to the receiver for frequency f is equal to the wavelength, the amplitude is reduced by a factor of $\exp(-2\pi) \approx 1/535.5$ (or 55 dB down). For this reason, CSEM surveys are designed such that distances of interest travelled from source to receiver are not much larger than one wavelength.

4.2.1 Space–Frequency Domain Solutions

The detailed derivation of the results in this section can be found in Appendix A. The scalar Green's functions of equations 4.54 and 4.55 are transformed back to the space–frequency domain by a two-dimensional inverse spatial Fourier transformation. The results are given by

$$\hat{G}_V(\mathbf{r}, \mathbf{r}_s, s) = \frac{\exp(-\gamma_v R_v)}{4\pi\lambda R_v}, \tag{4.67}$$

$$\hat{G}_H(\mathbf{r}, \mathbf{r}_s, s) = \frac{\exp(-\gamma R)}{4\pi R}, \tag{4.68}$$

where $R_v = \sqrt{x^2 + y^2 + (h_v^-)^2}$ is the TM mode distance from source to receiver, with $h_v^- = \lambda h^-$ being the scaled height difference between source and receiver, and $R = \sqrt{x^2 + y^2 + (h^-)^2}$ is the TE mode source–receiver distance.

The TE mode Green's function is given by

$$\hat{\mathbf{G}}_H^{ee}(0, 0, h^-, s) = -\begin{pmatrix} 1 & 0 & 0 \\ 0 & 1 & 0 \\ 0 & 0 & 0 \end{pmatrix} \zeta \frac{\exp(-\gamma|h^-|)}{8\pi|h^-|}, \tag{4.69}$$

when $r = 0$ while $h^- \neq 0$. When $r > 0$ it is given by

$$\hat{\mathbf{G}}_H^{ee}(\mathbf{r}, \mathbf{r}_s, s) = \begin{pmatrix} -\frac{y^2}{r^2} & \frac{xy}{r^2} & 0 \\ \frac{xy}{r^2} & -\frac{x^2}{r^2} & 0 \\ 0 & 0 & 0 \end{pmatrix} \zeta \frac{\exp(-\gamma R)}{4\pi R}$$

$$+\zeta\begin{pmatrix}\frac{x^2-y^2}{r^2} & 2\frac{xy}{r^2} & 0\\ 2\frac{xy}{r^2} & \frac{y^2-x^2}{r^2} & 0\\ 0 & 0 & 0\end{pmatrix}\frac{\exp(-\gamma R)}{4\pi\gamma r^2}. \tag{4.70}$$

We can write the space domain expression for TM–mode electric field Green's function for an electric current source as

$$\hat{\mathbf{G}}_V^{ee}(0,0,z,z_s,s) = -\left[\begin{pmatrix}1 & 0 & 0\\ 0 & 1 & 0\\ 0 & 0 & -2\lambda^2\end{pmatrix}\left(\frac{1}{|h^-|^2}+\frac{\gamma}{|h^-|}\right)\right.$$
$$\left.+\gamma^2\begin{pmatrix}\frac{1}{2} & 0 & 0\\ 0 & \frac{1}{2} & 0\\ 0 & 0 & 0\end{pmatrix}\right]\frac{\exp(-\gamma|h^-|)}{4\pi\sigma\lambda^2|h^-|}, \tag{4.71}$$

when $r = 0$ while $h^- \neq 0$. When $r > 0$, it is given by

$$\hat{\mathbf{G}}_V^{ee}(\mathbf{r},\mathbf{r}_s,s) = \left[\begin{pmatrix}\frac{3x^2}{R_v^2}-1 & \frac{3xy}{R_v^2} & \frac{3\lambda x h_v^-}{R_v^2}\\ \frac{3xy}{R_v^2} & \frac{3y^2}{R_v^2}-1 & \frac{3\lambda y h_v^-}{R_v^2}\\ \frac{3\lambda x h_v^-}{R_v^2} & \frac{3\lambda y h_v^-}{R_v^2} & \frac{3(\lambda h_v^-)^2}{R_v^2}-\lambda^2\end{pmatrix}\left(\frac{1}{\sigma_v R_v^2}+\frac{\gamma_v}{\sigma_v R_v}\right)\right.$$
$$\left.+\zeta\begin{pmatrix}\frac{x^2}{R_v^2}-\frac{x^2}{r^2} & \frac{xy}{R_v^2}-\frac{xy}{r^2} & \frac{\lambda x h_v^-}{R_v^2}\\ \frac{xy}{R_v^2}-\frac{xy}{r^2} & \frac{y^2}{R_v^2}-\frac{y^2}{r^2} & \frac{\lambda y h_v^-}{R_v^2}\\ \frac{\lambda x h_v^-}{R_v^2} & \frac{\lambda y h_v^-}{R_v^2} & \frac{(\lambda h_v^-)^2}{R_v^2}-\lambda^2\end{pmatrix}\right]\frac{\exp(-\gamma_v R_v)}{4\pi\lambda R_v}$$
$$-\zeta\begin{pmatrix}\frac{x^2-y^2}{r^2} & 2\frac{xy}{r^2} & 0\\ 2\frac{xy}{r^2} & \frac{y^2-x^2}{r^2} & 0\\ 0 & 0 & 0\end{pmatrix}\frac{\exp(-\gamma_v R_v)}{4\pi\gamma r^2}. \tag{4.72}$$

Note that the last term in the expression of the TM-mode Green's function cancels the second term in the expression of the TE-mode Green's function when the two modes are added to obtain the total electric field in an isotropic whole space.

The total electric field impulse response to an electric dipole is obtained by summing equations 4.70 and 4.72, which yields

$$\hat{\mathbf{G}}^{ee}(\mathbf{r},\mathbf{r}_s,s) = \left[\begin{pmatrix}\frac{3x^2}{R_v^2}-1 & \frac{3xy}{R_v^2} & \frac{3\lambda x h_v^-}{R_v^2}\\ \frac{3xy}{R_v^2} & \frac{3y^2}{R_v^2}-1 & \frac{3\lambda y h_v^-}{R_v^2}\\ \frac{3\lambda x h_v^-}{R_v^2} & \frac{3\lambda y h_v^-}{R_v^2} & \frac{3(\lambda h_v^-)^2}{R_v^2}-\lambda^2\end{pmatrix}\left(\frac{1}{\sigma_v R_v^2}+\frac{\gamma_v}{\sigma_v R_v}\right)\right.$$

$$+\zeta\begin{pmatrix}\frac{x^2}{R_v^2} & \frac{xy}{R_v^2} & \frac{\lambda x h_v^-}{R_v^2}\\ \frac{xy}{R_v^2} & \frac{y^2}{R_v^2} & \frac{\lambda y h_v^-}{R_v^2}\\ \frac{\lambda x h_v^-}{R_v^2} & \frac{\lambda y h_v^-}{R_v^2} & \frac{(\lambda h_v^-)^2}{R_v^2}-\lambda^2\end{pmatrix}\Bigg]\frac{\exp(-\gamma_v R_v)}{4\pi\lambda R_v}$$

$$-\zeta\left[\begin{pmatrix}\frac{x^2}{r^2} & \frac{xy}{r^2} & 0\\ \frac{xy}{r^2} & \frac{y^2}{r^2} & 0\\ 0 & 0 & 0\end{pmatrix}\frac{\exp(-\gamma_v R_v)}{4\pi\lambda R_v}+\begin{pmatrix}\frac{y^2}{r^2} & -\frac{xy}{r^2} & 0\\ -\frac{xy}{r^2} & \frac{x^2}{r^2} & 0\\ 0 & 0 & 0\end{pmatrix}\frac{\exp(-\gamma R)}{4\pi R}\right]$$

$$-\zeta\begin{pmatrix}\frac{x^2-y^2}{r^2} & \frac{2xy}{r^2} & 0\\ \frac{2xy}{r^2} & \frac{y^2-x^2}{r^2} & 0\\ 0 & 0 & 0\end{pmatrix}\frac{\exp(-\gamma_v R_v)-\exp(-\gamma R)}{4\pi\gamma r^2}. \tag{4.73}$$

The electric field impulse response to an electric dipole contains directional patterns contained in the matrices, which are all frequency-independent. Equation (4.73) has three grouped terms on the right-hand side. The terms in the first square brackets are all TM mode terms. The two scalar terms in the first expression on the right-hand side of equation 4.73 have the same radiation pattern. The scalar parts form the near-field term, proportional to inverse distance cubed, and the intermediate-field term, proportional to inverse distance squared. The second directional pattern corresponds to the far-field term and is proportional to inverse distance. The expression in the second square brackets is a far-field sum of the TM and TE modes. These show the effect of anisotropy and the splitting of the modes. This term shows that the x-component of the electric field generated by an x-directed electric dipole is a TM mode field for offsets in the line of the source–receiver direction, also known as the endfire acquisition setup, whereas for offsets perpendicular to the source–receiver direction it is a TE mode field, also known as the broadside acquisition setup. Notice that the latter is only true in the far field. The last expression on the right-hand side of equation 4.73 is also an effect of the VTI medium and would vanish for an isotropic whole space. Their directional patterns depend only on horizontal source–receiver distances, but their geometric spreading terms differ. Equation (4.73) is also valid when $r = 0$ while $h^- \neq 0$. The horizontal components of the total electric field impulse response to a horizontal electric dipole for $r = 0$ while $h^- \neq 0$ can be obtained by evaluating the first three terms of the Taylor series expansion of the exponentials around $r = 0$ for the terms divided by r^2, which is given by

$$-\lim_{r\to 0}\frac{\exp(-\gamma_v R_v)-\exp(-\gamma R)}{4\pi\gamma r^2}=\left(\frac{1}{\lambda^2}-1\right)\frac{\exp(-\gamma|h^-|)}{8\pi|h^-|}. \tag{4.74}$$

Keeping only the first term in the Taylor series expansion around $r = 0$ of the exponentials in the fourth term, adding all results and taking the limit of $r \to 0$, results in

$$\hat{G}^{ee}_{xx}(0,0,z,z_s,s) = -\left(\frac{2}{(h^-)^2} + \frac{2\gamma}{|h^-|} + \gamma^2(1+\lambda^2)\right)\frac{\exp(-\gamma|h^-|)}{8\pi\sigma\lambda^2|h^-|}, \tag{4.75}$$

$$\hat{G}^{ee}_{yy}(0,0,z,z_s,s) = \hat{G}^{ee}_{xx}(0,0,z,z_s,s), \tag{4.76}$$

$$\hat{G}^{ee}_{xy}(0,0,z,z_s,s) = \hat{G}^{ee}_{yx}(0,0,z,z_s,s) = 0. \tag{4.77}$$

With these expressions, the total electric field impulse response to an electric dipole in a VTI whole space has been determined and can be coded without numerical problems around $r = 0$ while $h^- \neq 0$.

In the isotropic limit the expressions for the electromagnetic impulse response to an electric dipole reduce to

$$\hat{\mathbf{G}}^{ee}(\mathbf{r},\mathbf{r}_s,s) = \begin{pmatrix} \frac{3x^2}{R^2}-1 & \frac{3xy}{R^2} & \frac{3xh^-}{R^2} \\ \frac{3xy}{R^2} & \frac{3y^2}{R^2}-1 & \frac{3yh^-}{R^2} \\ \frac{3xh^-}{R^2} & \frac{3yh^-}{R_v^2} & \frac{3(h^-)^2}{R^2}-1 \end{pmatrix}\left(\frac{1}{R^2}+\frac{\gamma}{R}\right)\frac{\exp(-\gamma R)}{4\pi\sigma R}$$
$$+\zeta\begin{pmatrix} \frac{x^2}{R^2}-1 & \frac{xy}{R^2} & \frac{xh^-}{R^2} \\ \frac{xy}{R^2} & \frac{y^2}{R^2}-1 & \frac{yh^-}{R^2} \\ \frac{xh^-}{R^2} & \frac{yh^-}{R^2} & \frac{(h^-)^2}{R^2}-1 \end{pmatrix}\frac{\exp(-\gamma R)}{4\pi R}. \tag{4.78}$$

This is the well-known expression of the electric field Green's functions for an electric dipole source.

For the notion of skin distance we need to look at the two scalar Green's functions, which are written in the frequency domain with $s = -\mathrm{i}\omega$ as

$$\hat{G}_H(\mathbf{r},\mathbf{r}_s,\omega) = \frac{\exp[-(1-\mathrm{i})R/\delta]}{4\pi R}, \tag{4.79}$$

$$\hat{G}_V(\mathbf{r},\mathbf{r}_s,\omega) = \frac{\exp[-(1-\mathrm{i})R_v/\delta_v]}{4\pi\lambda R_v}, \tag{4.80}$$

where we have used the same skin depth δ as in the plane wave domain and have introduced a new number $\delta_v = \delta\lambda$ to account for different behaviour in the horizontal and vertical directions that occurs in the TM mode. For the TE mode we can see that the skin depth applies in three-dimensional space in all directions and we should regard it as skin range. For the TM mode the situation is slightly more complicated. We introduce the dip angle θ relative to the (x, y)-plane and write $r = R\cos(\theta)$ and $h^- = R\sin(\theta)$. We then find

$$\hat{G}_V(\mathbf{r},\mathbf{r}_s,\omega) = \frac{\exp\left[-(1-\mathrm{i})(R/\delta)\sqrt{[\cos(\theta)]^2/\lambda^2+[\sin(\theta)]^2}\right]}{4\pi\lambda R_v}, \tag{4.81}$$

from which we can see that the skin range in the horizontal direction is a factor λ larger than the skin depth. If the vertical transverse isotropy is caused by small-scale horizontal layering, the horizontal conductivity will most likely be larger than the vertical conductivity. In this case, λ is larger than unity and the attenuation is less for propagation in the horizontal direction than in the vertical direction. The same argument applies to the phase of the field which increases less (and hence the field propagates faster) in the horizontal direction than in the vertical direction.

4.2.2 Space–Time Domain Solutions

The electromagnetic impulse response in space–time can be obtained by first writing the Green's functions for the electric field generated by an electric dipole in explicit frequency-dependent parts. These are obtained from equation 4.73. To separate frequency-independent from frequency-dependent terms, the following functions are introduced

$$\hat{F}^{(m)}(\tau, s) = s^{m/2} \exp(-2\sqrt{s\tau}) \tag{4.82}$$

to express the frequency-dependent parts of the electromagnetic fields. To use these functions for the Green's functions, we see that we need two diffusion times. The TE mode is characterised by $\tau = \sigma\mu_0 R^2/4$ and the TM mode is characterised by $\tau_v = \sigma_v \mu_0 R_v^2/4$. The functions $\hat{F}^{(m)}(\tau, s)$ have known time domain equivalents and they are given by the recurrence relation (see Abramowitz and Stegun (1972: 1026), combining formulae 29.3.86 and 29.3.87):

$$F^{(m)}(\tau, t) = \frac{\sqrt{\tau}}{t} F^{(m-1)}(\tau, t) - \frac{m}{2t} F^{(m-2)}(\tau, t). \tag{4.83}$$

From equation 4.83 it can be observed that all time functions are known if $F^{(-2)}(\tau, t)$ and $F^{(-1)}(\tau, t)$ are known. These are given by (see Abramowitz and Stegun (1972: 1026), formulae 29.3.83 for $m = -2$ and 29.3.84 for $m = -1$):

$$F^{(-2)}(\tau, t) = \mathrm{erfc}(\sqrt{\tau/t}) H(t), \tag{4.84}$$

$$F^{(-1)}(\tau, t) = \frac{\exp(-\tau/t)}{\sqrt{\pi t}} H(t), \tag{4.85}$$

in which erfc denotes the complementary error function defined by

$$\mathrm{erfc}(\sqrt{\tau/t}) = 1 - \mathrm{erf}(\sqrt{\tau/t}) = \frac{2}{\sqrt{\pi}} \int_{u=\sqrt{\tau/t}}^{\infty} \exp(-u^2) du, \tag{4.86}$$

and where erf is the error function given by

$$\mathrm{erf}(\sqrt{\tau/t}) = \frac{2}{\sqrt{\pi}} \int_{u=0}^{\sqrt{\tau/t}} \exp(-u^2) du. \tag{4.87}$$

The functions for $m = 0, 1, 2$ are found from these results and equation 4.83 as

$$F^{(0)}(\tau,t) = \sqrt{\frac{\tau}{\pi t^3}}\exp(-\tau/t)H(t), \tag{4.88}$$

$$F^{(1)}(\tau,t) = \left(\frac{\tau}{t} - \frac{1}{2}\right)\sqrt{\frac{1}{\pi t^3}}\exp(-\tau/t)H(t), \tag{4.89}$$

$$F^{(2)}(\tau,t) = \left(\frac{\tau}{t} - \frac{3}{2}\right)\sqrt{\frac{\tau}{\pi t^5}}\exp(-\tau/t)H(t). \tag{4.90}$$

To simplify the electromagnetic field expressions, we can use the relation

$$F^{(0)}(\tau,t) + 2\sqrt{\tau}F^{(1)}(\tau,t) = 2\sqrt{\frac{\tau^3}{\pi t^5}}\exp(-\tau/t)H(t). \tag{4.91}$$

This combination has the same late-time behaviour as $F^{(2)}(\tau,t)$.

For the electric field Green's function we take a closer look at the first TM mode part of the xx-component, which corresponds to the top left matrix elements in the first square bracketed term of equation 4.73. In this expression the factor $1/(\sigma_v R_v^2) + \gamma_v/(\sigma_v R_v)$ can be written as

$$\left(\frac{1}{\sigma_v R_v^2} + \frac{\gamma_v}{\sigma_v R_v}\right) = \frac{1 + 2\sqrt{s\tau_v}}{\sigma_v R_v^2}. \tag{4.92}$$

In the frequency domain we can now write the xx-component of the Green's function with the aid of equation 4.82 as

$$\begin{aligned}\hat{G}^{ee}_{xx}(\mathbf{r},\mathbf{r}_s,s) = {} & \left(\frac{3x^2}{R_v^2} - 1\right)\frac{\hat{F}^{(0)}(\tau_v,s) + 2\sqrt{\tau_v}\hat{F}^{(1)}(\tau_v,s)}{4\pi\lambda\sigma_v R_v^3} \\ & + \left(\frac{x^2}{R_v^2} - \frac{x^2}{r^2}\right)\frac{\mu_0\hat{F}^{(2)}(\tau_v,s)}{4\pi\lambda R_v} - \frac{y^2}{r^2}\frac{\mu_0\hat{F}^{(2)}(\tau,s)}{4\pi R} \\ & - \frac{x^2 - y^2}{r^2}\sqrt{\frac{\mu_0}{\sigma}}\frac{\hat{F}^{(1)}(\tau_v,s) - \hat{F}^{(1)}(\tau,s)}{4\pi r^2}.\end{aligned} \tag{4.93}$$

In the time domain this result can be simplified using equations 4.90 and 4.91 as

$$\begin{aligned}G^{ee}_{xx}(\mathbf{r},\mathbf{r}_s,t) = {} & \left(\frac{\sigma_v\mu_0 x^2}{2t} - 1\right)\left(\frac{\sigma_v\mu_0^3}{\pi^3 t^5}\right)^{1/2}\frac{\exp(-\tau_v/t)}{8\lambda} \\ & - \mu_0\left(\frac{x^2}{r^2}\frac{F^{(2)}(\tau_v,t)}{4\pi\lambda R_v} + \frac{y^2}{r^2}\frac{F^{(2)}(\tau,t)}{4\pi R}\right) \\ & - \frac{x^2 - y^2}{r^2}\sqrt{\frac{\mu_0}{\sigma}}\frac{\hat{F}^{(1)}(\tau_v,t) - \hat{F}^{(1)}(\tau,t)}{4\pi r^2}.\end{aligned} \tag{4.94}$$

Finally, in an isotropic whole space the expression simplifies further, because the two terms in the second line of equation 4.94 can be combined and merged with the expression in the first line and the expression in the third line vanishes. The result is given by

$$G^{ee}_{xx}(\mathbf{r},\mathbf{r}_s,t) = \frac{\mu_0 \exp(-\tau/t)}{4\pi t^{5/2}} \left(\frac{\sigma\mu_0}{\pi}\right)^{1/2} \left[1 - \frac{\sigma\mu_0(y^2+z^2)}{4t}\right]. \tag{4.95}$$

All other components can be obtained in a similar way. The electric field Green's function $[\Omega/(\mathrm{m}^2\mathrm{s})]$ can now be written as

$$\begin{aligned}
\mathbf{G}^{ee}(\mathbf{r},\mathbf{r}_s,t) = {} & \begin{pmatrix} \frac{\sigma_v\mu_0 x^2}{2t} - 1 & \frac{\sigma_v\mu_0 xy}{2t} & \frac{\sigma_v\mu_0\lambda x h_v^-}{2t} \\ \frac{\sigma_v\mu_0 xy}{2t} & \frac{\sigma_v\mu_0 y^2}{2t} - 1 & \frac{\sigma_v\mu_0\lambda y h_v^-}{2t} \\ \frac{\sigma_v\mu_0\lambda x h_v^-}{2t} & \frac{\sigma_v\mu_0\lambda y h_v^-}{2t} & \lambda^2\left(2 - \frac{\sigma_v\mu_0 r^2}{2t}\right) \end{pmatrix} \left(\frac{\sigma_v\mu_0^3}{\pi^3 t^5}\right)^{1/2} \\
& \times \frac{\exp(-\tau_v/t)}{8\lambda} \\
& - \frac{\mu_0}{4\pi}\left[\begin{pmatrix} \frac{x^2}{r^2} & \frac{xy}{r^2} & 0 \\ \frac{xy}{r^2} & \frac{y^2}{r^2} & 0 \\ 0 & 0 & 0 \end{pmatrix} \frac{F^{(2)}(\tau_v,t)}{\lambda R_v} + \begin{pmatrix} \frac{y^2}{r^2} & -\frac{xy}{r^2} & 0 \\ -\frac{xy}{r^2} & \frac{x^2}{r^2} & 0 \\ 0 & 0 & 0 \end{pmatrix} \frac{F^{(2)}(\tau,t)}{R}\right] \\
& - \begin{pmatrix} \frac{x^2-y^2}{r^2} & \frac{2xy}{r^2} & 0 \\ \frac{2xy}{r^2} & \frac{y^2-x^2}{r^2} & 0 \\ 0 & 0 & 0 \end{pmatrix} \sqrt{\frac{\mu_0}{\sigma}}\, \frac{F^{(1)}(\tau_v,t) - F^{(1)}(\tau,t)}{4\pi r^2},
\end{aligned} \tag{4.96}$$

where the $\times$-sign is not to be confused with a vector product, but just used to indicate that the first term on the right-hand side should be multiplied by the second term. The diffusion times are $\tau_v = \sigma_v\mu_0 R_v^2/4$ and $\tau = \sigma\mu_0 R^2/4$. The exponential part of the functions $F^{(m)}(\tau,t)$ in the electric field is a Gaussian function in space for fixed times, the centre of which lies at the source location and has a standard deviation of $\sqrt{2t/(\sigma\mu_0)}$, which determines the width. For a fixed position the behaviour of $F^{(m)}(\tau,t)$ as a function of time is more complicated. For early times, $t \ll \tau$, each function grows exponentially away from zero and for late times, $t \gg \tau$, each function behaves as an inverse polynomial in time with an odd half-power. With these $F^{(m)}(\tau,t)$-functions block-pulse source functions are easily modelled as well, because every step-function response can be obtained from the impulse response functions by using $m-2$ instead of m in $F^{(m)}(\tau,t)$. By convolving the Green's function of equation 4.96 with an electric current time behaviour, the electric field is obtained. Instead of using diffusion times τ we could have used diffusion distances D [m] such that $\tau/t = R^2/D^2$. With such expressions the Gaussian behaviour would become explicit and the standard deviation would be $D/\sqrt{2}$.

4.3 The Electromagnetic Field of a Source in a VTI Half-Space

We assign two half-spaces a constant magnetic permeability μ, a horizontal conductivity σ_n and a vertical conductivity $\sigma_{v;n}$ with $n = 0, 1$, separated by the boundary at $z_0 = 0$, as shown in Figure 4.2. Because here we are interested only in the electric field generated by an electric current source inside the conductive half-space, the incident field $\tilde{\mathbf{G}}^{ee;i}_{H,V}$ and the reflected field $\tilde{\mathbf{G}}^{ee;r}_{H,V}$ are given here, but not the transmitted fields $\tilde{\mathbf{G}}^{ee;t}_{H,V}$. The upper half-space is domain $\mathbb{D}_0$ for which $z < 0$ and the lower half-space is $\mathbb{D}_1$ for which $z > 0$. The source is located in the lower half-space at depth level z_s. In the diffusive approximation we take air as the upper half-space with $\sigma_{v;0} = \sigma_0 = 0$ and in the lower half-space we can omit the subscript and will use σ and σ_v for ease of notation. In the lower half space the scalar Green's functions for the incident and reflected fields are given by

$$\tilde{G}_V^{\pm}(k_x, k_y, z, z_s, s) = \frac{\exp(-\Gamma_v |h^{\pm}|)}{2\Gamma_v}, \tag{4.97}$$

$$\tilde{G}_H^{\pm}(k_x, k_y, z, z_s, s) = \frac{\exp(-\Gamma |h^{\pm}|)}{2\Gamma}, \tag{4.98}$$

where $\tilde{G}_H^-$ corresponds to the TE mode incident field or whole space Green's function given in equation 4.55, $\tilde{G}_H^+$ corresponds to the TE mode reflected field, $h^+ = z + z_s$ is the vertical source–receiver distance via the boundary and $h^- = z - z_s$, as defined below equation 4.63. Similar definitions apply to the TM mode Green's functions $\tilde{G}_V^{\pm}$. We write the final electromagnetic field in the lower half-space as an incident field and a reflected field as

$$\tilde{\mathbf{G}}_V^{ee} = \tilde{\mathbf{G}}_V^{ee;i} + \tilde{\mathbf{G}}_V^{ee;r}, \tag{4.99}$$

$$\tilde{\mathbf{G}}_H^{ee} = \tilde{\mathbf{G}}_H^{ee;i} + \tilde{\mathbf{G}}_H^{ee;r}, \tag{4.100}$$

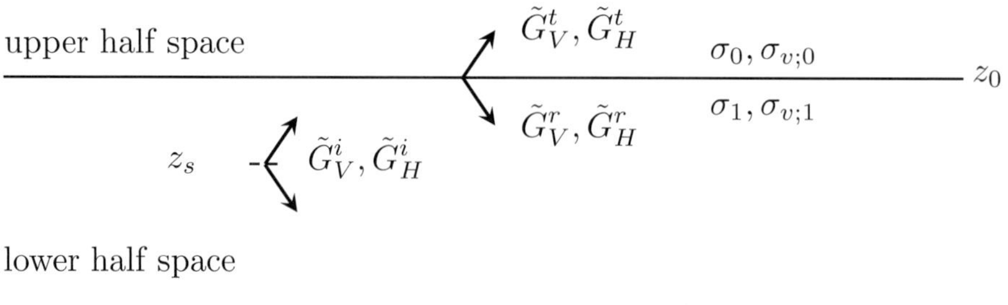

Figure 4.2 The diffusive fields in terms of the incident, reflected and transmitted field Green's functions, $\tilde{G}_M^{ee;i}$, $\tilde{G}_M^{ee;r}$, $\tilde{G}_M^{ee;t}$, respectively, in a half-space and where $M = H$ or $M = V$ to denote TE or TM modes, respectively.

where $\tilde{\mathbf{G}}_V^{ee;i}$ and $\tilde{\mathbf{G}}_H^{ee;i}$ are the incident field Green's functions given in equation 4.62 with $\tilde{G}_V$ replaced by $\tilde{G}_V^-$ and in equation 4.61 with $\tilde{G}_H$ replaced by $\tilde{G}_H^-$, respectively. The half-space problem is a special case of the layered medium problem for which solutions are derived in Appendix B. The reflected field Green's functions are obtained from equations B.50 and B.52 for the TM and TE modes by taking the Green's functions out of the expressions for $\tilde{\mathbf{E}}_{V,H}^{+-}$ with $\varsigma = 1$, and because there is only one conductive half-space we drop the subscript. Then we find that no multiple reflections can occur and $M_{V,H} = 1$ (cf. equations B.39 and B.46), the reflection against the surface is given by $r_V^- = 1$ and

$$r_H^- = (\kappa - \Gamma_1)/(\kappa + \Gamma_1) = -1 - \frac{2}{\gamma^2}(\kappa^2 - \kappa\Gamma), \tag{4.101}$$

and the factor $-\Gamma$ represents a vertical derivative acting on the downgoing diffusive field. Both expressions for the reflection coefficients are obtained using the diffusive approximation in equations B.19 and B.48. We further observe that we can write $\Gamma_v/(\sigma\kappa^2) = (1/\sigma_v + \zeta/\kappa^2)/\Gamma_v$. Using these substitutions in equations B.50 and B.52, the reflected field Green's functions are given by

$$\tilde{\mathbf{G}}_V^{ee;r} = -\left[\frac{1}{\sigma_v}\begin{pmatrix} k_x^2 & k_xk_y & \frac{-ik_x\partial}{\partial z} \\ k_xk_y & k_y^2 & \frac{-ik_y\partial}{\partial z} \\ \frac{-ik_x\partial}{\partial z} & \frac{-ik_y\partial}{\partial z} & \frac{\partial^2}{\partial z^2} - \gamma^2 \end{pmatrix} + \zeta \begin{pmatrix} \frac{k_x^2}{\kappa^2} & \frac{k_xk_y}{\kappa^2} & 0 \\ \frac{k_xk_y}{\kappa^2} & \frac{k_y^2}{\kappa^2} & 0 \\ 0 & 0 & 0 \end{pmatrix}\right]\tilde{G}_V^+, \tag{4.102}$$

$$\tilde{\mathbf{G}}_H^{ee;r} = \begin{pmatrix} -k_y^2 & k_xk_y & 0 \\ k_xk_y & -k_x^2 & 0 \\ 0 & 0 & 0 \end{pmatrix}\left(\frac{\zeta}{\kappa^2} + \frac{2}{\sigma} + \frac{2}{\sigma\kappa}\frac{\partial}{\partial z}\right)\tilde{G}_H^+. \tag{4.103}$$

The TM mode reflected field Green's function, apart from the height difference and the sign change in the zz-component, is the same as the incident field Green's function and does not present any new functions to transform back to the space domain. The first two terms in the TE mode reflected field Green's function are recognised as functions we have already transformed back to the space domain, and we can directly write down their solutions. The last term is new and needs to be evaluated.

4.3.1 Space–Frequency Domain Solutions

The TM mode reflected electric field can be obtained directly from the incident electric field. The TE mode reflected field can be obtained from the incident field and the result of equation B.78. Collecting all results and carrying out the differentiations yields the electric field Green's function for an electric current source in terms of its TM mode and TE mode Green's functions for any point in the lower half-space as

$$\hat{\mathbf{G}}^{ee}(\mathbf{r}, \mathbf{r}_s, s) = \hat{\mathbf{G}}_V^{ee}(\mathbf{r}, \mathbf{r}_s, s) + \hat{\mathbf{G}}_H^{ee}(\mathbf{r}, \mathbf{r}_s, s) + \hat{\mathbf{G}}_A^{ee}(\mathbf{r}, \mathbf{r}_s, s). \tag{4.104}$$

In this expression, $\hat{\mathbf{G}}_A^{ee}$ is purely TE mode and contains all the Bessel functions-related terms. The mode functions are written as

$$\hat{\mathbf{G}}_V^{ee}(\mathbf{r}, \mathbf{r}_s, s) = \sum_{m=0}^{2} \mathbf{g}_V^{m-} \hat{F}^{(m)}(\tau_v^-, s) + \mathbf{g}_V^{m+} \hat{F}^{(m)}(\tau_v^+, s), \tag{4.105}$$

$$\hat{\mathbf{G}}_H^{ee}(\mathbf{r}, \mathbf{r}_s, s) = \sum_{m=0}^{2} \mathbf{g}_H^{m-} \hat{F}^{(m)}(\tau^-, s) + \mathbf{g}_H^{m+} \hat{F}^{(m)}(\tau^+, s), \tag{4.106}$$

where all matrix coefficients are independent of s. We introduce diffusion times $\tau_v^\pm = \sigma_v \mu_0 (R_v^\pm)^2/4$ and $\tau^\pm = \sigma \mu_0 (R^\pm)^2/4$ that describe the diffusive time characteristic of the TM and TE modes, with $R_v^\pm = \sqrt{x^2 + y^2 + (h_v^\pm)^2}$, $h_v^\pm = \lambda h^\pm$ and $R^\pm = \sqrt{x^2 + y^2 + (h^\pm)^2}$. The direct distance from source to receiver is denoted R^-, and the distance from the source to the surface at $z = 0$ and then to the receiver is denoted R^+, which can be understood as a direct path coming to the receiver from a virtual source at $-z_s$. Parameters with a minus-sign in superscript are associated with the incident field and those with a plus-sign in superscript are associated with the reflected field. The diffusion functions $F^{(m)}$ are defined in equations 4.83–4.90. The TM mode matrix coefficients are given by

$$\mathbf{g}_V^{0\pm} = \begin{pmatrix} \frac{3x^2}{(R_v^\pm)^2} - 1 & \frac{3xy}{(R_v^\pm)^2} & \mp\frac{3\lambda x h_v^\pm}{(R_v^\pm)^2} \\ \frac{3xy}{(R_v^\pm)^2} & \frac{3y^2}{(R_v^\pm)^2} - 1 & \mp\frac{3\lambda y h_v^\pm}{(R_v^\pm)^2} \\ \frac{3\lambda x h_v^\pm}{(R_v^\pm)^2} & \frac{3\lambda y h_v^\pm}{(R_v^\pm)^2} & \mp\lambda^2\left(\frac{3(h_v^\pm)^2}{(R_v^\pm)^2} - 1\right) \end{pmatrix} \frac{1}{4\pi\lambda\sigma_v (R_v^\pm)^3}, \tag{4.107}$$

$$\mathbf{g}_V^{1\pm} = 2\sqrt{\tau_v^\pm}\,\mathbf{g}_V^{0\pm} - \begin{pmatrix} \frac{x^2-y^2}{4\pi r^4} & \frac{xy}{2\pi r^4} & 0 \\ \frac{xy}{2\pi r^4} & \frac{y^2-x^2}{4\pi r^4} & 0 \\ 0 & 0 & 0 \end{pmatrix} \sqrt{\frac{\mu_0}{\sigma}}, \tag{4.108}$$

$$\mathbf{g}_V^{2\pm} = \begin{pmatrix} \frac{x^2}{(R_v^\pm)^2} - \frac{x^2}{r^2} & \frac{xy}{(R_v^\pm)^2} - \frac{xy}{r^2} & \mp\frac{\lambda x h_v^\pm}{(R_v^\pm)^2} \\ \frac{xy}{(R_v^\pm)^2} - \frac{xy}{r^2} & \frac{y^2}{(R_v^\pm)^2} - \frac{y^2}{r^2} & \mp\frac{\lambda y h_v^\pm}{(R_v^\pm)^2} \\ \frac{\lambda x h_v^\pm}{(R_v^\pm)^2} & \frac{\lambda y h_v^\pm}{(R_v^\pm)^2} & \mp\lambda^2\left(\frac{(h_v^\pm)^2}{(R_v^\pm)^2} - 1\right) \end{pmatrix} \frac{\mu_0}{4\pi\lambda R_v^\pm}. \tag{4.109}$$

The TE mode coefficients are given by

$$\mathbf{g}_H^{0\pm} = \begin{pmatrix} \frac{3y^2}{(R^+)^2} - 1 & -\frac{3xy}{(R^+)^2} & 0 \\ -\frac{3xy}{(R^+)^2} & \frac{3x^2}{(R^+)^2} - 1 & 0 \\ 0 & 0 & 0 \end{pmatrix} \frac{1 \pm 1}{4\pi\sigma(R^+)^3}, \tag{4.110}$$

$$\mathbf{g}_H^{1\pm} = \begin{pmatrix} \frac{x^2-y^2}{4\pi r^4} & \frac{xy}{2\pi r^4} & 0 \\ \frac{xy}{2\pi r^4} & \frac{y^2-x^2}{4\pi r^4} & 0 \\ 0 & 0 & 0 \end{pmatrix} \sqrt{\frac{\mu_0}{\sigma}} + (1 \pm 1)\sqrt{\tau^+}\mathbf{g}_H^{0+}, \tag{4.111}$$

$$\mathbf{g}_H^{2\pm} = \begin{pmatrix} -y^2 & xy & 0 \\ xy & -x^2 & 0 \\ 0 & 0 & 0 \end{pmatrix} \left(\frac{1}{r^2} - \frac{1 \pm 1}{(R^\pm)^2} \right) \frac{\mu_0}{4\pi R^+}. \tag{4.112}$$

The incident field is of course the same field as the total field in the whole space. The reflected field, as far as it is described by the two mode's Green's functions with the coefficients $\mathbf{g}_H^{m+}, \mathbf{g}_V^{m+}$, has the geometrical interpretation of specular reflection against the Earth's surface from below signified by the distance functions R_v^+ and R^+, which can be seen as the direct distances from an image source at depth location $-z_s$. Both TM and TE modes have such reflected fields and only the TE mode has another term given by

$$\hat{\mathbf{G}}_A^{ee}(\mathbf{r}, \mathbf{r}_s, s) = \begin{pmatrix} -y^2 & xy & 0 \\ xy & -x^2 & 0 \\ 0 & 0 & 0 \end{pmatrix} \frac{\zeta\gamma h^+ \left[(3\mathrm{I}_0 + \mathrm{I}_2)\mathrm{K}_1 - \mathrm{I}_1(3\mathrm{K}_0 + \mathrm{K}_2)\right]}{16\pi(R^+)^3}$$
$$- \begin{pmatrix} \frac{3y^2}{(R^+)^2} - 1 & -\frac{3xy}{(R^+)^2} & 0 \\ -\frac{3xy}{(R^+)^2} & \frac{3x^2}{(R^+)^2} - 1 & 0 \\ 0 & 0 & 0 \end{pmatrix} \left(\frac{\zeta h^+(\mathrm{I}_0\mathrm{K}_0 - \mathrm{I}_1\mathrm{K}_1)}{4\pi(R^+)^2} + \frac{\xi^+\mathrm{I}_0\mathrm{K}_1 + \xi^-\mathrm{I}_1\mathrm{K}_0}{2\pi\sigma(R^+)^3} \right). \tag{4.113}$$

For improved numerical accuracy it is recommended to compute scaled Bessel functions given by

$$\bar{\mathrm{I}}_n(\xi^-) = \exp\left(-\Re\{\xi^-\}\right)\mathrm{I}_n(\xi^-), \tag{4.114}$$
$$\bar{\mathrm{K}}_n(\xi^+) = \exp(\xi^+)\mathrm{K}_n(\xi^+), \tag{4.115}$$

where $\Re\{\xi^-\}$ means the real part of ξ^-, and correct for the scaling in the product as

$$\mathrm{I}_n(\xi^-)\mathrm{K}_m(\xi^+) = \exp\left(-\Re\{\gamma\}h^+ - \mathrm{i}\Im\{\xi^+\}\right)\bar{\mathrm{I}}_n(\xi^-)\bar{\mathrm{K}}_m(\xi^+), \tag{4.116}$$

where $\Im\{\xi^+\}$ means the imaginary part of ξ^+.

The significance of this decomposition into TM and TE modes, and the deconstruction of the TE mode reflection response into a specular reflection response at the surface as if the signal was coming from a virtual source and the more complicated coupling with the air, lies in the physical interpretation of each term. The TM mode reflection response is entirely described by the response from the virtual source that has the same sign for the horizontal source components but the opposite sign for the vertical source. This is because we have used $\varepsilon_0 = 0$ in air, while a lossless air does not really warrant this approximation. As we discuss in Section 4.1.2, this approximation can be made, because the effect of non-zero permittivity is negligible, as we show in Chapter 5 with a numerical example. The TE mode virtual source has the same component as the original source, but also a component as if the x- and y-axes are interchanged. This second virtual source related near-field response is represented by $\mathbf{g}_H^{0+}$ and the corresponding intermediate and far-field behaviours can easily be found in $\mathbf{g}_H^{1+}$ and $\mathbf{g}_H^{2+}$ as the terms showing inverse proportionality to 3D distance and not to horizontal distance. Finally, the TE mode has a contribution in the form of a diffusive refracted wave from the wavefield that travels in the air and refracts back as a diffusive field into the half-space. We have seen that at large horizontal offsets this can be understood as the air wave, because then this contribution can be written as a field that diffuses vertically up to the surface as if it were a plane diffusive wave, propagates in the air along the surface as a wave in free space where it shows near-field geometrical spreading of inverse distance cubed, and then diffuses vertically down into the Earth as if it was a plane diffusive wave. For shorter offsets the behaviour is not easily interpreted geometrically because the Bessel functions have a more complicated behaviour. For a constant value of h^+ it is predictable because it scales with the square root of frequency, but for constant frequency there is no simple dependence on height versus offset where the approximation comes close to the true solution.

4.3.2 Space–Time Domain Solutions

Obtaining the space–time domain electric impulse response to an electric dipole source in a half-space is analogous to obtaining the whole space solution. We therefore write it as

$$\mathbf{G}^{ee}(\mathbf{r},\mathbf{r}_s,t) = \mathbf{G}_V^{ee}(\mathbf{r},\mathbf{r}_s,t) + \mathbf{G}_H^{ee}(\mathbf{r},\mathbf{r}_s,t) + \mathbf{G}_A^{ee}(\mathbf{r},\mathbf{r}_s,t), \tag{4.117}$$

with

$$\mathbf{G}_V^{ee}(\mathbf{r},\mathbf{r}_s,t) = \sum_{m=0}^{2} \mathbf{g}_V^{m-} F^{(m)}(\tau_v^-,t) + \mathbf{g}_V^{m+} F^{(m)}(\tau_v^+,t), \tag{4.118}$$

$$\mathbf{G}^{ee}_{H}(\mathbf{r},\mathbf{r}_s,t) = \sum_{m=0}^{2} \mathbf{g}^{m-}_{H} F^{(m)}(\tau^-,t) + \mathbf{g}^{m+}_{H} F^{(m)}(\tau^+,t), \tag{4.119}$$

where the coefficients are unchanged and the functions $F^{(m)}(\tau,t)$ are defined in Section 4.2.2. The Laplace transform pair exists for the product of the two Bessel functions of equal order (Oberhettinger and Badii, 1973: 346),

$$\mathrm{I}_0(\xi^-)\mathrm{K}_0(\xi^+) \overset{\mathcal{L}^{-1}}{\longleftrightarrow} \frac{\exp(-\tau_h^+/t - \tau_r/t)\mathrm{I}_0(\tau_r/t)}{2t} = \frac{\exp(-\tau_h^+/t)\bar{\mathrm{I}}_0(\tau_r/t)}{2t}, \tag{4.120}$$

with the new air wave-related diffusion times given by

$$\tau_h^+ = \sigma\mu_0(h^+)^2/4, \tag{4.121}$$

$$\tau_r = \sigma\mu_0 r^2/8. \tag{4.122}$$

Notice that for the air wave these new diffusion times do not describe diffusion in three-dimensional space, but rather diffusion in the vertical direction described by an exponential and polynomial behaviour in the horizontal direction described by the scaled modified Bessel function. This leads to the space–time domain expression of the air wave-related parts of the Green's function, as

$$\mathbf{G}^{ee}_{A}(\mathbf{r},\mathbf{r}_s,t) = \frac{\sigma\mu_0^2 h^+ \exp(-\tau_h^+/t)}{32\pi t^3}\left[\begin{pmatrix}1&0&0\\0&1&0\\0&0&0\end{pmatrix}\left[\bar{\mathrm{I}}_0(\tau_r/t)-\bar{\mathrm{I}}_1(\tau_r/t)\right]\right.$$
$$\left. -2\begin{pmatrix}-\frac{y^2}{r^2}&\frac{xy}{r^2}&0\\\frac{xy}{r^2}&-\frac{x^2}{r^2}&0\\0&0&0\end{pmatrix}\left\{\bar{\mathrm{I}}_1(\tau_r/t)-2\frac{\tau_r}{t}\left[\bar{\mathrm{I}}_0(\tau_r/t)-\bar{\mathrm{I}}_1(\tau_r/t)\right]\right\}\right]. \tag{4.123}$$

Late-time behaviour of the impulse responses can be analysed by first looking at the late-time asymptote of the Bessel functions occurring in the air wave, which is given by

$$\lim_{\tau_r/t\downarrow 0} \bar{\mathrm{I}}_n(\tau_r/t) = \frac{\tau_r^n}{n!\,(2t)^n}. \tag{4.124}$$

Comparing this result with the parts of the Green's functions that are expressed in terms of ordinary functions, we see that the air wave has a late-time response proportional to t^{-3}, while the remainder has a late-time response proportional to $t^{-5/2}$ and dominates over the air wave response. For early times, however, the air wave is expected to be important for large enough offsets because the travel time through the air is zero.

Step responses are obtained for the functions $F^{(m)}$ by replacing m by $m-2$, but for the air wave no analytical result is found. We can find the unit step response numerically using the Gaver–Stehfest method (Gaver, 1966; Stehfest, 1970), which computes the inverse Laplace transformation for the unit step response with discrete real values of s for a value of t as

$$\mathbf{G}^{ee}_{AS}(\mathbf{r}, \mathbf{r}_s, t) = \sum_{k=1}^{2K} D_k \hat{\mathbf{G}}^{ee}_{A}(\mathbf{r}, \mathbf{r}_s, s_k), \tag{4.125}$$

where $\mathbf{G}^{ee}_{AS}$ is the half-space step response; $s_k = k\log(2)/t$ and the coefficients D_k are independent of time and given by

$$D_k = \frac{(-1)^{k+K}}{K!\,k} \sum_{j=(k+1)/2}^{\min(k,K)} j^{K+1} \begin{pmatrix} K \\ j \end{pmatrix} \begin{pmatrix} 2j \\ j \end{pmatrix} \begin{pmatrix} j \\ k-j \end{pmatrix}. \tag{4.126}$$

Note that the half-space impulse response $\hat{\mathbf{G}}^{ee}_{A}$ is used and the division by s is taken inside the coefficients D_k. The maximum number of points to be taken depends on machine precision and in double precision arithmetic usually $K = 7$ or 8. Because $(k+1)/2$ is not always an integer, the starting point for the lower bound in the summation of equation 4.126 is ambiguous. It should be taken as the smaller integer value at even values of k.

4.3.3 Horizontal Dipoles at the Interface

For a source on the interface between air and the VTI half-space, we look at the horizontal electric dipole as source and as receiver. For any point in the subsurface the previous expressions for the electric field generated by a horizontal electric dipole source at the surface can be used by taking the source depth level at the interface, $z_s = 0$. If the receiver is also at the interface, only the horizontal electric dipole receiver can be used and the expressions can be obtained from the above results by taking the receiver depth level at the surface, $z = 0$, as well. The expressions simplify considerably and are given in the frequency domain by

$$\begin{aligned}\hat{G}^{ee}_{xx}(\mathbf{r}, \mathbf{0}, s) = {} & \left[\left(3\frac{y^2}{r^2} - 1\right) + \frac{y^2}{r^2}\gamma r\right] \frac{\exp(-\gamma r)}{2\pi\sigma r^3} - \left(3\frac{y^2}{r^2} - 1\right)\frac{1}{2\pi\sigma r^3} \\ & + \left[\left(3\frac{x^2}{r^2} - 1\right)\lambda + \frac{x^2}{r^2}\gamma r\right] \frac{\exp(-\gamma_v r)}{2\pi\sigma r^3},\end{aligned} \tag{4.127}$$

in which the first two terms are TE mode contributions and the third is the TM mode contribution. The horizontal cross-component is found as

$$\hat{G}^{ee}_{xy}(\mathbf{r}, \mathbf{0}, s) = \frac{xy}{r^2}\left[(3\lambda + \gamma r)\frac{\exp(-\gamma_v r)}{2\pi\sigma r^3} - (3 + \gamma r)\frac{\exp(-\gamma r)}{2\pi\sigma r^3} + \frac{3}{2\pi\sigma r^3}\right]. \tag{4.128}$$

Horizontal symmetry is preserved in the Green's function for a VTI half-space and, therefore,

$$\hat{G}^{ee}_{yx}(\mathbf{r},\mathbf{0},s) = \hat{G}^{ee}_{xy}(\mathbf{r},\mathbf{0},s). \tag{4.129}$$

The Green's function $\hat{G}^{ee}_{yy}(\mathbf{r},\mathbf{0},s)$ is the same as $\hat{G}^{ee}_{xx}(\mathbf{r},\mathbf{0},s)$ when x and y are interchanged.

In the limit of an isotropic half-space, $\lambda = 1$, equations 4.127 and 4.128 reduce to

$$\hat{G}^{ee}_{xx}(\mathbf{r},\mathbf{0},s) = \frac{(1+\gamma r)\exp(-\gamma r) - (3y^2/r^2 - 1)}{2\pi\sigma r^3}, \tag{4.130}$$

$$\hat{G}^{ee}_{xy}(\mathbf{r},\mathbf{0},s) = \frac{3xy}{2\pi\sigma r^5}, \tag{4.131}$$

as found by Weir (1980).

We can write these electric Green's functions as

$$\hat{G}^{ee}_{xx}(\mathbf{r},\mathbf{0},s) = -a_0 + \sum_{m=0}^{1} a_m \hat{F}^{(m)}(s,\tau) + \sum_{m=0}^{1} b_m \hat{F}^{(m)}(s,\tau/\lambda^2), \tag{4.132}$$

$$\hat{G}^{ee}_{xy}(\mathbf{r},\mathbf{0},s) = c_0 - \sum_{m=0}^{1} c_m \hat{F}^{(m)}(s,\tau) + \sum_{m=0}^{1} d_m \hat{F}^{(m)}(s,\tau/\lambda^2), \tag{4.133}$$

with the coefficients given by

$$a_0 = \left(3\frac{y^2}{r^2} - 1\right)\frac{1}{2\pi\sigma r^3}, \qquad a_1 = \frac{y^2\sqrt{\mu_0}}{2\pi\sqrt{\sigma}r^4}, \tag{4.134}$$

$$b_0 = \left(3\frac{x^2}{r^2} - 1\right)\frac{\lambda}{2\pi\sigma r^3}, \qquad b_1 = \frac{x^2\sqrt{\mu_0}}{2\pi\sqrt{\sigma}r^4}, \tag{4.135}$$

$$c_0 = \frac{3xy}{2\pi\sigma r^5}, \qquad c_1 = \frac{xy\sqrt{\mu_0}}{2\pi\sqrt{\sigma}r^4}, \tag{4.136}$$

$$d_0 = \lambda c_0, \qquad d_1 = c_1. \tag{4.137}$$

For an in-line source and receiver we have

$$\hat{G}^{ee}_{xx}(r,0,0,s) = \frac{1}{2\pi\sigma r^3}\left(1 - \exp(-\gamma r) + (2\lambda + \gamma r)\exp(-\gamma_v r)\right), \tag{4.138}$$

with $r = |x - x_s|$. If the dipoles have a source length L_s and distance between the receiver electrodes L_r with the midpoints of the source and receiver located a distance r apart such that $r > (L_r + L_s)/2$, then we can find the elongated dipole response to the elongated source as the electric potential difference given by

$$\Delta V(r,0,0,s) = \hat{I}^e_x(s)\int_{x'=-L_s/2}^{L_s/2}\int_{x''=-L_r/2}^{L_r/2} \hat{G}^{ee}_{xx}(r + x'' - x', 0, 0, s)\,dx''\,dx', \tag{4.139}$$

in which $\hat{I}^e_x(s)$ is the current strength of the source as a function of frequency.

The time domain expressions corresponding to equations 4.132 and 4.133 can be obtained by replacing the functions $\hat{F}^{(m)}(s,\tau)$ by $F^{(m)}(t,\tau)$ and for step response functions by $F^{(m-2)}(t,\tau)$ we find the impulse and step responses at the receiver location as

$$G^{ee}_{xx}(r,0,0,t) = \frac{\delta(t) + \sqrt{\frac{\tau}{\pi t^3}}\left[\left(2\frac{\tau_v}{t}+1\right)\exp(-\tau_v/t) - \exp(-\tau/t)\right]}{2\pi\sigma r^3}H(t), \quad (4.140)$$

$$IG^{ee}_{xx}(r,0,0,t) = \frac{\mathrm{erf}(\sqrt{\tau/t}) + 2\lambda\mathrm{erfc}(\sqrt{\tau_v/t}) + 2\sqrt{\frac{\tau}{\pi t}}\exp(-\tau_v/t)}{2\pi\sigma r^3}H(t), \quad (4.141)$$

where IG stands for the step response given by

$$IG^{ee}_{xx}(r,0,0,t) = \int_{t'=0}^{t} G^{ee}_{xx}(r,0,0,t')dt'. \quad (4.142)$$

The electric potential difference is obtained for an impulsive time source as $\Delta V^{\delta}(r,0,0,t)$ and for a step switch-on time source as $\Delta V^{H}(r,0,0,t)$ given by

$$\Delta V^{\delta}(r,0,0,t) = I^{\delta}\int_{x'=-L_s/2}^{L_s/2}\int_{x''=-L_r/2}^{L_r/2} G^{ee}_{xx}(r+x''-x',0,0,t)dx''dx', \quad (4.143)$$

$$\Delta V^{H}(r,0,0,t) = I^{H}\int_{x'=-L_s/2}^{L_s/2}\int_{x''=-L_r/2}^{L_r/2} IG^{ee}_{xx}(r+x''-x',0,0,t)dx''dx', \quad (4.144)$$

where I^{δ} is the strength of the impulsive source and has unit [A s] and I^{H} is the strength of the unit step source and has unit [A], such that in both situations the electric potential difference has the expected units of [V].

The step response Green's function has an early-time value of the first term in equation 4.141 when the error function has the value 1 and a late-time value of the second term when the complementary error function has the value 1. This means that the field starts in early times as TE mode and ends in late times as TM mode. We can find the coefficient of anisotropy as

$$\lambda = \frac{\Delta V^{H}(r,0,0,t\to\infty)}{2\Delta V^{H}(r,0,0,t\downarrow 0)}, \quad (4.145)$$

as found by Werthmüller (2009). The step switch-off response is obtained by subtracting equation 4.141 from the late-time result of the same equation, as given by Kaufman and Keller (1983). When the source current is switched off, the field is proportional to $t^{-3/2}$. Isotropic half-space impulse and step switch-on responses are obtained by taking $\lambda = 1$ in these results. This leads to

$$G^{ee}_{xx}(r,0,0,t) = \frac{\delta(t) + 2\sqrt{\frac{\tau^3}{\pi t^5}}\exp(-\tau/t)}{2\pi\sigma r^3}H(t), \quad (4.146)$$

$$IG^{ee}_{xx}(r,0,0,t) = \frac{1+\mathrm{erfc}(\sqrt{\tau/t})+2\sqrt{\frac{\tau}{\pi t}}\exp(-\tau/t)}{2\pi\sigma r^3}H(t). \tag{4.147}$$

For every in-line impulse response measurement there is an equivalent half-space impulse response that has a peak at the same time as the data. This equivalent half-space has an apparent conductivity that can be obtained at the time the signal has a peak. By differentiating equation 4.146 to t and putting the result to zero, we find that the time to the peak, t_p occurs at $t_p = 0.4\tau$ from which we find an expression for the apparent conductivity as

$$\sigma\mathrm{app} = \frac{10t_p}{\mu_0 r^2}, \tag{4.148}$$

as found by Ziolkowski et al. (2007).

By inserting the expression for the Green's function of equation 4.146 in equation 4.143 the half-space impulse response can be found. The first term in equation 4.146 is a polynomial term and can be integrated as

$$\int_{x'=-L_s/2}^{L_s/2}\int_{x''=-L_r/2}^{L_r/2}\frac{1}{(r+x''-x')^3}dx''dx' = \frac{1}{2}\sum_{m,n=1}^{2}\frac{(-1)^{m-n}}{r_{mn}}, \tag{4.149}$$

with the distances given by

$$r_{mn} = r + (-1)^m L_r/2 + (-1)^n L_s/2. \tag{4.150}$$

The second term in equation 4.146 is an exponential term and can be integrated as

$$\begin{aligned}\int_{x'=-L_s/2}^{L_s/2}\int_{x''=-L_r/2}^{L_r/2} & \exp[-(r+x''-x')^2/D^2]dx''dx' \\ &= \frac{\sqrt{\pi}D}{2}\int_{x''=-L_r/2}^{L_r/2}\mathrm{erfc}[(r-L_s/2+x'')/D]dx'' \\ &- \frac{\sqrt{\pi}D}{2}\int_{x''=-L_r/2}^{L_r/2}\mathrm{erfc}[(r+L_s/2+x'')/D]dx'',\end{aligned} \tag{4.151}$$

in which the diffusion distance is given by $D = 2\sqrt{t/(\sigma\mu_0)}$. The integral of the complementary error function is given by

$$\int_{x'=\pm L_r/2}^{\infty}\mathrm{erfc}[(r\pm L_s/2+x')/D]dx' = D\mathrm{ierfc}[(r\pm L_r/2\pm L_s/2)/D]. \tag{4.152}$$

With these results we can write the potential for each electrode pair as

$$V^{\delta}_{mn}(t) = I^{\delta}\left[\frac{\delta(t)}{4\pi\sigma r_{mn}} + \left(\frac{\mu_0}{\sigma}\right)^{1/2}\frac{\mathrm{ierfc}(r_{mn}/D)}{4\pi t^{3/2}}\right], \tag{4.153}$$

where the integrated error function ierfc is given by

$$\text{ierfc}(r_{mn}/D) = \frac{\exp(-r_{mn}^2/D^2)}{\sqrt{\pi}} - \frac{r_{mn}}{D}\text{erfc}(r_{mn}/D) \tag{4.154}$$

and the measured electric potential difference is given by

$$\Delta V^{\delta}(r, L_r, L_s, t) = V_{11}^{\delta}(t) + V_{22}^{\delta}(t) - V_{21}^{\delta}(t) - V_{12}^{\delta}(t). \tag{4.155}$$

This is the half-space impulse response for an in-line dipole–dipole configuration. The time to the peak with this equation is different from the value corresponding to the point-source–point-receiver approximation. The accuracy of the point-device approximation improves with increasing offset. The error in the obtained apparent conductivity is proportional to the error in the time to the peak for the point-device model compared with the value found in the data. When the half-space conductivity is $\sigma = 1/20$ S m^{-1}, the distance between the midpoints of the array is 3 km, the source is 1 km long and the electrode distance for the potential difference measurement is 20 m, the error in the time to the peak and therefore that of the estimated apparent conductivity is 10%.

The half-space step response can easily be obtained by integrating in time. The impulse of the airwave changes to a unit step function and the time integral of the integrated error function can be evaluated as

$$\int_{t'=0}^{t} (t')^{-3/2}\text{ierfc}(r_{mn}/D)dt' = \frac{4}{\sqrt{\sigma\mu_0}r_{mn}}\int_{\chi=r_{mn}/D}^{\infty} \text{ierfc}(\chi)d\chi, \tag{4.156}$$

which is the next order integrated error function given by

$$\int_{\chi=r_{mn}/D}^{\infty} \text{ierfc}(\chi)d\chi = \frac{1}{4}\text{erfc}(r_{mn}/D) - \frac{r_{mn}}{2D}\text{ierfc}(r_{mn}/D). \tag{4.157}$$

Collecting the results for each potential, we find

$$V_{mn}^{H}(t) = I^{H}\left[\frac{H(t) + \text{erfc}(r_{mn}/D)}{4\pi\sigma r_{mn}} - \left(\frac{\mu_0}{\sigma}\right)^{1/2}\frac{\text{ierfc}(r_{mn}/D)}{4\pi t^{1/2}}\right]. \tag{4.158}$$

The measured potential difference is expressed as

$$\Delta V^{H}(r, L_r, L_s, t) = V_{11}^{H}(t) + V_{22}^{H}(t) - V_{21}^{H}(t) - V_{12}^{H}(t). \tag{4.159}$$

Equation (4.158) shows that when the current is switched on, half the end value is directly obtained from the air wave. The half-space response slowly contributes the other half through the complementary error function. The integrated error function goes to zero for time to zero and to infinity. In the late-time limit the potential difference attains the value that would be measured with the DC electric resistivity dipole–dipole configuration. It is given by

$$\lim_{t\to\infty} \Delta V^H(r, L_r, L_s, t) = \frac{I^H}{2\pi\sigma}\left[\frac{1}{r_{11}} + \frac{1}{r_{22}} - \frac{1}{r_{12}} - \frac{1}{r_{21}}\right]. \tag{4.160}$$

When the half-space conductivity is $\sigma = 1/20$ S m^{-1}, the distance between the midpoints of the array is 3 km, the source is 1 km long and the electrode distance for the potential difference measurement is 20 m, 98% of the DC value is obtained in the half-space step response within 1 s after the current is switched on. For this configuration the point-dipole approximation would result in an error of 5% at early and late times and has a maximum error of 8% at $t = 58$ ms.

4.4 The Electromagnetic Field for Marine CSEM

In this section we give the closed form expressions for the x-component of the electric field generated by an x-directed electric dipole point source located inside the top layer of a VTI layered medium. The expression is exact in the horizontal wavenumber–frequency domain. Space–frequency and space–time domain results must be obtained numerically. The detailed derivations are given in Appendix B. Let us assume a layered model with $N + 1$ interfaces at depth levels z_n with $n = 0, 1, 2, \cdots, N-1, N$. Because zero depth can be chosen anywhere, we take $z_0 = 0$. Each layer $\mathbb{D}_n$ is characterised by a constant magnetic permeability of μ_0, a horizontal conductivity σ_n, a vertical conductivity $\sigma_{v;n}$, and layer thickness $d_n = z_n - z_{n-1}$, with the understanding that the upper and lower half-spaces, $\mathbb{D}_0$ and $\mathbb{D}_{N+1}$, respectively, have an infinite thickness. Let the source be located at z_s with $0 < z_s < z_1$. The marine application of the CSEM method has two acquisition modes of operation. Both modes use a horizontal electric dipole source towed behind a vessel at a certain depth. In one configuration the receivers are stations positioned at fixed locations on the seafloor, whereas the other configuration uses a streamer containing the receiver electrodes. The seabed stations usually measure all components of the magnetic flux and the two horizontal components of the electric field, whereas a streamer measures only the in-line electric field component. We take the receiver depth level z_r as $z_s < z_r < z_1$. Figure 4.3 sketches the situation for a horizontal electric dipole source and an in-line electric field receiver.

Then the electric field can be written in terms of TE and TM mode fields that diffuse in upgoing and downgoing directions. The fields are separated in terms of direction in which they leave the source and at which they arrive at the receiver. These are denoted $\tilde{E}^{\pm\pm}_{xx,M}(k_x, k_y, z_r, z_s, s)$, where the first superscript denotes the direction in which the field diffuses towards the receiver and the second superscript

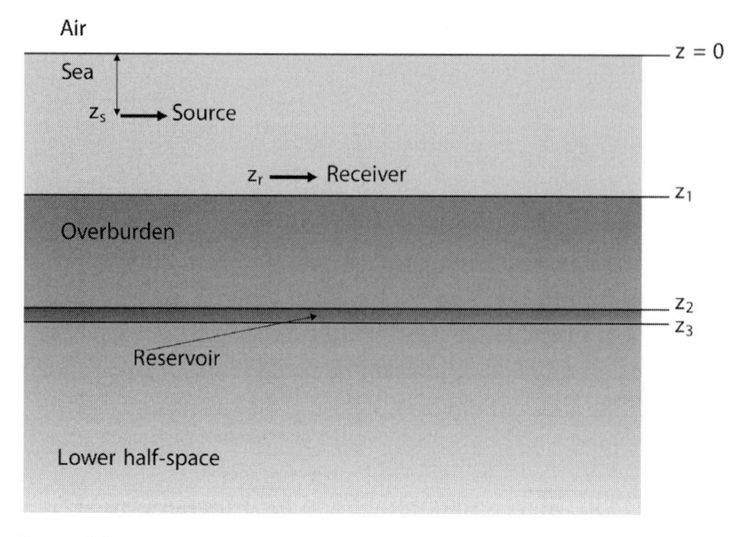

Figure 4.3 Sketch of the marine CSEM setting in a very simple model.

denotes the direction in which the field diffuses away from the source. For both, the plus-sign indicates that the field diffuses in the downward direction and the minus-sign indicates that the field diffuses in the upward direction. This gives the following expression for the field inside the layer that contains the source:

$$\tilde{E}_{xx} = \tilde{E}^{++}_{xx;H} + \tilde{E}^{-+}_{xx;H} + \tilde{E}^{+-}_{xx;H} + \tilde{E}^{--}_{xx;H} + \tilde{E}^{++}_{xx;V} + \tilde{E}^{-+}_{xx;V} + \tilde{E}^{+-}_{xx;V} + \tilde{E}^{--}_{xx;V}. \quad (4.161)$$

We give them for each mode separately because they are of interest in the detection of a buried resistive layer. They are given by

$$\tilde{E}^{\pm\pm}_{xx;V} = \frac{(\mathrm{i}k_x)^2\Gamma_{v;1}}{\kappa^2\sigma_1}\hat{I}^e_x\tilde{G}^{\pm\pm}_{V;1}, \quad (4.162)$$

$$\tilde{E}^{\pm\pm}_{xx;H} = \pm\pm\frac{(\mathrm{i}k_y)^2\zeta_0}{\kappa^2\Gamma_1}\hat{I}^e_x\tilde{G}^{\pm\pm}_{H;1}, \quad (4.163)$$

in which the TM mode directional scalar Green's functions are given in equations B.35–B.38 and the reverberation operator and global and local reflection coefficients are given in equation B.39 and equations B.18 and B.19, respectively. The TE mode equivalents are given in equations B.42–B.48.

4.4.1 Space–Frequency Domain Solutions

The usual way to find space domain solutions is to add all contributions together and then perform the inverse Fourier–Bessel transformation. It is interesting, however, to transform each upgoing and downgoing contribution to the space domain separately to investigate its contribution to the total field. In the space–frequency domain equation 4.161 is given by

$$\hat{E}_{xx}(\mathbf{r}_r, \mathbf{r}_s, \omega) = \hat{E}^{++}_{xx;V} + \hat{E}^{-+}_{xx;V} + \hat{E}^{+-}_{xx;V} + \hat{E}^{--}_{xx;V} + \hat{E}^{++}_{xx;H} + \hat{E}^{-+}_{xx;H} + \hat{E}^{+-}_{xx;H} + \hat{E}^{--}_{xx;H}. \quad (4.164)$$

For zero offset the TM mode downgoing electric field from the downgoing part of the electric dipole source is given by

$$\hat{E}^{++}_{xx;V}(0,0,z_r,z_s,\omega) = -\frac{\hat{I}^e_x}{8\pi\sigma_1}\int_{\kappa=0}^{\infty}\frac{\Gamma_{v;1}}{M_{V;1}}\exp(-\Gamma_{v;1}|h^-|)\kappa\, d\kappa, \quad (4.165)$$

which presents no numerical difficulties. For all non-zero offsets we already know from the analysis of the whole space response that the separate mode responses will contain an artefact that we will remove by evaluating

$$\begin{aligned}\hat{E}^{++}_{xx;V}(\mathbf{r}_r, \mathbf{r}_s, \omega) &= -\frac{\hat{I}^e_x x^2}{4\pi\sigma_1 r^2}\int_{\kappa=0}^{\infty}\frac{\Gamma_{v;1}}{M_{V;1}}\exp(-\Gamma_{v;1}|h^-|)\mathrm{J}_0(\kappa r)\kappa\, d\kappa\\ &+\frac{\hat{I}^e_x(x^2-y^2)}{4\pi\sigma_1 r^3}\int_{\kappa=0}^{\infty}\left(\frac{\Gamma_{v;1}}{M_{V;1}}-\frac{\gamma_1^2}{M_{V;1}(\kappa=0)\Gamma_{v;1}}\right)\exp(-\Gamma_{v;1}|h^-|)\mathrm{J}_1(\kappa r)d\kappa\\ &-\frac{\zeta_1\hat{I}^e_x(x^2-y^2)}{4\pi\gamma_1 r^4}\frac{\exp(-\gamma_{v;1}R_v)}{M_{V;1}(\kappa=0)}.\end{aligned} \quad (4.166)$$

To arrive at this result we have removed the part of the exponent that depends only on vertical distance. This makes sense because the same term is found in the TE mode with opposite sign just as in the homogeneous space, because $M_{V;1}(\kappa = 0) = M_{H;1}(\kappa = 0)$. Now both integrals give physically meaningful results together with the third correction factor. The TE mode equivalent can be written at zero offset as

$$\hat{E}^{++}_{xx;H}(0,0,z_r,z_s,\omega) = -\frac{\zeta_1\hat{I}^e_x}{8\pi}\int_{\kappa=0}^{\infty}\frac{\exp(-\Gamma_1|h^-|)}{\Gamma_1 M_{H;1}}\kappa\, d\kappa, \quad (4.167)$$

and for non-zero offsets as

$$\begin{aligned}\hat{E}^{++}_{xx;H}(\mathbf{r}_r, \mathbf{r}_s, \omega) &= -\frac{\zeta_1\hat{I}^e_x y^2}{4\pi r^2}\int_{\kappa=0}^{\infty}\frac{\exp(-\Gamma_1|h^-|)}{\Gamma_1 M_{H;1}}\mathrm{J}_0(\kappa r)\kappa\, d\kappa\\ &-\frac{\zeta_1\hat{I}^e_x(x^2-y^2)}{4\pi r^3}\int_{\kappa=0}^{\infty}\left(\frac{1}{M_{H;1}}-\frac{1}{M_{H;1}(\kappa=0)}\right)\frac{\exp(-\Gamma_1|h^-|)}{\Gamma_1}\mathrm{J}_1(\kappa r)d\kappa\\ &+\frac{\zeta_1\hat{I}^e_x(x^2-y^2)}{4\pi\gamma_1 r^4}\frac{\exp(-\gamma_1 R)}{M_{H;1}(\kappa=0)}.\end{aligned} \quad (4.168)$$

Equations 4.165–4.168 are the same as equations B.70–B.73. Because the only difference in the other up–down combinations is a possible sign change, we can write down all other combinations at zero offset as

$$\hat{E}^{-+}_{xx;V}(0,0,z_r,z_s,\omega) = -\frac{\hat{I}^e_x}{8\pi\sigma_1}\int_{\kappa=0}^{\infty}\frac{\Gamma_{v;1}R^+_{V;1}}{M_{V;1}}\exp(-\Gamma_{v;1}h^{-+})\kappa d\kappa, \tag{4.169}$$

$$\hat{E}^{-+}_{xx;H}(0,0,z_r,z_s,\omega) = \frac{\zeta_1\hat{I}^e_x}{8\pi}\int_{\kappa=0}^{\infty}\frac{R^+_{H;1}\exp(-\Gamma_1 h^{-+})}{\Gamma_1 M_{H;1}}\kappa d\kappa, \tag{4.170}$$

$$\hat{E}^{+-}_{xx;V}(0,0,z_r,z_s,\omega) = -\frac{\hat{I}^e_x}{8\pi\sigma_1}\int_{\kappa=0}^{\infty}\frac{\Gamma_{v;1}R^-_{V;1}}{M_{V;1}}\exp(-\Gamma_{v;1}h^{+-})\kappa d\kappa, \tag{4.171}$$

$$\hat{E}^{+-}_{xx;H}(0,0,z_r,z_s,\omega) = \frac{\zeta_1\hat{I}^e_x}{8\pi}\int_{\kappa=0}^{\infty}\frac{R^-_{H;1}\exp(-\Gamma_1 h^{+-})}{\Gamma_1 M_{H;1}}\kappa d\kappa, \tag{4.172}$$

$$\hat{E}^{--}_{xx;V}(0,0,z_r,z_s,\omega) = -\frac{\hat{I}^e_x}{8\pi\sigma_1}\int_{\kappa=0}^{\infty}\frac{\Gamma_{v;1}R^-_{V;1}R^+_{V;1}}{M_{V;1}}\exp(-\Gamma_{v;1}h^{--})\kappa d\kappa, \tag{4.173}$$

$$\hat{E}^{--}_{xx;H}(0,0,z_r,z_s,\omega) = -\frac{\zeta_1\hat{I}^e_x}{8\pi}\int_{\kappa=0}^{\infty}\frac{R^-_{H;1}R^+_{H;1}\exp(-\Gamma_1 h^{--})}{\Gamma_1 M_{H;1}}\kappa d\kappa, \tag{4.174}$$

in which $h^{-+} = 2z_1 - z_r - z_s$, $h^{+-} = z_r + z_s$ and $h^{--} = 2d_1 - z_r + z_s$. At non-zero offsets we find

$$\begin{aligned}\hat{E}^{-+}_{xx;V}(\mathbf{r}_r,\mathbf{r}_s,\omega) &= -\frac{\hat{I}^e_x x^2}{4\pi\sigma_1 r^2}\int_{\kappa=0}^{\infty}\frac{\Gamma_{v;1}R^+_{V;1}}{M_{V;1}}\exp(-\Gamma_{v;1}h^{-+})\mathrm{J}_0(\kappa r)\kappa d\kappa \\ &+\frac{\hat{I}^e_x(x^2-y^2)}{4\pi\sigma_1 r^3}\int_{\kappa=0}^{\infty}\left(\frac{\Gamma^2_{v;1}R^+_{V;1}}{M_{V;1}} - \frac{\gamma_1^2 R^+_{V;1}(\kappa=0)}{M_{V;1}(\kappa=0)}\right)\frac{\exp(-\Gamma_{v;1}|h^{-+}|)}{\Gamma_{v;1}}\mathrm{J}_1(\kappa r)d\kappa \\ &-\frac{\zeta_1\hat{I}^e_x(x^2-y^2)}{4\pi\gamma_1 r^4}\frac{R^+_{V;1}(\kappa=0)}{M_{V;1}(\kappa=0)}\exp(-\gamma_{v;1}R^{-+}_v),\end{aligned} \tag{4.175}$$

in which $R^{-+}_v = \sqrt{r^2 + (\lambda_1 h^{-+})^2}$.

$$\begin{aligned}\hat{E}^{-+}_{xx;H}(\mathbf{r}_r,\mathbf{r}_s,\omega) &= \frac{\zeta_1\hat{I}^e_x y^2}{4\pi r^2}\int_{\kappa=0}^{\infty}\frac{R^+_{H;1}\exp(-\Gamma_1 h^{-+})}{\Gamma_1 M_{H;1}}\mathrm{J}_0(\kappa r)\kappa d\kappa \\ &+\frac{\zeta\hat{I}^e_x(x^2-y^2)}{4\pi r^3}\int_{\kappa=0}^{\infty}\left(\frac{R^+_{H;1}}{M_{H;1}} - \frac{R^+_{H;1}(\kappa=0)}{M_{H;1}(\kappa=0)}\right)\frac{\exp(-\Gamma_1 h^{-+})}{\Gamma_1}\mathrm{J}_1(\kappa r)d\kappa \\ &-\frac{\zeta_1\hat{I}^e_x(x^2-y^2)}{4\pi\gamma_1 r^4}\frac{R^+_{H;1}(\kappa=0)}{M_{H;1}(\kappa=0)}\exp(-\gamma_1 R^{-+}),\end{aligned} \tag{4.176}$$

where $R^{-+} = \sqrt{r^2 + (h^{-+})^2}$. Note that the sign-change in the TE mode still removes the non-physical event because $R^{\pm}_{V;1}(\kappa = 0) = -R^{\pm}_{H;1}(\kappa = 0)$. For the upgoing parts of the source, the downgoing field at the receiver is given by

$$\hat{E}^{+-}_{xx;V}(\mathbf{r}_r, \mathbf{r}_s, \omega) = -\frac{\hat{I}^e_x x^2}{4\pi\sigma_1 r^2}\int_{\kappa=0}^{\infty}\frac{\Gamma_{v;1}R^-_{V;1}}{M_{V;1}}\exp(-\Gamma_{v;1}h^{+-})\mathrm{J}_0(\kappa r)\kappa d\kappa$$
$$+\frac{\hat{I}^e_x(x^2-y^2)}{4\pi\sigma_1 r^3}\int_{\kappa=0}^{\infty}\left(\frac{\Gamma^2_{v;1}R^-_{V;1}}{M_{V;1}} - \frac{\gamma_1^2 R^-_{V;1}(\kappa=0)}{M_{V;1}(\kappa=0)}\right)\frac{\exp(-\Gamma_{v;1}|h^{+-}|)}{\Gamma_{v;1}}\mathrm{J}_1(\kappa r)d\kappa$$
$$-\frac{\zeta_1\hat{I}^e_x(x^2-y^2)}{4\pi\gamma_1 r^4}\frac{R^-_{V;1}(\kappa=0)}{M_{V;1}(\kappa=0)}\exp(-\gamma_{v;1}R^{+-}_v), \tag{4.177}$$

in which $R^{+-}_v = \sqrt{r^2 + (\lambda_1 h^{+-})^2}$, and

$$\hat{E}^{+-}_{xx;H}(\mathbf{r}_r, \mathbf{r}_s, \omega) = \frac{\zeta_1\hat{I}^e_x y^2}{4\pi r^2}\int_{\kappa=0}^{\infty}\frac{R^-_{H;1}\exp(-\Gamma_1 h^{+-})}{\Gamma_1 M_{H;1}}\mathrm{J}_0(\kappa r)\kappa d\kappa$$
$$+\frac{\zeta\hat{I}^e_x(x^2-y^2)}{4\pi r^3}\int_{\kappa=0}^{\infty}\left(\frac{R^-_{H;1}}{M_{H;1}} - \frac{R^-_{H;1}(\kappa=0)}{M_{H;1}(\kappa=0)}\right)\frac{\exp(-\Gamma_1 h^{+-})}{\Gamma_1}\mathrm{J}_1(\kappa r)d\kappa$$
$$-\frac{\zeta_1\hat{I}^e_x(x^2-y^2)}{4\pi\gamma_1 r^4}\frac{R^-_{H;1}(\kappa=0)}{M_{H;1}(\kappa=0)}\exp(-\gamma_1 R^{+-}), \tag{4.178}$$

where $R^{+-} = \sqrt{r^2 + (h^{+-})^2}$. The upgoing TM mode field at the receiver is given by

$$\hat{E}^{--}_{xx;V}(\mathbf{r}_r, \mathbf{r}_s, \omega) = -\frac{\hat{I}^e_x x^2}{4\pi\sigma_1 r^2}\int_{\kappa=0}^{\infty}\frac{\Gamma_{v;1}R^-_{V;1}R^+_{V;1}}{M_{V;1}}\exp(-\Gamma_{v;1}h^{--})\mathrm{J}_0(\kappa r)\kappa d\kappa$$
$$+\frac{\hat{I}^e_x(x^2-y^2)}{4\pi\sigma_1 r^3}\int_{\kappa=0}^{\infty}\left(\frac{\Gamma^2_{v;1}R^-_{V;1}R^+_{V;1}}{M_{V;1}} - \gamma_1^2 F_V\right)\frac{\exp(-\Gamma_{v;1}|h^{--}|)}{\Gamma_{v;1}}\mathrm{J}_1(\kappa r)d\kappa$$
$$-\frac{\zeta_1\hat{I}^e_x(x^2-y^2)}{4\pi\gamma_1 r^4}F_V\exp(-\gamma_{v;1}R^{--}_v), \tag{4.179}$$

in which $R^{--}_v = \sqrt{r^2 + (\lambda_1 h^{--})^2}$, and

$$F_V = \frac{R^-_{V;1}(\kappa=0)R^+_{V;1}(\kappa=0)}{M_{V;1}(\kappa=0)}. \tag{4.180}$$

Finally, the upgoing TE mode field at the receiver is found as

$$\hat{E}^{--}_{xx;H}(\mathbf{r}_r, \mathbf{r}_s, \omega) = -\frac{\zeta_1\hat{I}^e_x y^2}{4\pi r^2}\int_{\kappa=0}^{\infty}\frac{R^-_{H;1}R^+_{H;1}\exp(-\Gamma_1 h^{--})}{\Gamma_1 M_{H;1}}\mathrm{J}_0(\kappa r)\kappa d\kappa$$

$$- \frac{\zeta \hat{I}^e_x (x^2 - y^2)}{4\pi r^3} \int_{\kappa=0}^{\infty} \left(\frac{R^-_{H;1} R^+_{H;1}}{M_{H;1}} - F_H \right) \frac{\exp(-\Gamma_1 h^{--})}{\Gamma_1} \mathrm{J}_1(\kappa r) d\kappa$$
$$+ \frac{\zeta_1 \hat{I}^e_x (x^2 - y^2)}{4\pi \gamma_1 r^4} F_H \exp(-\gamma_1 R^{--}), \tag{4.181}$$

with $R_v^{--} = \sqrt{r^2 + (h^{--})^2}$, and

$$F_H = \frac{R^-_{H;1}(\kappa = 0) R^+_{H;1}(\kappa = 0)}{M_{H;1}(\kappa = 0)}. \tag{4.182}$$

With these expressions the total electric has been defined. These integrals can all be evaluated numerically by quadrature integration or with the aid of filter tables. Using filter tables is known as the fast Hankel transform method, which was introduced by Ghosh (1970) for transforming electric potential fields in layered media. He published his work in 1971 (Ghosh, 1971) and many authors followed to improve the filters. Koefoed et al. (1972) were possibly the first to extend the method for transforming electromagnetic fields and also here many authors followed to improve the filters. Recent advances can be found in the work of Kong (2007) and Key (2012a). For zero-offset results the filters do not work and the quadrature method should be used. For non-zero offsets we use the 201-point Bessel function transformation filters, commonly known as Hankel filters, provided by Key (2012a). As an example we show how they can be used for one component of one mode of the electric field. Let us take the integrals in equation 4.166; we want to evaluate them for one particular receiver location $\mathbf{r}_r = (x, y, z_r)$ and the source is located at $\mathbf{r}_s = (0, 0, z_s)$. The Bessel function filters are designed for the Bessel function argument, hence $\bar{\kappa} = \kappa r$ where $r = \sqrt{x^2 + y^2}$. These are provided in the first column of the table called 'kk201Hankel.txt' and we can call them $\bar{\kappa}_m$, for $m = 1, 2, \cdots, M$, with $M = 201$ in this case, but filters with other lengths can be used as well. The zero-order Bessel function filter coefficients are given in the second column of that table and those for the first-order Bessel function are given in the third column. We can simply call them $\mathrm{J0}_m$ and $\mathrm{J1}_m$, for $m = 1, 2, \cdots, M$. The discretisation involves treating $d\kappa$ as $1/r d\bar{\kappa}$ and $d\bar{\kappa}$ is incorporated in the filter coefficients. Because the separate factor κ occurs in the expressions with J_0, we use $\mathrm{J}_0(\bar{\kappa})\bar{\kappa}/r$. This means that in equation 4.166 the part with the J_0 term gets an extra factor $1/r^2$ and the term involving J_1 a factor $1/r$ when we move from the integral representation to the discrete representation. Then the discrete expression of equation 4.166 can be written as

$$\hat{E}^{++}_{xx;V}(\mathbf{r}_r, \mathbf{r}_s, \omega) = - \frac{\hat{I}^e_x x^2}{4\pi \sigma_1 r^4} \sum_{m=1}^{M} \frac{\bar{\Gamma}_{v;m}}{\bar{M}_{V;m}} \exp(-\bar{\Gamma}_{v;m} |h^-|) \mathrm{J0}_m \bar{\kappa}_m$$

$$+\frac{\hat{I}^e_x(x^2-y^2)}{4\pi\sigma_1 r^4}\sum_{m=1}^{M}\left(\frac{\bar{\Gamma}_{v;m}}{\bar{M}_{V;m}}-\frac{\gamma_1^2}{M_{V;1}(0)\bar{\Gamma}_{v;m}}\right)\exp(-\bar{\Gamma}_{v;m}|h^-|)\mathrm{J1}_m$$
$$-\frac{\zeta_1\hat{I}^e_x(x^2-y^2)}{4\pi\gamma_1 r^4}\frac{\exp(-\gamma_{v;1}R_v)}{M_{V;1}(0)}, \tag{4.183}$$

where the discrete parameters are computed using the actual expressions with $\kappa = \bar{\kappa}_m/r$. Notice that in the summed expressions the factors $\mathrm{J0}_m\bar{\kappa}_m$ and $\mathrm{J1}_m$ do not depend on the horizontal distance and the terms in front of the summation signs do not depend on the discrete horizontal wavenumber. This can be exploited when the electric field is computed for many offsets. All other components can be computed in a similar way.

4.4.2 Space–Time Domain Solutions

The space–frequency domain expressions given in the previous section can be transformed back to time with a filter as well. To see that, it is useful to consider that in space–time each function is a real-valued function and exists only for $t > 0$. For any such real and causal time function $f(t)$, we can find its frequency domain counterpart by the Fourier transformation defined in Chapter 3 as

$$\hat{f}(\omega) = \hat{f}_r(\omega) + \mathrm{i}\hat{f}_i(\omega) = \int_{t=0}^{\infty} f(t)\exp(\mathrm{i}\omega t)dt, \tag{4.184}$$

in which $\hat{f}_r$ is the real part of $\hat{f}$ and $\hat{f}_i$ is the imaginary part of $\hat{f}$. We can write these separately as

$$\hat{f}_r(\omega) = \int_{t=0}^{\infty} f(t)\cos(\omega t)dt = \frac{1}{2}\int_{t=-\infty}^{\infty}[f(t)+f(-t)]\cos(\omega t)dt, \tag{4.185}$$

where the fact has been used that for a fixed value of ω the cosine is an even function of time. Note that $\hat{f}_r$ is even in frequency, $\hat{f}_r(-\omega) = \hat{f}_r(\omega)$. Similarly, we can write the imaginary part of the Fourier transform as

$$\hat{f}_i(\omega) = \int_{t=0}^{\infty} f(t)\sin(\omega t)dt = \frac{1}{2}\int_{t=-\infty}^{\infty}[f(t)-f(-t)]\sin(\omega t)dt, \tag{4.186}$$

where now the fact has been used that for a fixed value of ω the sine is an odd function of time and the whole integrand is an even function of time. Note that $\hat{f}_i$ is odd in frequency, $\hat{f}_i(-\omega) = -\hat{f}_i(\omega)$. If we now look at the inverse Fourier transformation we obtain

$$f(t) = \frac{1}{2\pi}\int_{\omega=-\infty}^{\infty}\hat{f}(\omega)\exp(-\mathrm{i}\omega t)d\omega. \tag{4.187}$$

We can use the fact that $\hat{f}_r$ is even and $\hat{f}_i$ is odd in frequency to obtain

$$f(t) + f(-t) = \frac{1}{\pi} \int_{\omega=-\infty}^{\infty} \hat{f}_r(\omega) \cos(\omega t) d\omega, \tag{4.188}$$

and a similar result can be obtained using the sine function in the transformation, given by

$$f(t) - f(-t) = \frac{1}{\pi} \int_{\omega=-\infty}^{\infty} \hat{f}_i(\omega) \sin(\omega t) d\omega. \tag{4.189}$$

For almost all functions it is redundant, but we need it to compute electric field step switch-on responses. We exploit the fact that only the even part of the whole integrand contributes to the integral and only evaluate at positive frequencies and only for positive times. We can write

$$f(t) = \frac{2}{\pi} \int_{\omega=0}^{\infty} \hat{f}_r(\omega) \cos(\omega t) d\omega H(t), \tag{4.190}$$

$$f(t) = \frac{2}{\pi} \int_{\omega=0}^{\infty} \hat{f}_i(\omega) \sin(\omega t) d\omega H(t). \tag{4.191}$$

For diffusive fields these integrals can be approximated, similar to what was done for the Fourier–Bessel integral, with a filter known as the fast cosine or fast sine transform. Possibly the first to work on these filters was Anderson (1973), whose filters became popular. Kong (2012) showed how to compute the digital filter coefficients that perform the integrals of equations 4.190 and 4.191 with a different approach. Once the coefficients are found, the time domain function is found from

$$f(t) = \frac{2}{\pi t} \sum_{n=1}^{N} \hat{f}_r(\bar{\omega}_n / t) c_n, \tag{4.192}$$

$$f(t) = \frac{2}{\pi t} \sum_{n=1}^{N} \hat{f}_i(\bar{\omega}_n / t) s_n. \tag{4.193}$$

The discrete frequencies are given by $\omega_n = \bar{\omega}_n / t$ and with $N = 201$ the values of $\bar{\omega}_n$ are provided in the first column of the filter table file called 'kk201CosSin.txt' (Key, 2012a). The filter coefficients c_n, s_n are given in the second and third columns, respectively, of this file. Notice that similar to the Hankel filters, the cosine and sine filter coefficients do not depend on time, which can be exploited when the time domain result is desired at many time points. Equation (4.192) can be used to compute impulse and step switch-off responses, where in the latter case minus the real part of the frequency domain step response should be taken as $\hat{f}_r$. Equation (4.193) can be used to compute the impulse response for $t > 0$, and the step switch-on response, in which case the imaginary part of the frequency domain step response should be taken as $\hat{f}_i$. Note that if $\hat{f}$ denotes an impulse response function in the frequency domain it should be divided by $-\mathrm{i}\omega$ to compute the step response in time.

4.5 The Electromagnetic Field for Land CSEM

When we put the source and receivers on the Earth's surface we mimic a land CSEM survey. In this situation we need to consider only horizontal sources and receivers. For these combinations the equations can easily be obtained by taking the source in the upper half-space first, hence we take $\varsigma = 0$ in equations B.49–B.52. In the air the source generates only upgoing fields. Hence, the only upgoing field at the receivers comes from the layered medium below the source and we find that $\tilde{\mathbf{E}}^{\pm-}_{V;0} = 0$ and $M_{V;\varsigma} = 0$. The same applies to the TE mode and we take the limit of receiver and source going to the surface. The result is

$$\begin{pmatrix} \tilde{E}_{xx} & \tilde{E}_{xy} \\ \tilde{E}_{yx} & \tilde{E}_{yy} \end{pmatrix} = \begin{pmatrix} (\mathrm{i}k_x)^2 & \mathrm{i}k_x\mathrm{i}k_y \\ \mathrm{i}k_x\mathrm{i}k_y & (\mathrm{i}k_y)^2 \end{pmatrix} \frac{\Gamma_{v;0}}{2\kappa^2\sigma_0}[1 + R^+_{V;0}] + \begin{pmatrix} (\mathrm{i}k_y)^2 & -\mathrm{i}k_x\mathrm{i}k_y \\ -\mathrm{i}k_x\mathrm{i}k_y & (\mathrm{i}k_x)^2 \end{pmatrix} \frac{\zeta_0}{2\kappa^2\Gamma_0}[1 - R^+_{H;0}]. \tag{4.194}$$

We then write

$$\frac{\Gamma_{v;0}}{2\kappa^2\sigma_0}[1 + R^+_{V;0}] = \frac{\Gamma_{v;0}}{2\kappa^2\sigma_0} \frac{(1 + r^+_{V;0})[1 + R^+_{V;1}\exp(-\Gamma_{v;1}d_1)]}{1 + r_{V;0}R^+_{V;1}\exp(-\Gamma_{v;1}d_1)}, \tag{4.195}$$

$$\frac{\zeta_0}{2\kappa^2\Gamma_0}[1 - R^+_{H;0}] = \frac{\zeta_0}{2\kappa^2\Gamma_0} \frac{(1 - r^+_{H;0})[1 - R^+_{H;1}\exp(-\Gamma_1 d_1)]}{1 + r_{H;0}R^+_{H;1}\exp(-\Gamma_1 d_1)}. \tag{4.196}$$

We use

$$\frac{\Gamma_{v;0}}{2\kappa^2\sigma_0}(1 + r^+_{V;0}) = \frac{\Gamma_{v;1}\Gamma_{v;0}}{\kappa^2(\sigma_0\Gamma_{v;1} + \sigma_1\Gamma_{v;0})}, \tag{4.197}$$

$$\frac{\zeta_0}{2\kappa^2\Gamma_0}(1 - r^+_{H;0}) = \frac{\zeta_0}{\kappa^2(\Gamma_0 + \Gamma_1)} = \frac{\Gamma_1 - \Gamma_0}{\kappa^2\sigma_1}. \tag{4.198}$$

We can now substitute these results into equation 4.194 and invoke the diffusive approximation in the results. This gives

$$\begin{pmatrix} \tilde{E}_{xx} & \tilde{E}_{xy} \\ \tilde{E}_{yx} & \tilde{E}_{yy} \end{pmatrix} = \begin{pmatrix} (\mathrm{i}k_x)^2 & \mathrm{i}k_x\mathrm{i}k_y \\ \mathrm{i}k_x\mathrm{i}k_y & (\mathrm{i}k_y)^2 \end{pmatrix} \frac{\Gamma_{v;1}}{\kappa^2\sigma_1} \frac{1 + R^+_{V;1}\exp(-2\Gamma_{v;1}d_1)}{1 - R^+_{V;1}\exp(-2\Gamma_{v;1}d_1)} + \begin{pmatrix} (\mathrm{i}k_y)^2 & -\mathrm{i}k_x\mathrm{i}k_y \\ -\mathrm{i}k_x\mathrm{i}k_y & (\mathrm{i}k_x)^2 \end{pmatrix} \frac{\Gamma_1 - \kappa}{\kappa^2\sigma_1} \frac{1 - R^+_{H;1}\exp(-2\Gamma_1 d_1)}{1 + r^+_{H;0}R^+_{H;1}\exp(-2\Gamma_1 d_1)}. \tag{4.199}$$

With these expressions the horizontal source–receiver component combinations are determined.

The space–frequency domain electric fields are obtained in a similar way, as explained in Appendix A for the whole space Green's functions. Here the resulting Fourier–Bessel transforms are given by

$$\begin{pmatrix} \hat{E}_{xx} & \hat{E}_{xy} \\ \hat{E}_{yx} & \hat{E}_{yy} \end{pmatrix}(x,y,0,0,\omega) = -\begin{pmatrix} \frac{x^2}{r^2} & \frac{xy}{r^2} \\ \frac{xy}{r^2} & \frac{y^2}{r^2} \end{pmatrix}\mathcal{I}_{V;0} - \begin{pmatrix} \frac{y^2}{r^2} & -\frac{xy}{r^2} \\ -\frac{xy}{r^2} & \frac{x^2}{r^2} \end{pmatrix}\mathcal{I}_{H;0} + \begin{pmatrix} \frac{x^2-y^2}{r^3} & \frac{2xy}{r^3} \\ \frac{2xy}{r^3} & \frac{y^2-x^2}{r^3} \end{pmatrix}\mathcal{I}_1, \tag{4.200}$$

and the three integrals are given by

$$\mathcal{I}_{V;0} = \frac{1}{2\pi\sigma_1}\int_{\kappa=0}^{\infty} \frac{1+R^{+}_{V;1}\exp(-2\Gamma_{v;1}d_1)}{1-R^{+}_{V;1}\exp(-2\Gamma_{v;1}d_1)}\Gamma_{v;1}\mathrm{J}_0(\kappa r)\kappa d\kappa, \tag{4.201}$$

$$\mathcal{I}_{H;0} = \frac{1}{2\pi\sigma_1}\int_{\kappa=0}^{\infty} \frac{1-R^{+}_{H;1}\exp(-2\Gamma_1 d_1)}{1+r^{+}_{H;0}R^{+}_{H;1}\exp(-2\Gamma_1 d_1)}(\Gamma_1-\kappa)\mathrm{J}_0(\kappa r)\kappa d\kappa, \tag{4.202}$$

$$\mathcal{I}_1 = \frac{1}{2\pi\sigma_1}\int_{\kappa=0}^{\infty}\left[\frac{1+R^{+}_{V;1}\exp(-2\Gamma_{v;1}d_1)}{1-R^{+}_{V;1}\exp(-2\Gamma_{v;1}d_1)}\Gamma_{v;1} - \frac{1-R^{+}_{H;1}\exp(-2\Gamma_1 d_1)}{1+r^{+}_{H;0}R^{+}_{H;1}\exp(-2\Gamma_1 d_1)}(\Gamma_1-\kappa)\right]\mathrm{J}_1(\kappa r)d\kappa. \tag{4.203}$$

These integrals can be evaluated in the same way as explained in Section 4.4.1.

The electric field components given in equation 4.200 are impulse response functions in the frequency domain. Time domain impulse response functions and unit step switch-off response functions can then be obtained with the fast cosine transform method and unit step switch-on response functions can be obtained with the fast sine transform method, as explained in Section 4.4.2.

5
Numerical Examples

In this chapter we present numerical examples of the electric field in a whole space and a half-space, in the frequency domain and in the time domain. We begin with the field separated into the transverse electric (TE) and transverse magnetic (TM) modes. We also show the total vector magnitude in the horizontal and vertical planes that contain the electric current source vector. We then present numerical frequency domain and transient time domain results in the marine and land CSEM settings, where we restrict ourselves to in-line components and receiver positions at various offsets. In the marine setting we start by showing results at a single frequency for a towed source and sea-bottom receivers. We show the electric field in the presence and absence of a buried resistive layer in both amplitude and phase as a function of offset. We then identify the components of fields that diffuse in the upwards or downwards direction away from the source and are received in the upwards or downwards direction at the sea floor. This distinction is made for the TE and TM modes separately so that the most important contributions to the total electric field can be identified and understood physically. We repeat this analysis in a shallow sea configuration, where the results for a source towed close to the sea floor and sea-bottom receivers are compared with results for a shallow towed source and towed receivers in the middle of the sea. We continue by showing total electric fields and normalised amplitudes of differences in the fields in the presence and absence of a buried resistive layer. This is done for impulse responses in the frequency domain and in the time domain as a function of offset. In the land setting we investigate a half-space in which a buried resistive layer is present at 1 km depth and at 3 km depth. We end the chapter with a land example of a conductive layer buried at 2 km depth. All results presented here are coded by the authors. These results can be reproduced with an independent code from Werthmüller, and the corresponding information on the Python code is presented by Werthmüller (2017).[1]

[1] A ZIP file containing Jupyter Notebooks with Python code for reproducing figures in the book can be found at www.cambridge.org/csem. The ZIP file also contains the current version of the modeller empymod,

5.1 The Electric Field in a VTI Whole Space

The whole space electric field response to an impulsive electric dipole is described in terms of the TE and TM mode Green's function matrices in equations 4.69–4.72. The total electric field is the sum of the modes and is given in equation 4.73. These are all expressions in the frequency domain and can be used to visualise the field when the source operates at a single frequency f by taking $s = -2i\pi f$ in these expressions.

We show the TE mode and TM mode electric field and the total electric field magnitudes generated by an x-directed electric dipole in the (x, z)-plane at $y = 0$ and in the (x, y)-plane at $z = z_s$, avoiding the source location in the plots. An isotropic whole space is modelled with $\lambda = 1$ and a VTI whole space is modelled with $\lambda = 2$. In both models $\sigma = 1$ S/m and the source frequency of operation is $f = 1$ Hz. The results for the separate modes are shown in Figure 5.1 for an isotropic whole space, where the greyscale bar indicates the logarithmic values of the magnitudes of the electric field vector in the planes. This colour scheme is used for all the greyscale plots in this section. The left column of Figure 5.1 shows that the TE and TM modes have low-amplitude zones that are perpendicular to each other in the horizontal plane, whereas in the vertical plane, shown in the right column, there is a common factor that shows up as a light (high-amplitude) bar around $x = 0$. It can be seen by comparing equations 4.70 and 4.72 that they have terms that cancel each other for small x-distances. This is the effect of separating the two modes mathematically. Figure 5.2 shows the total electric vector magnitude in the (x, y)-plane (left) and the (x, z)-plane (right) and it can be seen that the two graphs are equal. This is because, in an isotropic whole space, the electric field vector magnitude is a body of revolution around the axis containing the source vector, which is the x-axis in this example. At 1 Hz and with $\sigma = 1$ S/m, the skin distance is $\delta \approx 503$ m and the wavelength is 3.16 km. This means that for most of the plotted response $R \gg \delta$. The amplitude behaviour in both plots of Figure 5.2 shows that the field at distances more than 6 km from the source is at least 5.5 orders of magnitude smaller than the largest amplitude shown.

Figure 5.3 shows the mode results and Figure 5.4 shows the total electric field vector magnitudes for $\lambda = 2$. The TE mode field results are the same as in the isotropic model, because this mode does not depend on σ_v. The TM mode field has a much stronger magnitude now that $\lambda = 2$, because the exponential damping in the horizontal plane is reduced by the factor λ. In the bottom plot of the right column

and instructions for running the programs. Evolving versions of the codes can also be accessed at https://github.com/empymod/csem-ziolkowski-and-slob. These codes are provided courtesy of Dr Dieter Werthmüller and are made available under an Apache License (Version 2.0). Full details of the license and permitted usage of the codes are also included in the ZIP file. For further information regarding the modeller empymod, please see: Werthmüller, D., 2017, An open-source full 3D electromagnetic modeler for 1D VTI media in Python: empymod: *Geophysics*, **82**(6), WB9–WB19.

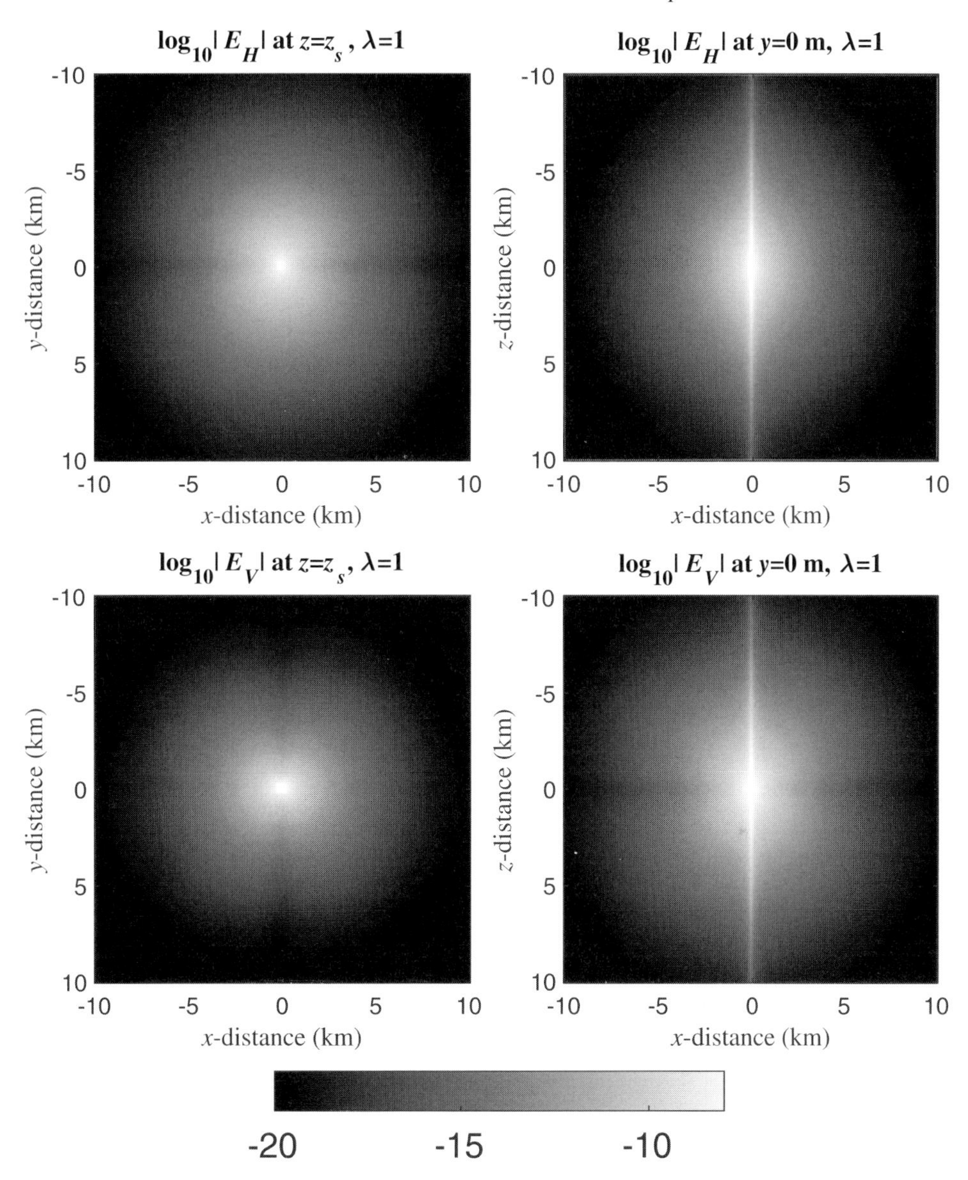

Figure 5.1 The TE mode (top) and TM mode (bottom) electric field vector magnitude in the (x, y)- (left) and the (x, z)-planes (right) for an x-directed electric current dipole operating at 1 Hz in an isotropic whole space.

it can be seen that the TM mode attenuation in the vertical direction is the same as in an isotropic medium. The field has strong magnitudes laterally, but attenuates rapidly in the vertical direction. The strong field visible at large vertical distances at small horizontal offsets is also visible in the TE mode and this cancels with the corresponding phenomenon in the TM mode in the total field, as can be seen in the right-hand plot of Figure 5.4.

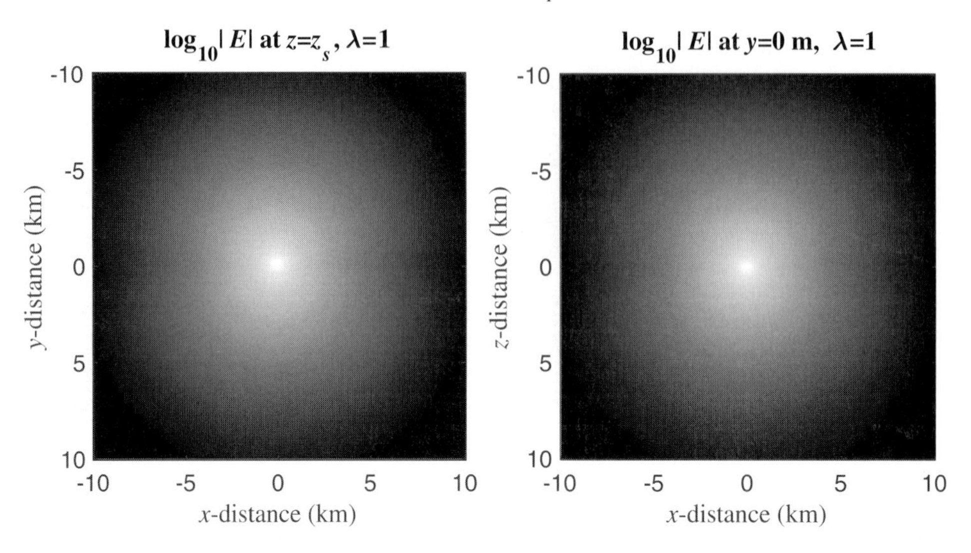

Figure 5.2 The total electric field vector magnitude in the (x, y)-plane (left) and the (x, z)-plane (right) for an x-directed electric current dipole at 1 Hz in an isotropic whole space.

For the impulse response electric fields we use the same example as for the frequency domain results and show the TE mode and TM mode electric field amplitudes generated by an impulsive x-directed electric dipole in the (x, z)-plane at $y = 0$ and in the (x, y)-plane at $z = z_s$, avoiding the source location in the plots. An isotropic whole space is modelled with $\lambda = 1$ and a VTI space is modelled with $\lambda = 2$. In both models $\sigma = 1$ S/m and the time instant at which the fields are shown is chosen as the value of τ at $R = 1262$ m at 1 Hz, which is at $t = 0.5$ s. The TM mode part of the electric field in the time domain can be found by taking all parts that have an exponential term depending on τ_v in equation 4.96. The TE mode part of the electric field in the time domain can be found by taking all parts that have an exponential term depending on τ in equation 4.96. The separate mode results are shown in Figure 5.5 for an isotropic whole space. The left column of Figure 5.5 shows again that the TE and TM modes have low amplitude zones that are perpendicular to each other in the horizontal plane. In the vertical plane, shown in the right column, there is a common factor. It can be seen from equation 4.96 that the last two terms vanish for isotropic media. Figure 5.6 shows the total electric vector magnitude in the (x, y)- (left) and the (x, z)-planes (right) and it can be seen that the two graphs are equal. As in the frequency domain, this is because, in an isotropic space, the electric field vector magnitude is a body of revolution around the axis containing the source vector. In the isotropic model the time instant $t = 0.5$ s means that for distances $R \gg 1262$ m, $\tau \gg t$, and the plots show the

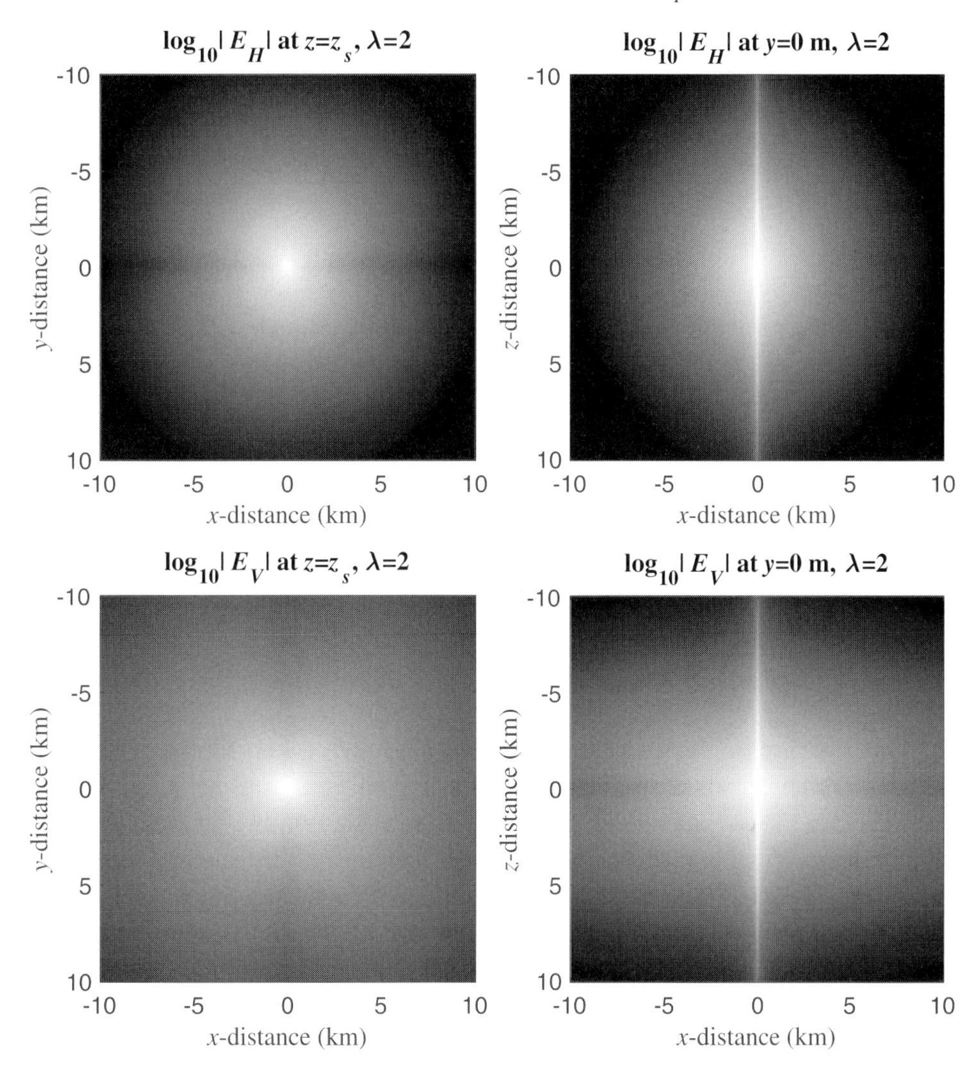

Figure 5.3 The TE mode (left) and TM mode (right) electric field vector magnitude in the (x, y)-plane (top) and the (x, z)-plane (bottom) for an x-directed electric current dipole operating at 1 Hz in a VTI whole space with $\lambda = 2$.

electric field in the early-time regime, whereas for smaller distances $R \ll 1262$ m the plots show the electric field in the late-time regime, which is only near the source. This means that the plots show the behaviour of the electric field in the transition zone between early and late times for almost all offsets. The black parts in the plots have an amplitude that is at least ten orders of magnitude smaller than the maximum amplitude shown. It can be concluded from the amplitude behaviour shown in the plots that the impulse response electric field strength is extremely

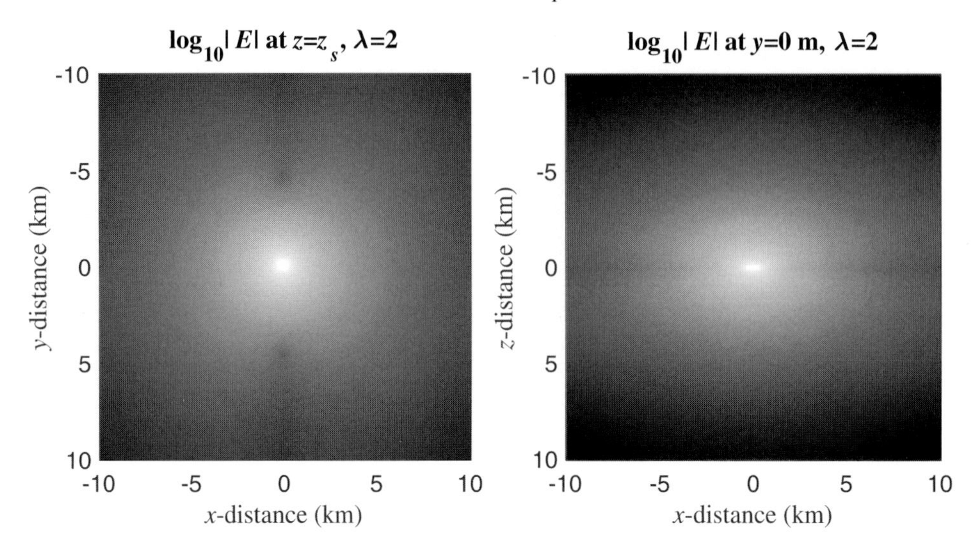

Figure 5.4 The total electric field vector magnitude in the (x, y)-plane (left) and the (x, z)-plane (right) for an x-directed electric current dipole operating at 1 Hz in a VTI whole space with $\lambda = 2$.

small for distances larger than 5 km, which means for $R > 10\delta$ at 1 Hz and that 10δ is just 1.6 times the wavelength at 1 Hz.

Figure 5.7 shows the mode results and Figure 5.8 shows the total electric field vector magnitudes for $\lambda = 2$. The TE mode field results are the same as in the isotropic model, because this mode does not depend on σ_v. The TM mode field has a much stronger magnitude now that $\lambda = 2$, because the exponential damping in the horizontal plane is reduced by the factor λ and the TM mode of the diffusive field spreads much faster in the horizontal directions than in depth. In the bottom plot in the right column it can be seen that the TM mode attenuation in the vertical direction is the same as in an isotropic medium. The field has strong magnitude laterally, but attenuates rapidly in the vertical direction. The strong field visible at large vertical distances at small horizontal offsets is also visible in the TE mode and this cancels with the corresponding phenomenon in the TM mode in the total field, as can be seen in the right-hand plot of Figure 5.8. For the TM mode that propagates in the vertical direction there is no change compared with the isotropic situation. For the TM mode that propagates in the horizontal plane, the time instant $t = 0.5$ s is seen as a diffusion time corresponding to a horizontal skin distance $\delta = 2500$ m, because $\lambda = 2$. This means that at $z = z_s$ for horizontal distances $r \gg 2500$ m, $\tau_v \gg t$, and the plots show the electric field in the early-time regime, whereas for distances $r \ll 2500$ m the plots show the electric field in the late-time regime, which is only near the source. This means that the plots show the behaviour

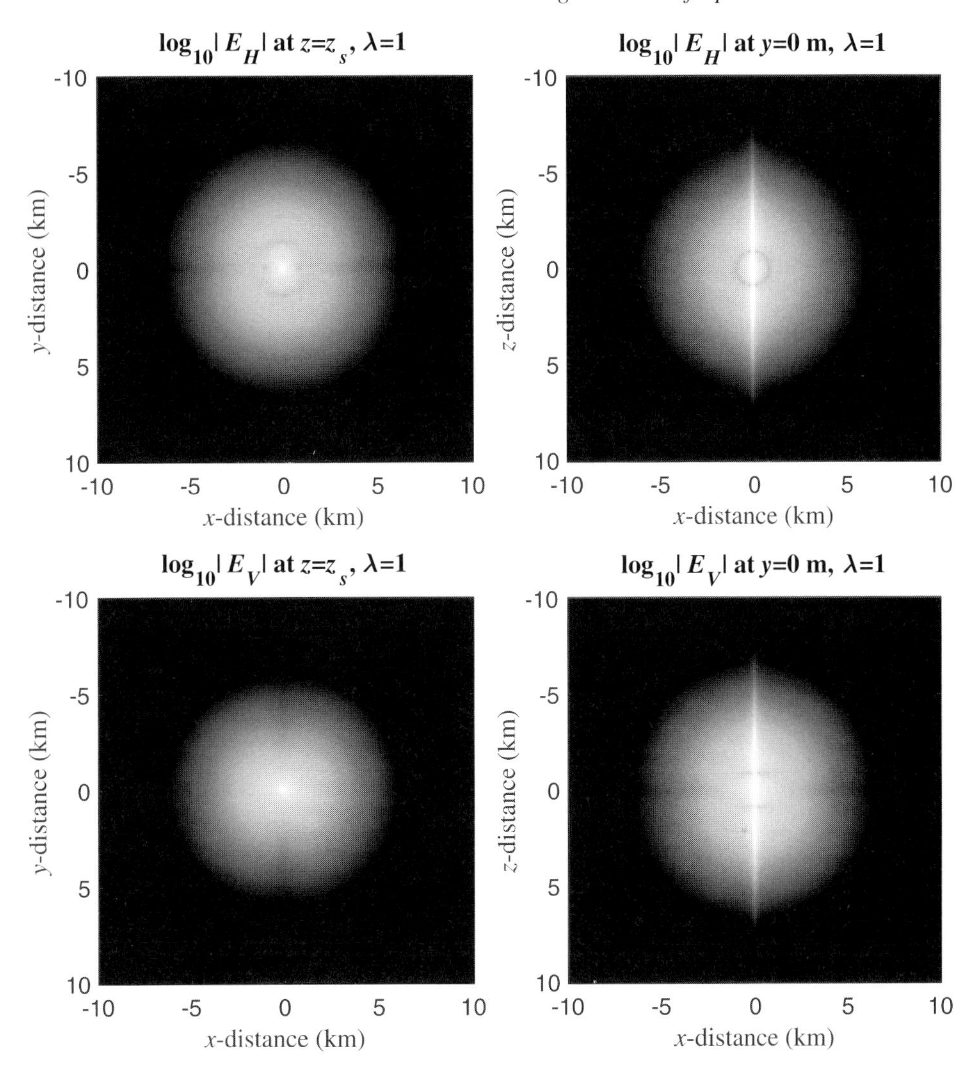

Figure 5.5 The TE mode (top) and TM mode (bottom) electric field vector magnitude at $t = 0.5$ s in the (x, y)-plane (left) and the (x, z)-plane (right) for an x-directed electric current dipole in an isotropic whole space.

of the electric field in the transition zone between early and late times for almost all offsets.

5.2 The Electric Field in a Homogeneous Half-Space

For the homogeneous half-space example, we take the situation of a sea as lower half-space with air as upper half-space, and their interface at $z = 0$ m. The sea is an isotropic layer with conductivity $\sigma = 3.2$ S/m. We place the x-directed electric

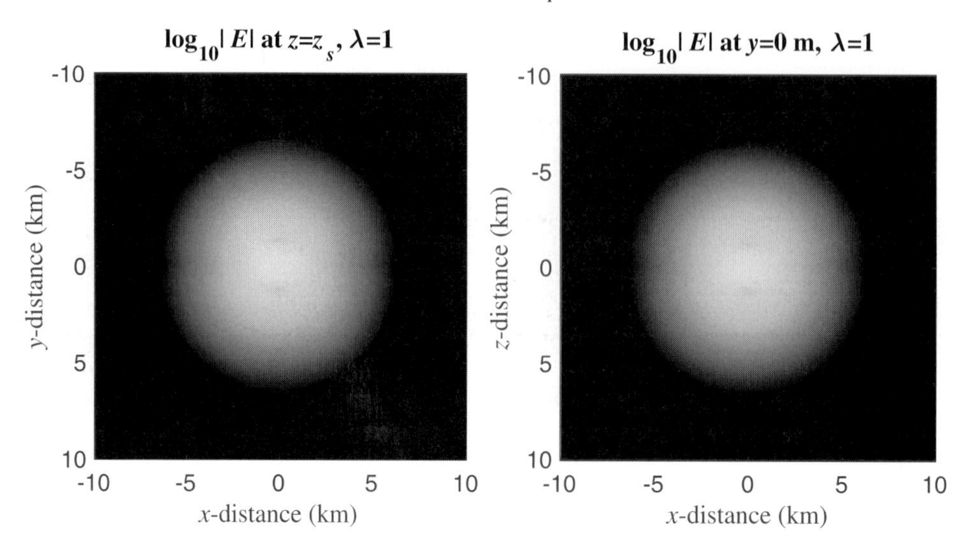

Figure 5.6 The total electric field vector magnitude at $t = 0.5$ s in the (x, y)-plane (left) and the (x, z)-plane (right) for an x-directed electric current dipole in an isotropic whole space.

dipole at two depths, at 10 m below the sea surface and at 975 m below the sea surface. The in-line and vertical components of the electric field are shown together with the total electric field magnitude in the (x, z)-plane in Figure 5.9. The left column of the figure shows results for a source at 975 m depth and the right column shows the results for a source at 10 m depth. The x-component of the electric field is shown in the top row, the z-component is shown in the middle row, and the total electric field magnitude is shown in the bottom row. Comparing the two graphs in the top row, it can be seen that the air wave is very strong for the shallow source and its influence extends over the whole offset range and down to beyond 1 km depth. The vertical field shown in the middle row is forced to zero at the surface. For this reason the field amplitude for the deep source is slightly asymmetric, whereas for the shallow source the surface amplifies the vertical field component towards depth. The total electric field magnitude below the source depth is dominated by the direct field for all offsets for the deep source, whereas it is dominated by the air wave for most offsets for the shallow source, as can be seen in the graphs in the bottom row. It is difficult to see in these graphs that at relatively large offsets the air wave contribution in the subsurface is a vertically downward-diffusing field. This is because the path in the air still has near-field geometrical spreading causing the amplitude to vary inversely as distance cubed. By looking at the phase of the field we expect lines of constant phase to be horizontal when the air wave behaves as a horizontal plane wave diffusing vertically down. This is shown in Figure 5.10

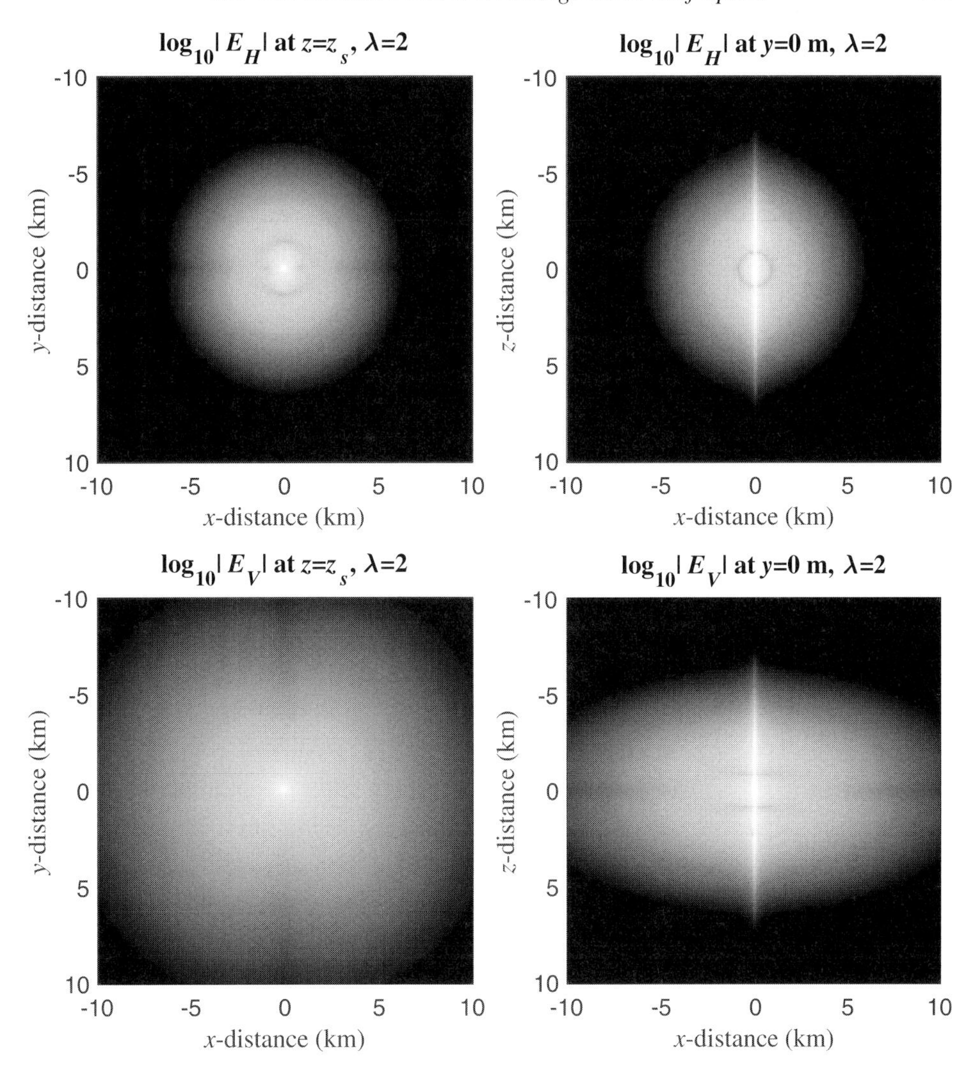

Figure 5.7 The TE (left) and TM mode (right) electric field vector magnitude at $t = 0.5$ s in the (x, y)-plane (top) and the (x, z)-plane (bottom) for an x-directed electric current dipole in a VTI whole space with $\lambda = 2$.

for the deep (left) and shallow (right) source. For the deep source the lines of constant phase become horizontal at the surface from 1.5 km offset onwards, and this increases to 4 km offset onwards at 2 km depth. For the shallow source this is the case for offsets larger than approximately 2 km at all depths.

Snapshots of the electric field impulse response for the same two source depth levels are shown in Figure 5.11 for $t = 0.5$ s. The left column of the figure shows results for a source at 975 m depth and the right column shows results for a source

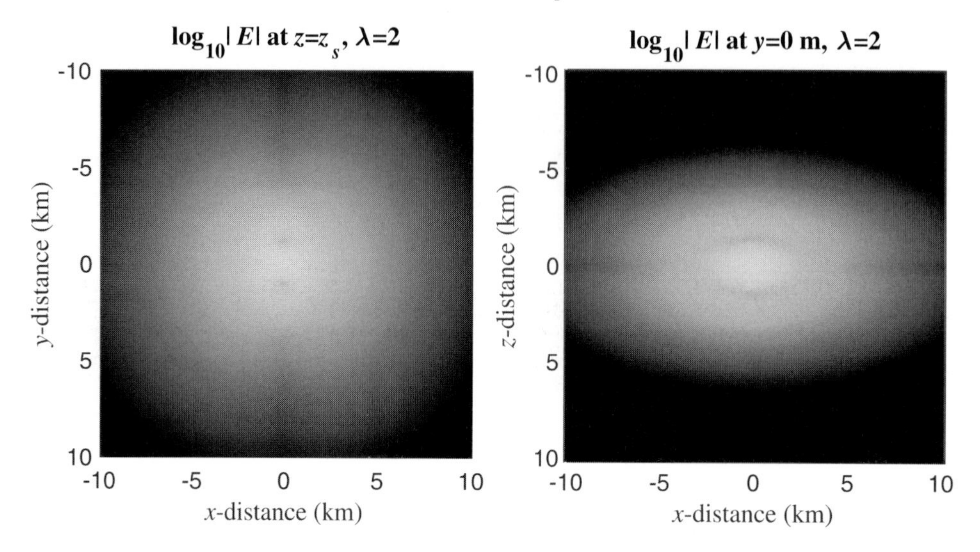

Figure 5.8 The total electric field vector magnitude at $t = 0.5$ s in the (x, y)-plane (left) and the (x, z)-plane (right) for an x-directed electric current dipole in a VTI whole space with $\lambda = 2$.

at 10 m depth. The x-component of the electric field is shown in the top row, the z-component is shown in the middle row and the total electric field magnitude is shown in the bottom row. Comparing the two graphs in the top row, it can be seen that the air wave is very strong for the shallow source and its influence extends over the whole offset range and down to almost 2 km depth for offsets larger than 2 km. The vertical field shown in the middle row is forced to zero at the surface. For this reason the field amplitude for the deep source is slightly asymmetric, whereas for the shallow source the surface amplifies the vertical field component towards depth. The total electric field magnitude below the source depth is dominated by the direct field for all offsets for the deep source, whereas it is dominated by the air wave for most offsets for the shallow source, as can be seen in the graphs in the bottom row.

Instead of looking at cross-section snapshots, we can also look at the space–time behaviour along a receiver line. Figure 5.12 shows the electric field components along a receiver line as a function of time, with time on a logarithmic axis from 0.1 ms to 100 s for the deep (left column) and shallow (right column) sources. The in-line electric field component has been decomposed into the air-wave (top row) and the direct wave plus specular reflected field (second row). The third row shows the vertical component and the last row shows the total electric field magnitude. The air wave arrives late and is weak for the deep source, whereas it is strong and arrives early for the shallow source. It has no visible time delay but there is amplitude loss as a function of offset. The direct and specular reflected parts of the

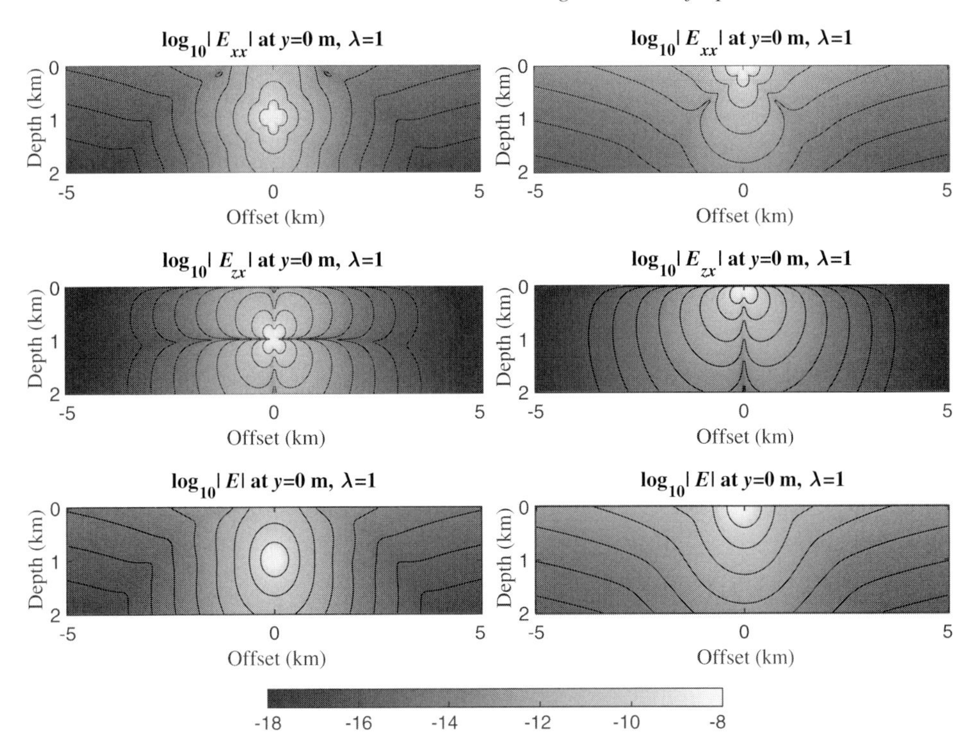

Figure 5.9 The amplitudes of the x-component (top) and z-component (middle) of the electric field, and total electric field magnitude (bottom) in the (x, z)-plane for an x-directed electric dipole source operating at 1 Hz at 975 m (left) and 10 m (right) below surface of a conductive half-space with $\sigma = 3.2$ S/m. The colour bar indicates the amplitude in logarithmic scale. For the colour version, please refer to the plate section.

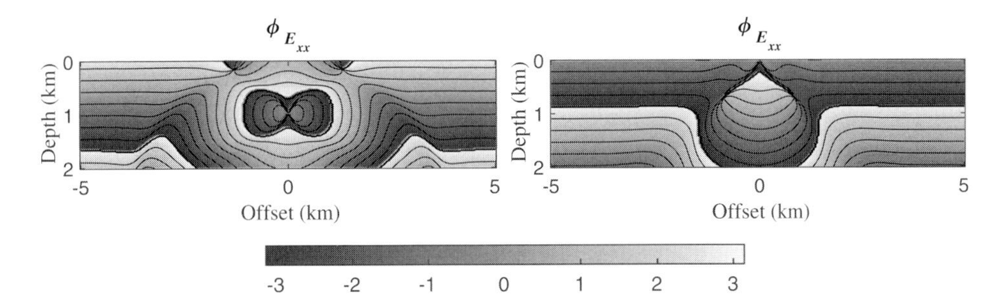

Figure 5.10 The phases of the x-component of the electric field in the (x, z)-plane for a source operating at 1 Hz at 975 m (left) and 10 m (right) below the surface of a conductive half-space with $\sigma = 3.2$ S/m. For the colour version, please refer to the plate section.

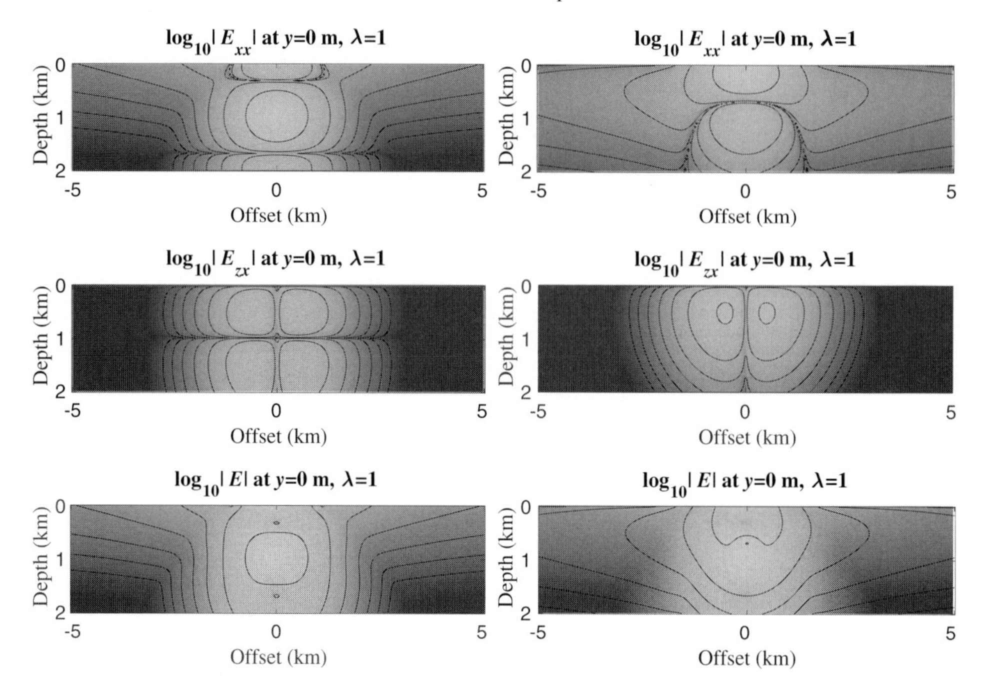

Figure 5.11 Snapshots at $t = 0.5$ s showing the amplitudes of the x-component (top), z-component (middle) of the electric field, and total electric field magnitude (bottom) in the (x, z)-plane for an impulsive source at 975 m (left) and 10 m (right) below the surface of a conductive half-space with $\sigma = 3.2$ S/m. For the colour version, please refer to the plate section.

inline component are quite similar to the vertical components for both deep and shallow sources. It can be seen that the shallow source has a small amplitude dip just after 10 ms that is absent for the deep source. This is caused by the difference in the vertical source–receiver distance and is caused by the direct field. The vertical component generated by the shallow source is visibly stronger for larger distances and times than that from the deep source, which is because of the mirror effect of the sea surface for the vertical electric field component. The total field magnitude from the deep source is mostly affected by the air wave for times later than 1 s and for offsets larger than 2 km. For the shallow source the air wave is strong for almost all offsets and for times before 10 ms, while from 2 km offset onward, the direct and reflected fields are clearly the strongest parts after 1 s.

Figure 5.13 shows the step switch-off response corresponding to the impulse responses shown in Figure 5.12. It can be seen that the time windows of the air wave and the direct and specular reflected fields overlap completely. By comparing the top and second-row graphs it can be seen that the air wave has a faster time decay

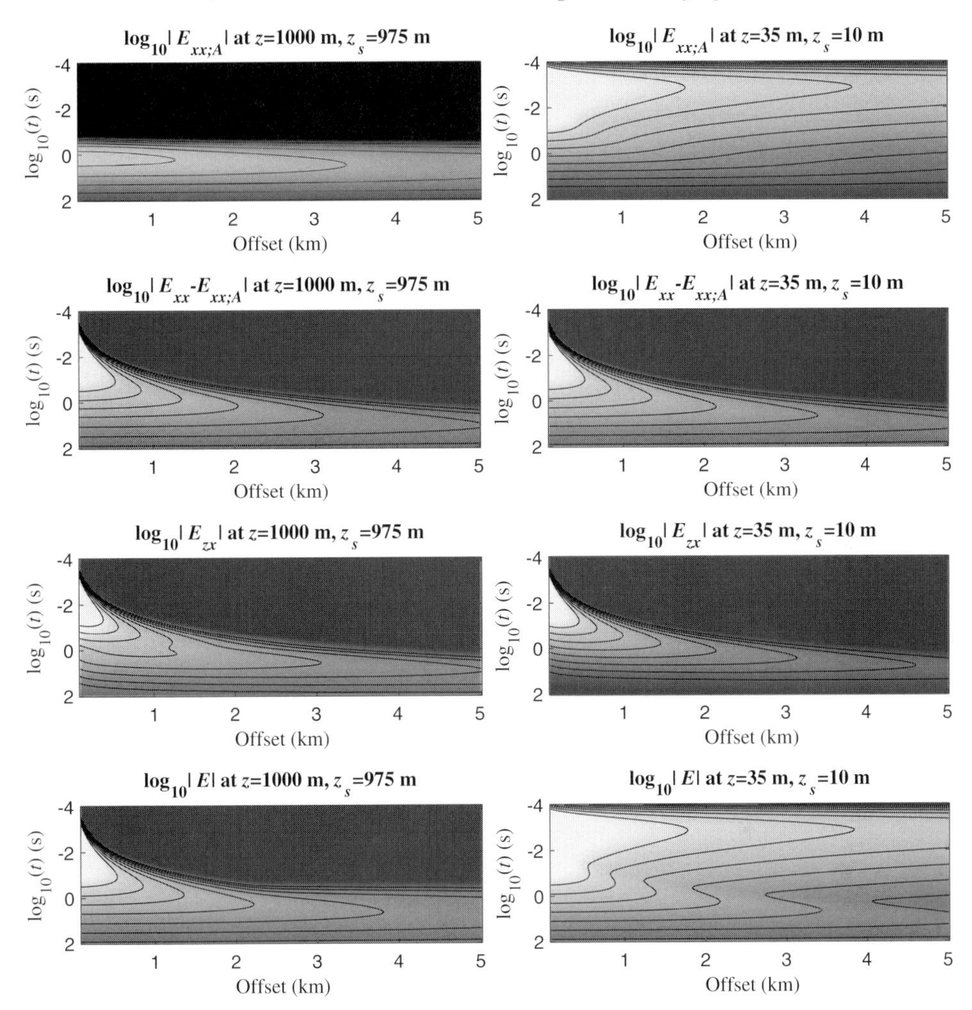

Figure 5.12 Amplitudes of the electric field impulse response 25 m below the source decomposed into the air wave (top), direct plus specular reflection (second) in the x-component, and z-component (third), and total electric field magnitude (bottom) for a source at 975 m (left) and 10 m (right) below the surface of a conductive half-space with $\sigma = 3.2$ S/m. For the colour version, please refer to the plate section.

than the direct and reflected fields. In the third row the vertical electric field component is shown and from these graphs it can be seen that the direct field and specular reflection interfere destructively, and therefore the shallow source response attenuates much faster than the deep source response. The total field is dominated by the in-line component of the direct field for the deep source whereas the shallow source total field response is a mixture of air wave and direct and specular reflected fields.

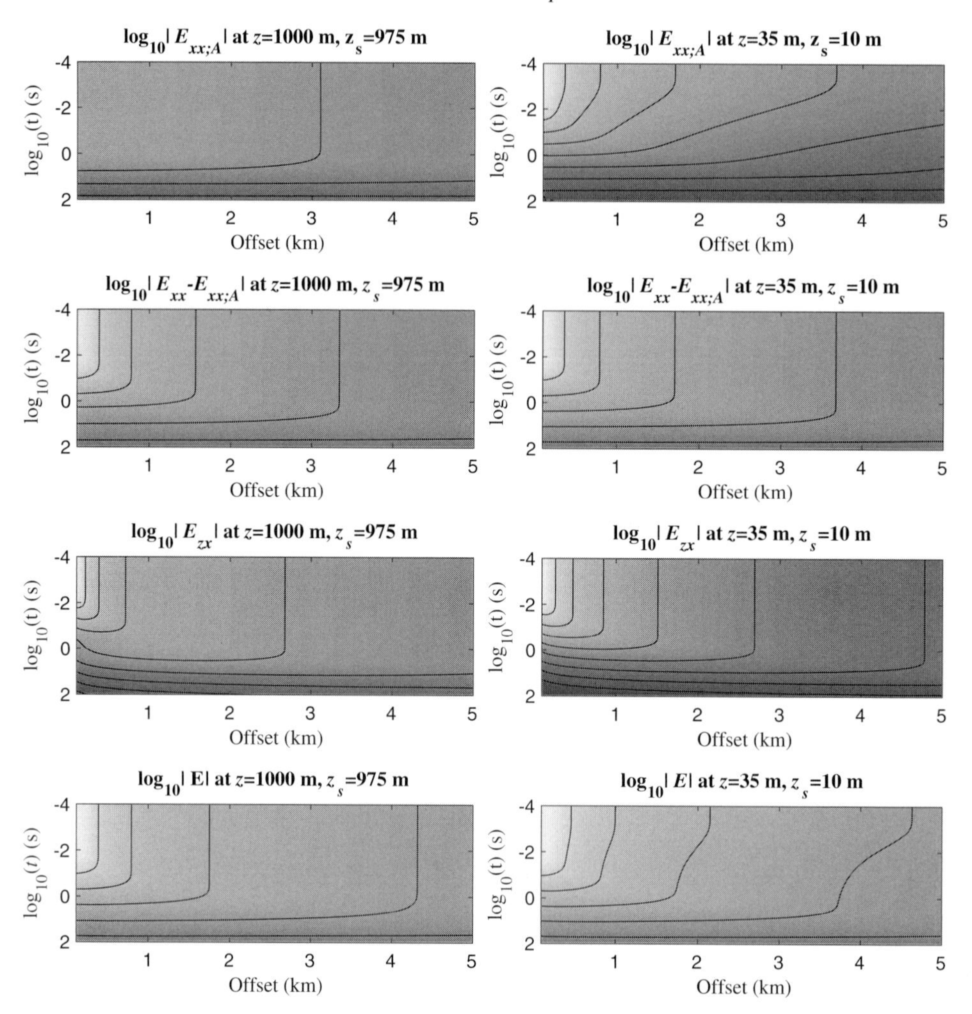

Figure 5.13 Amplitudes of the electric field step switch-off response 25 m below the source decomposed in the air wave (top) and direct plus specular reflection (second) in the x-component, and z-component (third), and total electric field magnitude (bottom) for a source at 975 m (left) and 10 m (right) below the surface of a conductive half-space with $\sigma = 3.2$ S/m. For the colour version, please refer to the plate section.

To understand the similarities and differences of the responses from a point dipole and an elongated dipole, we show the unit impulse response for a 1 km long source dipole measured with 100 m between the receiver electrodes and compare them with the equivalent point-source–point-receiver responses at 1 km, 3 km and 5 km between the midpoints of the source and receiver dipoles on the surface of a homogeneous isotropic half-space with $\sigma = 1$ S/m. The left plot of Figure 5.14

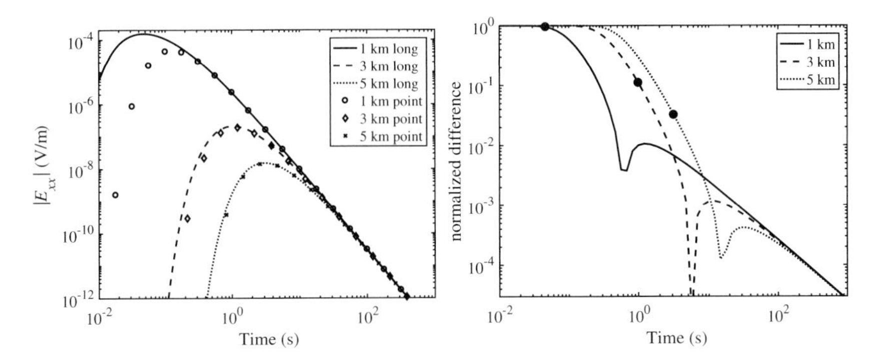

Figure 5.14 Homogeneous isotropic half-space impulse response (left plot) with a 1 km source and a 100 m electric dipole at 1, 3 and 5 km offsets (solid, dashed and dotted lines, respectively) at the surface compared with the corresponding point-source–point-receiver responses (circles, diamonds and crosses, respectively); normalised difference between responses with elongated devices and point devices with the location of the maximum amplitudes indicated by dots (right plot).

shows the electric field impulse response computed at the surface using equation 4.143 in solid, dashed and dotted lines for long dipoles at 1 km, 3 km and 5 km offsets, respectively, and computed from equation 4.140, corrected for the dipole lengths in circles, diamonds and crosses for point devices and the same offsets, respectively. The plot suggests that the error made by assuming the source to be a point is much larger for short offsets than for larger offsets. The right plot shows that the actual normalised error is indeed very large for early times at 1 km offset, but also decreases at earlier times than for the larger offsets. The dots in the right plot show the time instants at which the electric field responses have their maximum value. The error made by computing the responses with point devices is less than 1% after 0.5 s at 1 km offset, which is where the electric field is more than an order of magnitude below its maximum value, whereas for 3 km this occurs after 2.7 s and for 5 km offset after 5 s. Only for offsets of 10 km and greater is the error as low as 1% at or before the maximum field strength has occurred.

5.3 The Electric Field in a Marine CSEM Setting

5.3.1 Isotropic Model Results in the Frequency Domain

The effect of the sea surface and of the sea floor

In Section 4.3 we derive closed form expressions for the electric field in a VTI half-space assuming the diffusive approximation can be used in air, by taking $\sigma = 0$ and $\varepsilon_0 = 0$. Because of this assumption the TM mode reflection response from the half-space boundary can be found explicitly in the space–frequency and space–

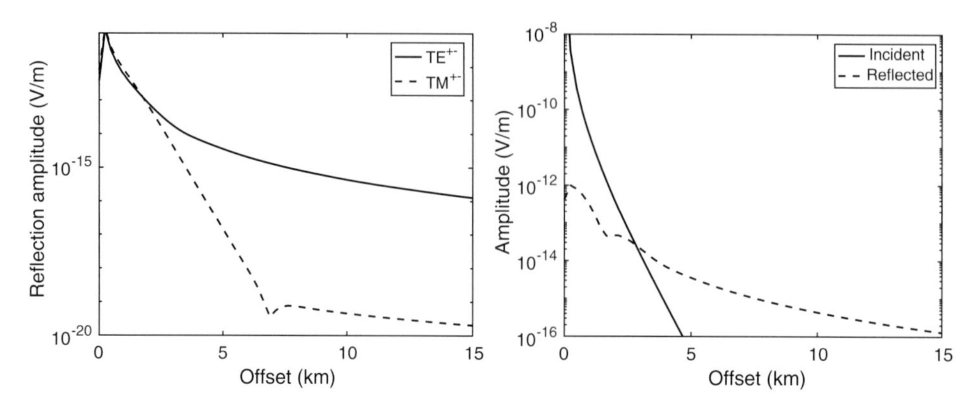

Figure 5.15 Surface reflections at $f = 0.5$ Hz in TE and TM modes (left) and incident and reflected electric fields (right).

time domains. Here we use our integral expressions of Section 4.4 without making this assumption and keeping ε_0 as its free-space value. We model a homogeneous half-space with a conductivity of $\sigma = 3$ S/m, the x-directed electric dipole at 975 m operating at $f = 0.5$ Hz, and we compute the x-component of the electric field at 1 km below the surface. We compute the TE and TM mode surface reflections separately using the free-space value of ε_0 in the air. Figure 5.15 shows in the left plot the reflection of the surface in the TE mode (solid line) and in the TM mode (dashed line). It can be seen that the TM mode also has an air wave whose presence becomes clearly visible at 7 km offset. It can also be seen that the TM mode reflection is almost four orders of magnitude smaller than the TE mode at those offsets. For the streamer model with a source 10 m below the surface and the receiver at 100 m below surface, the amplitudes are of course much stronger, but the four orders of magnitude difference between the reflections in both modes stays intact. We can therefore conclude that the diffusion approximation in the air works well. In the right plot of Figure 5.15 we can see that the incident field decays rapidly and the surface reflection becomes the stronger part of the signal around 3 km offset. This is the effect of the air wave in the TE mode, as can be seen from the separate mode responses in the left plot.

We also model an infinite half-space of sea with $\sigma_1 = 3$ S/m above an infinite half-space of ground with $\sigma_2 = 1$ S/m. Both half-spaces are isotropic. We put the source 25 m above the sea floor and the receivers on the sea floor. The results are shown in Figure 5.16. Now that both half-spaces are conductive, the TM mode reflection response is much stronger than the TE mode reflection response for offsets larger than a few kilometres, as can be seen in the left plot. In the right plot we see that the incident field has a slope at large offsets that is proportional to the conductivity of the upper half-space, whereas the reflected field at large

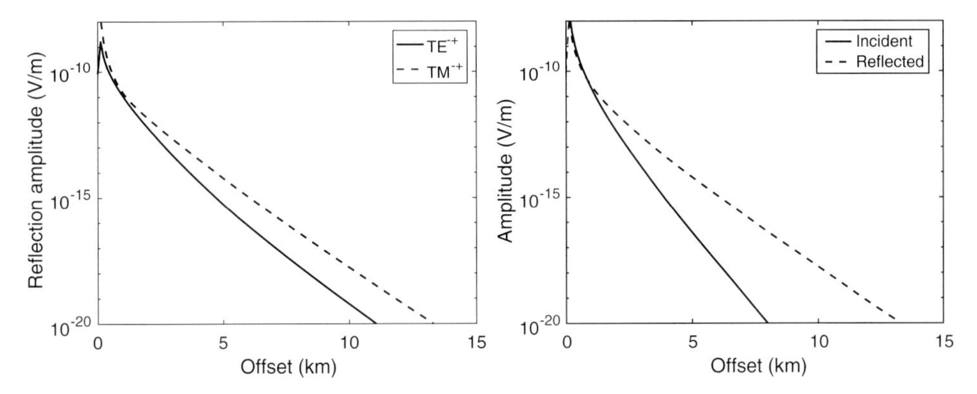

Figure 5.16 Sea-floor reflections at $f = 0.5$ Hz in TE and TM modes (left) and incident and reflected electric fields (right).

offsets has a slope proportional to the conductivity of the lower half-space. This can be physically understood as a signal that travels and attenuates just below the boundary and radiates upwards, where it is detected by the receivers. We could call this a diffusive head wave coming from the diffusive refracted wave just below the boundary. This situation is similar to that of the previous example with air as the upper medium, but with two differences. The first is the obvious one that in the previous example the source and receivers were below the boundary, whereas now they are above. Hence the reflected field originated from an upwards-diffusing field and arrived at the receivers as a downwards-diffusing field. Here the situation is opposite. The second difference is that in the previous situation with the non-conductive air, the TE mode reflection response contained the strongest head wave in the air wave response, whereas here it is the TM mode response that contains the strongest head wave component.

Presence and absence of a buried resistive layer

Deep sea model

We now use a model with a 1 km sea layer between a half-space of air at the top and a half-space below. In one situation the half-space is homogeneous and in the other situation the half-space contains a 40 m thick resistive layer whose top is located 1 km below the sea floor. The conductivity values are $\sigma = 3$ S/m for the sea, $\sigma = 1$ S/m for the half-space and $\sigma = 1/70$ S/m for the resistive layer. The source is 25 m above the sea floor, the receivers are on the sea floor, and $f = 0.5$ Hz. Figure 5.17 shows the electric field amplitude with and without the resistive layer in the left plot and the amplitude of the normalised difference in the right plot. Because we have shown that the electric field can be understood as a sum of events, it makes sense

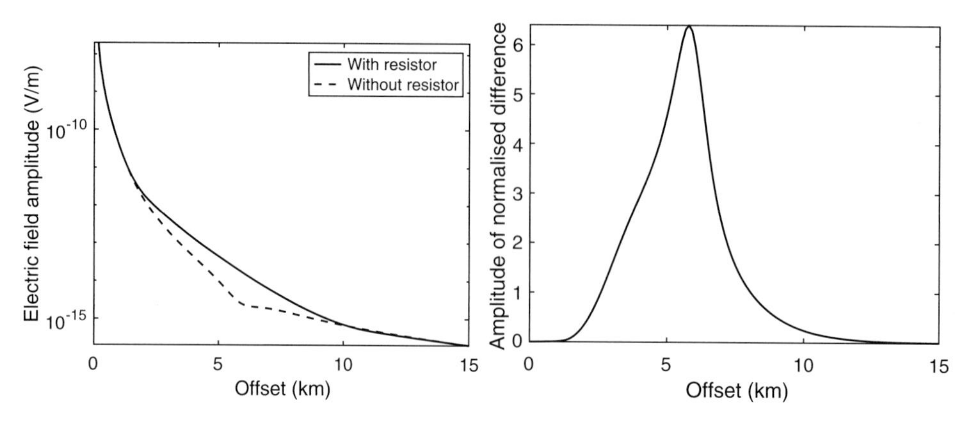

Figure 5.17 Electric field amplitudes (left) and amplitude of the normalised difference (right) for $f = 0.5$ Hz.

to look at the amplitude of the difference and not the difference in amplitudes. The physical signal at one location and for one frequency is a complex number that is the sum of several complex numbers, each of which represents a physical event. The direct field from source to receiver, the direct air wave response and the direct sea-floor response occur in both models. By subtracting these complex numbers only the other responses are left, and we take the amplitude of those. The normalisation is to the amplitude of the model without the resistive layer.

Comparing the reflected field (dashed line) in the right plot of Figure 5.16 with the model response of the model without resistor (dashed line) in the left plot of Figure 5.17, we can see that they are almost identical between 3 km and 5 km offset. This means that for those offsets the sea-floor reflection dominates the signal in the presence of the sea surface. For shorter offsets the incident field is also measurably present and for larger offsets the air wave starts to become visible. This can be seen from the change of slope that starts at 5 km offset, where it first becomes larger, and after 6 km it becomes much smaller due to the absence of attenuation in the air. Between approximately 2 km and 10 km offset there is a difference in the responses from the two models, as can be seen in the left plot of Figure 5.17. This can be due only to the following events. Multiple upward- and downward-diffusing fields are generated, because the sea is now bounded by air and by the lower half-space, and their attenuation is governed by the conductivity of the sea. The resistive layer reflects and generates internal multiples, the upward-diffusing parts of which are measured by the receivers. These create reverberations between the resistive layer and the sea floor and between the resistive layer and the sea surface, and of course all possible combinations of these occur as well. We discuss these in detail in the following when we look at the different directional contributions in the total signal.

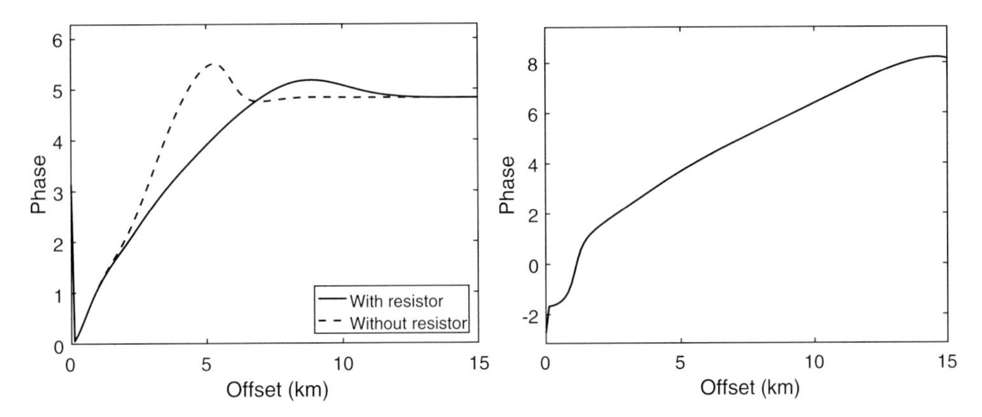

Figure 5.18 Electric field phases (left) and the phase of the difference in the fields (right).

We can see in the right plot that the amplitude of the normalised difference has a maximum of just over 6 at 6 km offset. The range of offsets for which the difference is larger than 100% is between 2.5 km and 8 km.

Another interesting property is the phase of the signals, shown in the left plot of Figure 5.18. The phase of the model without resistor advances rapidly, meaning the signal attenuates quickly. The flat phase at large offsets we have already seen as being caused by the air wave, which is the strongest event in the field. We can subtract the phases to highlight where the signals are the same and where they are different in the two models. The phase of a sum of events in a signal is non-linear in the events, whereas each event has its own distinctive phase. Because some events in both models are the same, the difference of the phases does not give a good indication of the events that are in one response and not in the other. Therefore we should look at the phase of the difference between the two electric fields. This is what we show in the right plot of Figure 5.18. We can see that the phase is nearly linear with increasing offset for a large offset range, suggesting that the difference signal is dominated by one event.

Figure 5.19 shows the contributions from the different upgoing and downgoing components at the source and receiver sides for the model with the resistive layer separated into modes, with the TE mode in the left plot and the TM mode in the right plot. In the TE mode the field that diffuses away from the source in the upward direction and arrives as a downward-diffusing field at the receivers (indicated as TE^{+-} in the figure) is the strongest signal from around 5 km offset onward. In the TM mode the field that diffuses away from the source in the downward direction and arrives as an upward-diffusing field at the receivers (indicated as TM^{-+} in the figure) is the strongest signal at all offsets. The final question is: In which part does

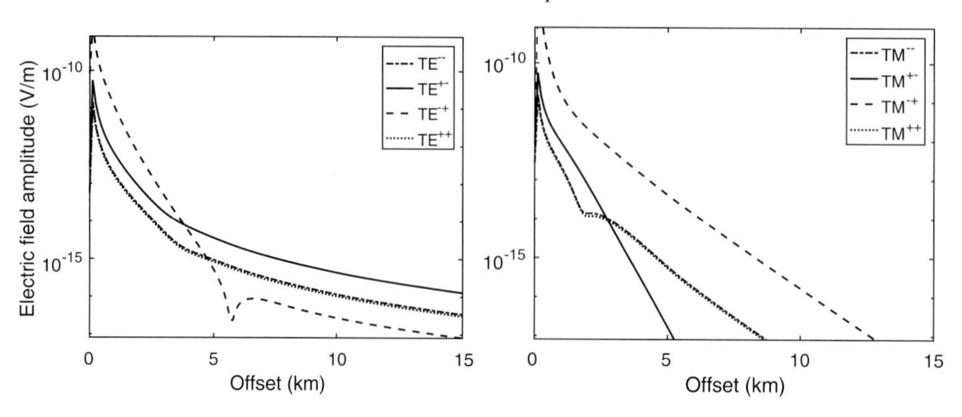

Figure 5.19 Reflections in TE (left) and TM (right) modes.

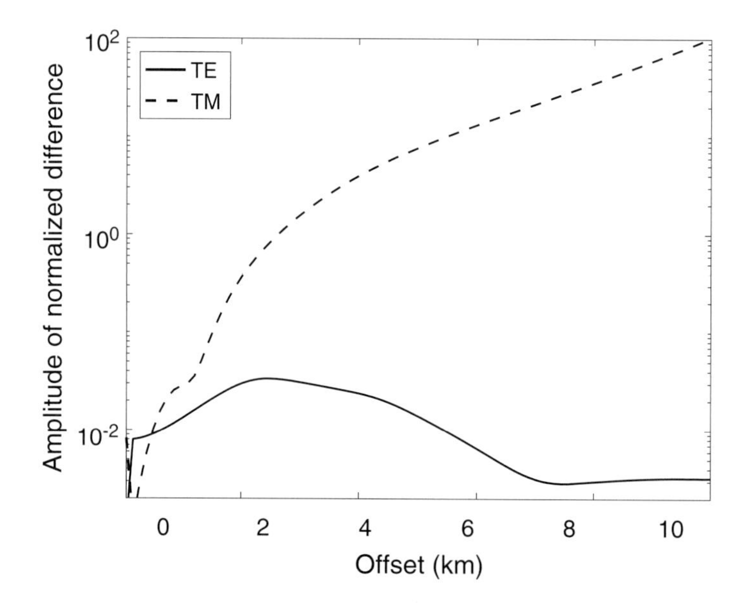

Figure 5.20 Amplitude of normalised difference of the reflected field in the TE (solid) and TM (dashed) modes.

the presence of the resistive layer show up? to answer this we sum all four directional contributions in each mode and take the amplitude of the difference between the reflected field with and without the resistor present in the model and normalise to the reflected field without the resistor. The result is shown in Figure 5.20, where the TE mode is shown with the solid line and the TM mode with the dashed line. It can be seen that the information on the presence or absence of the reservoir is in the TM mode part of the reflected field. The amplitude of the difference in the TE mode reflected field is, at its maximum, on the order of a few per cent, whereas the amplitude of the difference in the TM mode reflected field grows with offset. Of course, we do not measure the modes separately and the normalised difference between the total electric fields has a maximum amplitude shown in Figure 5.17, because the TE mode containing the air wave becomes the strongest part of the signal at large offsets.

Shallow sea model

Instead of using a 1000 m deep sea, we now use a 100 m deep sea and leave all other medium parameters the same. In this shallow sea we can investigate two acquisition geometries. The first is similar to the geometry used in the deep sea model, with a source at 25 m above the sea floor and receivers on the sea floor. The second uses a shallow towed source 10 m below the sea surface and streamer data are recorded at 50 m depth.

Figure 5.21 shows the electric field amplitude with and without the resistive layer in the left column, with sea-floor data in the top plot and streamer data in the bottom plot. The amplitude of the normalised difference is shown in the right column with the sea-floor data result in the top plot and the streamer data result in the bottom plot. The normalisation is again to the amplitude of the model without the resistive layer. The data plots show that now the amplitude of the data from the model without the resistor is larger than that of the model with the resistive layer at offsets between 3 km and 7.5 km. This can be understood from the interplay of different events in the two models, which we discuss when we look at the contributions from the directional fields in the two modes. The two plots in the right column of Figure 5.21 show that the maximum amplitude of the difference in the data from the models with and without the resistive layer is larger with the sea floor acquisition than with the streamer acquisition geometry. By comparing the plots shown in Figure 5.21 with the equivalents of the plots shown in Figure 5.17, it can be seen that the amplitude differences between presence and absence of a resistive layer in the shallow sea model is an order of magnitude smaller than in the configuration with a deep sea. It can also be seen that the maximum amplitude difference has shifted from 6 km offset in the deep sea to 3 km offset in the shallow sea. At that offset also the normalised difference in the deep sea model is small. Because the only parameter we changed in the model is the depth of the sea layer, the first reflection response of the resistive layer that is in the field that diffuses down from the source to the resistive layer and then diffuses upward to the receivers is not changed. Only events that involve reflection at the sea surface have changed for the sea-floor data acquisition configuration.

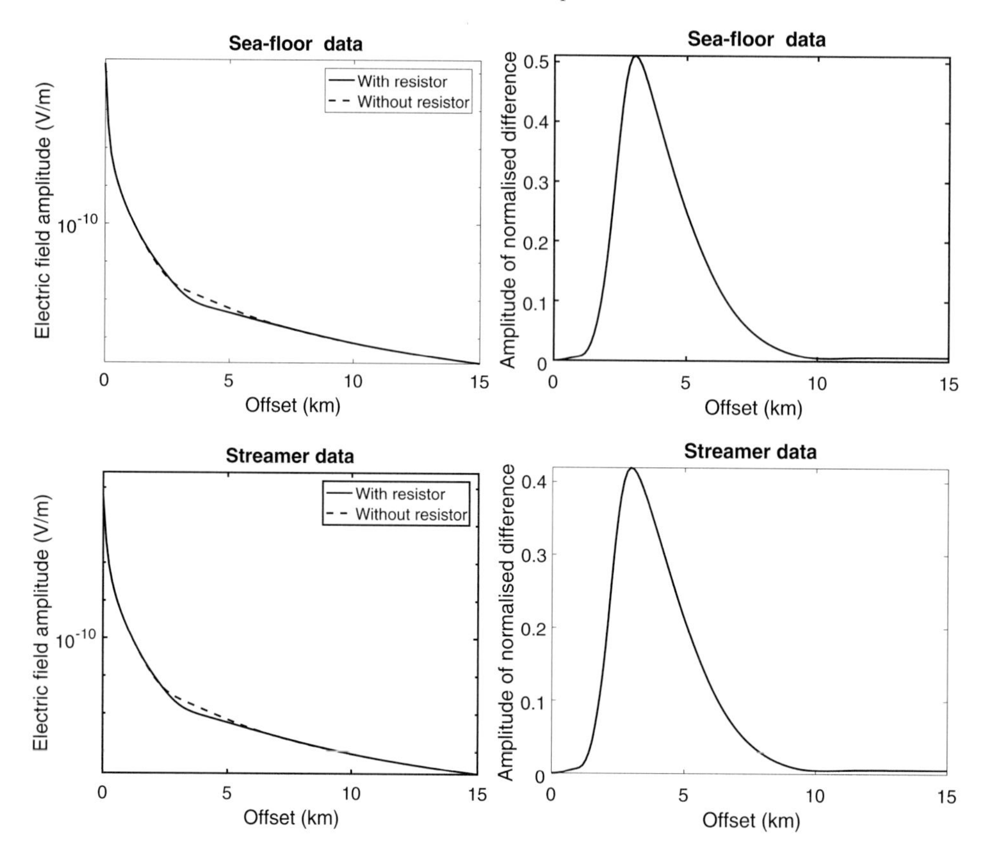

Figure 5.21 Electric field amplitudes (left) and amplitude of the normalised difference (right) for the deep source and sea floor receivers (top) and shallow sources and towed receivers (bottom); $f = 0.5$ Hz.

The phases are shown in Figure 5.22, where again the phases of the two models are shown in the left column and the phase of the electric field difference between the two models is shown in the right column, with the results for the sea-floor data in the top row and those for the streamer data in bottom row. In the left column it can be seen that for offsets between 1.5 km and 4 km the phases differ more in the sea-floor data than in the streamer data. When we subtract the electric fields and look at the phase of the resulting difference field, the resemblance of the two dataset results is striking. Between 2.5 km and 8.5 km offset the phase is almost linear, showing that there is a single strong event in the difference data. For both configurations the phase of the difference field becomes almost horizontal, suggesting that the air wave component becomes the strongest field in the difference field. This must be a secondary air wave effect that does not exist when the resistive layer is absent. The sea layer has a skin depth of 411 m at 0.5 Hz, and 100 m is a thin layer with respect

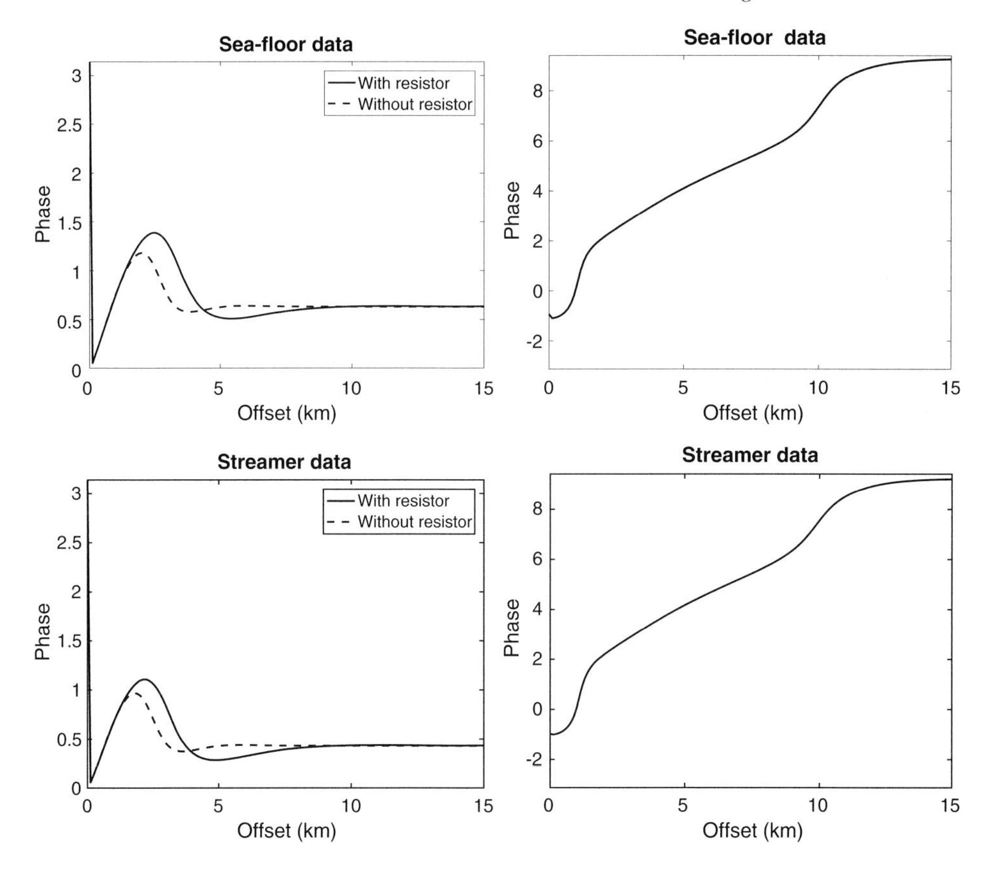

Figure 5.22 Electric field phases (left) and the phase of the difference in the fields (right) for the deep source and sea-floor receivers (top) and shallow sources and towed receivers (bottom).

to damping. The reflection from the resistive layer that diffuses upwards and reflects at the sea surface producing an air wave that diffuses down to the receiver exists in the difference field. The reciprocal field that diffuses upwards to the sea surface and then diffuses down to the reservoir and diffuses up to the receivers is also present in the difference field. This first event is part of what is indicated as TE^{++} and the last event is part of what is indicated as TE^{--} in Figure 5.23. The plots in the left column show that these are less than one order of magnitude smaller than the field containing the direct air wave, which is indicated as TE^{+-}. In the right column it can be seen that the contribution of the resistive layer is present in all directional parts of the TM mode field. In the sea-floor data they differ more because the source is close to the sea floor and the receivers are on the sea floor, whereas in the streamer data the source is close to the sea surface and the receivers are in the middle of the sea layer.

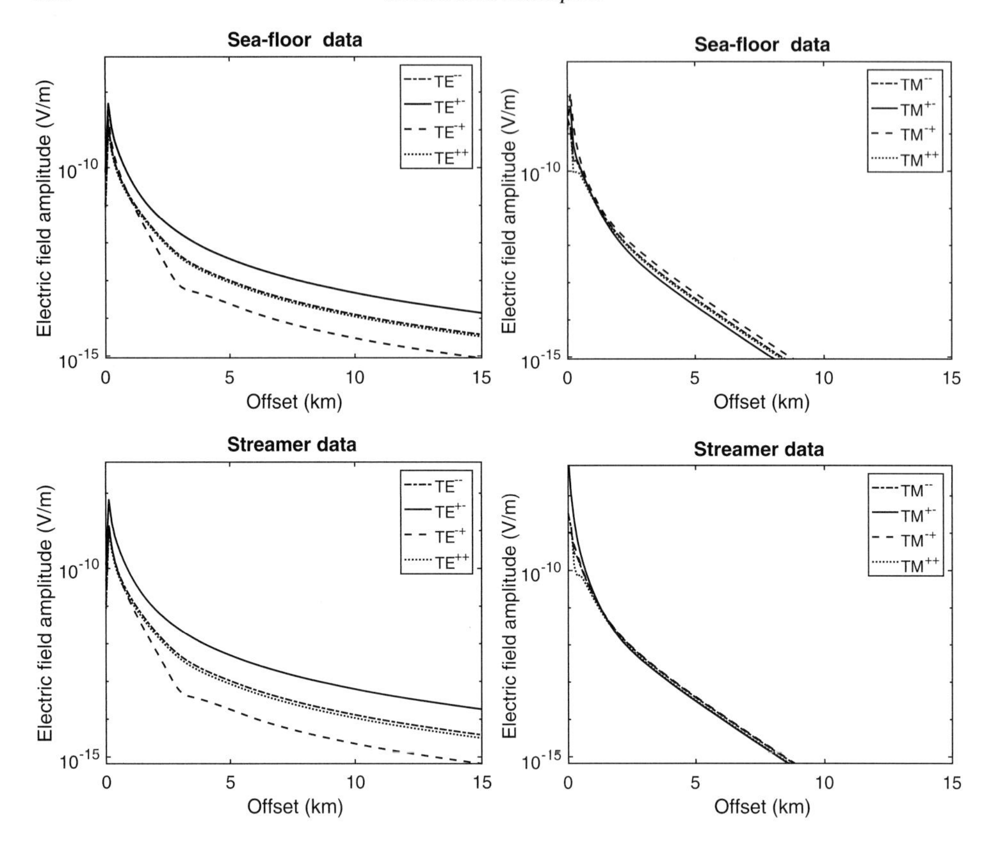

Figure 5.23 Reflections in TE (left) and TM (right) modes for the deep source and sea-floor receivers (top) and shallow sources and towed receivers (bottom).

The differences in the TE mode and TM mode parts of the fields are shown in Figure 5.24 for the sea-floor data in the left plot and for the streamer data in the right plot. The TE mode difference between absence and presence of the resistive layer is very similar to the difference in the TE mode field in the 1 km sea layer model. Compared with the TE mode normalised amplitude of the difference in Figure 5.20 in the shallow sea model, the maximum has reduced by a factor of 3 and the offset at which it occurs has increased slightly, whereas the large offset difference has increased by a factor of 2. This is because there is a stronger contribution from the air wave reflecting off the resistive layer in the shallow sea than in the deep sea. This is not relevant for measurements because the TM mode still dominates the signal at short offsets and the direct air wave dominates at middle and large offsets. Changes of less than 1% cannot be expected to be detectable in field data at such large offsets where signal strength is low and possibly below the noise level. This is because from those offsets onwards the TM mode air wave becomes so strong that it

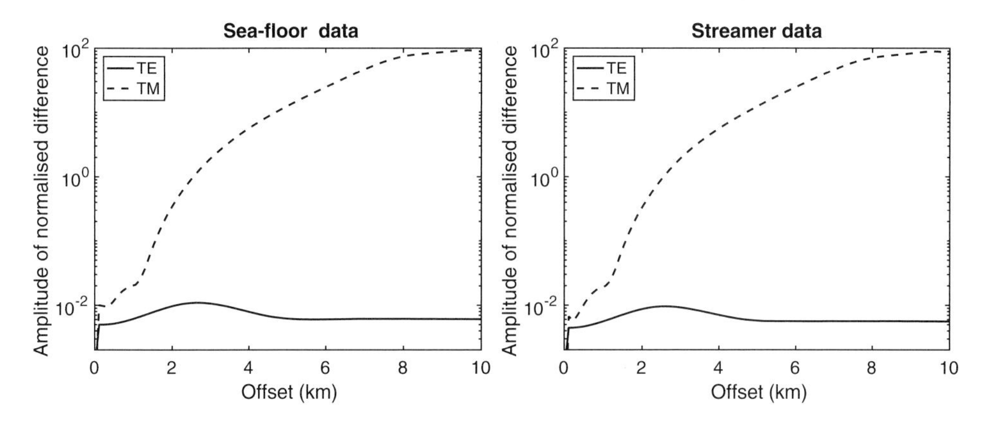

Figure 5.24 Amplitude of normalised difference of the reflected field in the TE (solid) and TM (dashed) modes for the deep source and sea-floor receivers (left) and shallow sources and towed receivers (right).

is visible in the difference. From 10 km onwards the TM mode air wave dominates and the two fields become more and more equal, reducing the difference further. In this model the difference at 15 km is still 100%. Notice that this behaviour is only visible because we model the air with the true electric permittivity and the TM air wave shows up. Also this effect is irrelevant for field measurements because the TE mode part of the field is so much stronger than the TM mode field at those large offsets, as shown by the results in Figure 5.23, that the TM mode part can most probably not be detected in the data.

5.3.2 Anisotropic Model Results in the Frequency Domain

In a VTI model the effect of the sea surface and sea floor do not change in an essential way and we focus on the effect of a VTI half-space and embedded resistive layer. All examples are computed for a source operating at $f = 0.5$ Hz.

Deep sea model

The deep sea model is the same as in the isotropic example, but now we introduce reduced vertical conductivity. The sea is an isotropic layer, the half-space is characterised by vertical conductivity $\sigma_v = 0.5$ S/m and the resistive layer has vertical conductivity given by $\sigma_v = 1/140$ S/m. This means that $\lambda = \sqrt{2}$ for all anisotropic layers. Figure 5.25 shows the electric field amplitudes at the sea floor of the model in the left plot and the normalised amplitude of the difference between presence and absence of the resistive layer in the right plot. At 2.5 km offset the amplitude of the electric field with the resistive layer is visibly larger than in the model without

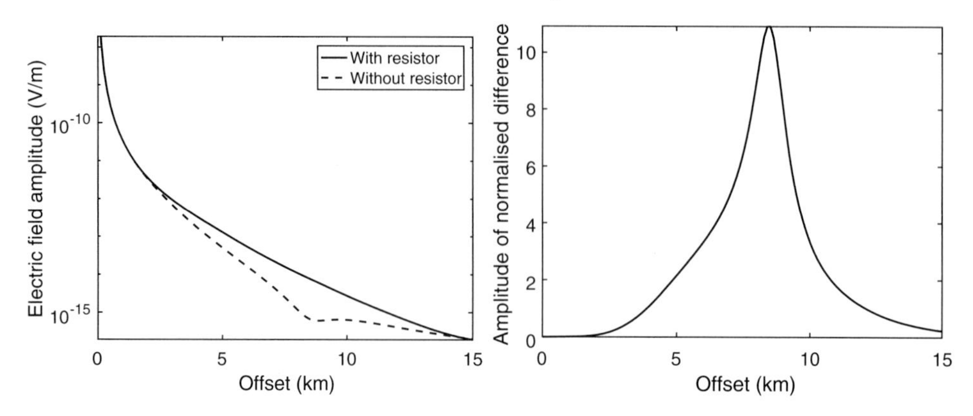

Figure 5.25 Electric field amplitudes (left) and amplitude of the normalised difference (right) in the VTI model with $\lambda = \sqrt{2}$ except for the sea layer that is isotropic.

the resistive layer. At 14 km the two responses have almost converged, indicating the air wave starts to dominate in both models. This can be seen in the right plot, where the normalised amplitude of the difference shows a wide offset range with significant difference between the two models. The maximum difference is just over an order of magnitude and is located at 8.5 km offset. Comparing these results with the results of the isotropic model shown in Figure 5.17, it can be seen that in the VTI model the offset range where the presence of the resistive layer creates a significant difference has increased and shifted to larger offsets. The shift to larger offsets occurs because the half space response is now stronger as its amplitude behaviour is controlled by the vertical conductivity that has decreased in the VTI model by a factor of 2 compared with that in the isotropic model. For this reason the direct air wave becomes the strongest event at offsets larger in the VTI model than in the isotropic model. In Figure 5.20 we can see that the TM mode contribution from the resistive layer reflection increases with offset. In the VTI model it also increases, and because now the maximum difference is shifted to larger offsets, the maximum difference has a higher value in the VTI model than in the isotropic model. It can be concluded that a lower vertical resistivity than horizontal resistivity increases the ability to detect a buried resistive layer in a VTI half-space compared with an isotropic half-space. Of course, in this model $\lambda = \sqrt{2}$ in the half-space and in the resistive layer. If the resistive layer were isotropic, the maximum difference would reduce to a factor of 5 and occur at approximately the same offset. The reduction of the maximum difference in amplitude between a VTI model with all layers having the same λ-value and the model where the resistive layer is isotropic is approximately a factor $\exp(-1/\lambda)$. This is because the electric field diffuses almost entirely horizontally through the resistive layer and its amplitude decay is therefore governed almost entirely by the vertical conductivity. By taking it as an isotropic layer, its vertical conductivity is increased by a factor λ leading to a reduction of $\exp(-1/\lambda)$ compared with the response of the resistive layer when it is also anisotropic. It should be understood that this is an approximate analysis that cannot be taken to be generally valid for every value of λ. For example, if we would model the half space with $\lambda = 2$, keeping the horizontal conductivity the same, the maximum normalised amplitude difference between the presence and absence of the resistive layer would be 22, located at 12 km offset if the resistive layer would also have $\lambda = 2$ and reduce to 4.3 at the same offset if the resistive layer were isotropic. This is not a reduction of a factor $\exp(-1/\lambda)$ but rather a reduction of a factor $\exp(-1/\lambda)/3$, indicating the response of the resistive layer is not simply determined by the horizontally diffusing field through the resistive layer, but also by the two-way transmission into and out of the resistive layer influences the strength of the event, and reverberations inside the resistive layer also play a role.

Figure 5.26 shows the phase of the electric field in the presence (solid line) and absence (dashed line) of the resistive layer in the left plot and the phase of the difference in electric fields in the right plot. It can be seen that the phase changes strongly from 2.5 km offset onwards and the right plot shows a linear phase as a function of offset, indicating that a single event dominates the difference field and the additional path with increasing offset is mostly horizontal. This can be understood as the response of the resistive layer.

Shallow sea model

In the shallow sea model we use the same parameters as in the deep sea model, but the sea layer has a depth of 100 m. Figure 5.27 shows the electric field amplitudes in

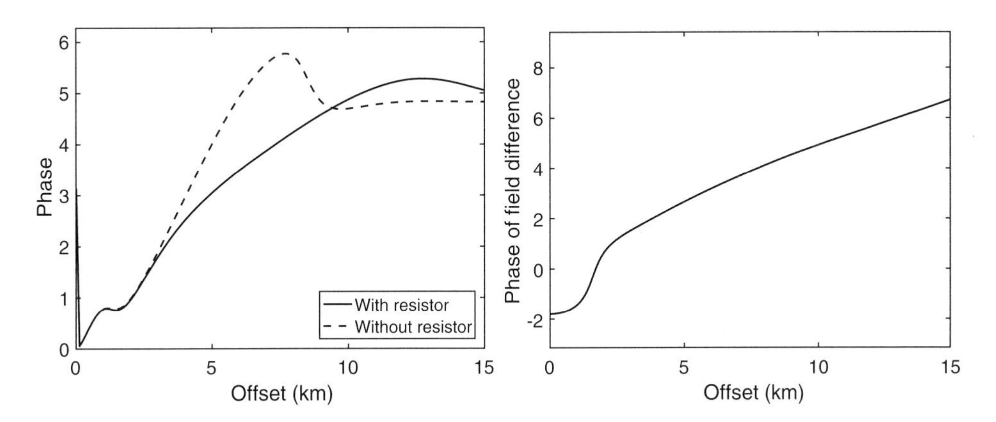

Figure 5.26 Electric field phases (left) and the phase of the difference in the fields (right).

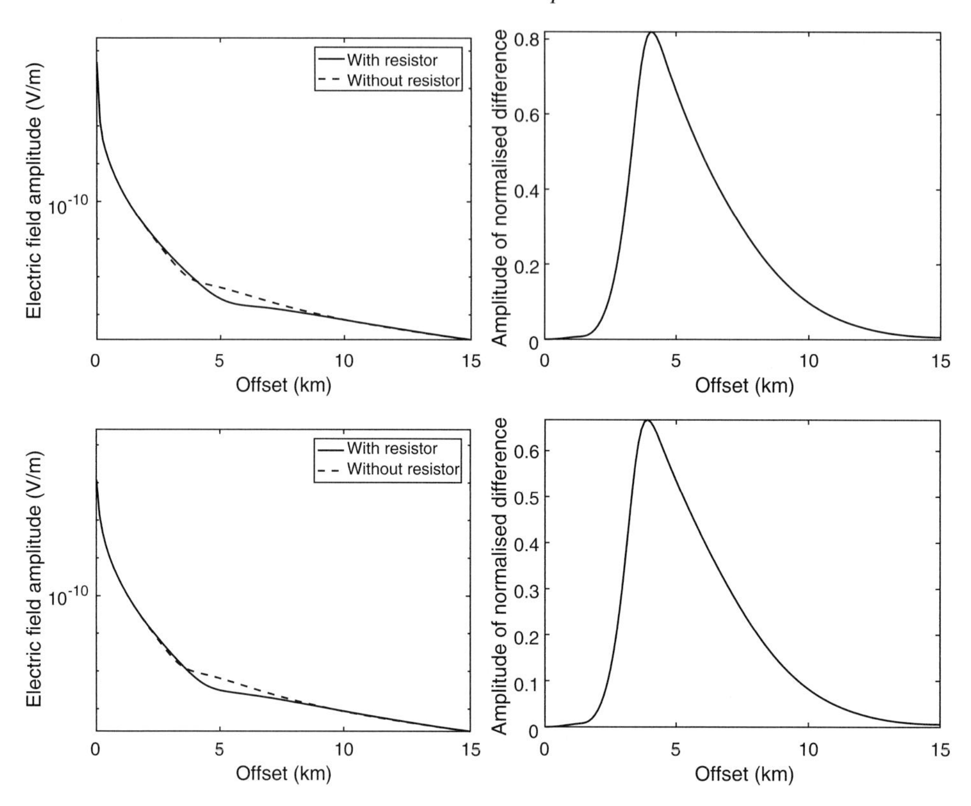

Figure 5.27 Electric field amplitudes (left) and amplitude of the normalised difference (right) for the deep source and sea-floor receivers (top) and shallow sources and towed receivers (bottom) in the VTI model with $\lambda = \sqrt{2}$ except for the sea layer that is isotropic.

the presence and absence of the resistive layer in the left column and the normalised amplitude of the difference field in the right column for the sea-floor and streamer data in the top and bottom rows, respectively. By comparing the results with those from the isotropic model in Figure 5.21, it can be seen that the effect of anisotropy is to increase the detectability of the resistive layer by more than 60% compared with the isotropic situation. The maximum difference now occurs at larger offsets than in the deep sea model. The increased detectability occurs because the resistive layer has reduced vertical conductivity that determines the attenuation of the horizontally diffusing field, while the attenuation of the direct air wave has remained unchanged. For this reason the effect of the air wave sets in at larger offsets.

Figure 5.28 shows the phases of the data from the models with (solid lines) and without (dashed lines) the resistive layer in the left column and the phase of the difference field in the right column for the sea-floor data (top row) and streamer data (bottom row). Comparing these results with those in Figure 5.22, it can be seen

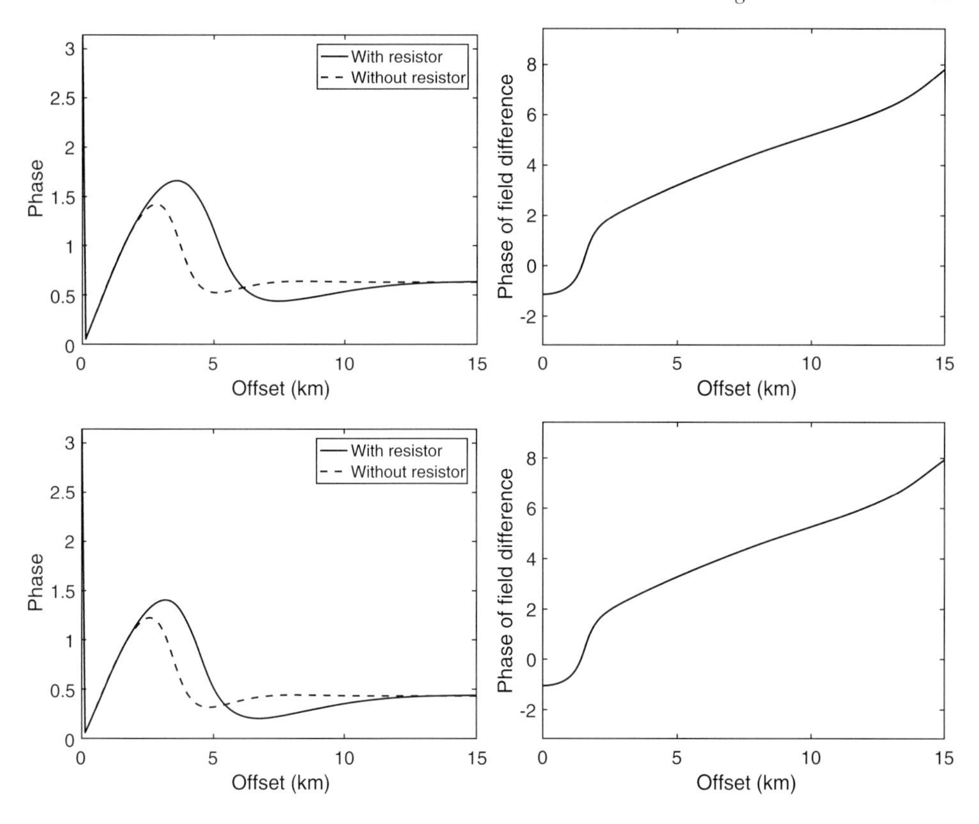

Figure 5.28 Electric field phases (left) and the phase of the difference in the fields (right) for the deep source and sea-floor receivers (top) and shallow sources and towed receivers (bottom).

that in the shallow sea the effect of anisotropy is to extend the range of detectability of the resistive layer and to shift the offset where the resistive layer first becomes detectable to larger offsets.

5.3.3 Isotropic Model Results in the Time Domain

Time domain results are given as impulse responses in the same models as we use in the sections where we look at single-frequency results.

Deep sea model

The electric field impulse responses are given in Figure 5.29 for the models with (left) and without (right) the resistive layer present. The colour bar indicates the electric field amplitudes in logarithmic scale for both plots. At small offsets and early times there is no visible difference in the impulse responses, but in the left

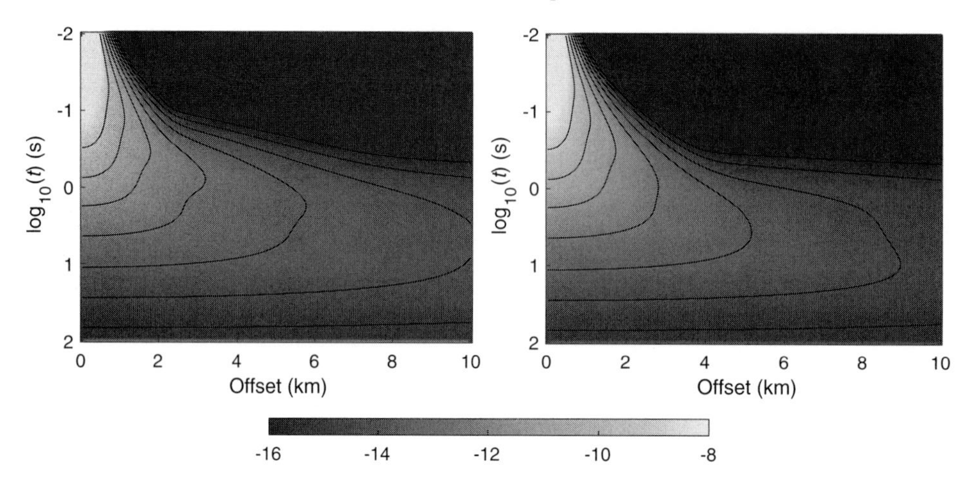

Figure 5.29 Electric field impulse response as a function of offset for the deep sea model with (left) and without (right) a 40 m thick buried resistive layer 1000 m below the sea floor. The colour bar indicates the amplitudes in logarithmic scale. For the colour version, please refer to the plate section.

plot that corresponds to the model with the resistive layer there is a clear change in amplitude slope in the onset of the signal from 2 km onwards compared with the right plot that corresponds to the model without the resistive layer. This is better visualised in Figure 5.30, where the normalised amplitude of the difference in the electric fields is shown as a function of offset and time in logarithmic scale.

The colour bar indicates the amplitude in logarithmic scale, showing that the presence of the resistive layer causes 100% difference in green to two orders of magnitude in yellow, which is the saturation level in the colour bar. The left plot is the result of the normalised difference taken from the data shown in Figure 5.29 and shows extremely high values in time–offset ranges, where the electric field in the earth's impulse response in the model without buried resistor is very small. This can be circumvented by introducing some random noise in the data. We assume the data are measured with 1% uncertainty by introducing a multiplicative random error with an average of 1% of the field at each location. We assume an average noise level in the electric field of 10^{-15} V/m and add it to the electric fields. This is done with different random number realisations for each error type and for the data with and without the resistive layer. The resulting maximum difference is just over a factor of 55 at 4 km offset and at 0.4 s with the current noise settings 10^{-15} V/m, and this would increase to 150 if 10^{-16} V/m were used as the noise level. In the offset range of approximately 3 km to 7 km the difference is very large between roughly 0.15 s and 1 s. In the offset range from 6 km to 10 km the difference is

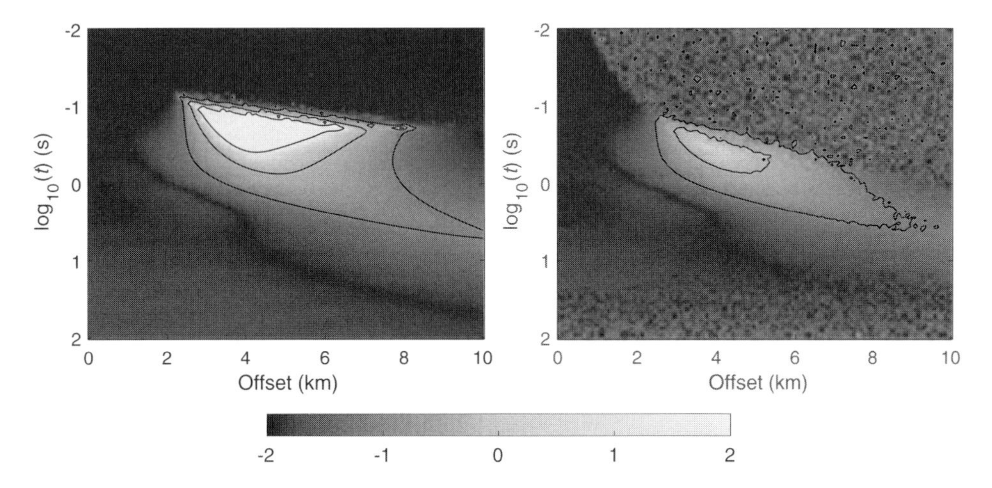

Figure 5.30 Electric field impulse response difference between models with and without resistive layer shown in Figure 5.29 as a function of offset for the deep sea model for noise-free (left) and noise (right) models. The colour bar indicates the normalised difference in logarithmic scale. For the colour version, please refer to the plate section.

still 100% to a factor of 3 in the time range from 1 s to 10 s. This offset–time range overlaps with the air wave that is present in both models, but still the presence of the resistive layer is clearly visible. The electric field with added and multiplicative noise is shown in Figure 5.31 as a function of offset for two fixed time instants and as a function of time for two fixed offsets from the data with noise. The fields corresponding to the model with the resistive layer are shown in solid lines and the fields for the model without the resistive layer in dashed lines. In the top row the left plot shows the field strength as a function of offset at $t = 0.4$ s and in the right plot at $t = 1$ s. In the bottom row the left plot shows the field strength as a function of time at 4 km offset and in the right plot at 6.5 km offset. The top-left plot shows that the electric field for the model without resistive layer at $t = 0.4$ s goes to zero very rapidly from 2.5 km offset onwards, whereas the electric field for the model with the resistive layer decays much more slowly and reaches the noise level around 7 km offset. At 4 km offset the maximum normalised difference reached is a factor of 55. In the top-right plot it can be seen that at 1 s the fields in the models with and without the resistive layer decay at a lower rate than at 0.4 s. The visible difference is between 3 km and 9 km offset and the maximum normalised difference is around a factor of 5 at 5 km offset. The plots in the bottom rows show the distinctive difference in arrival time of the resistive layer response. In the left plot it can be seen that higher amplitudes occur in the model with the resistive layer than in the model without this layer between 0.2 s and 2 s at 4 km offset. A similar

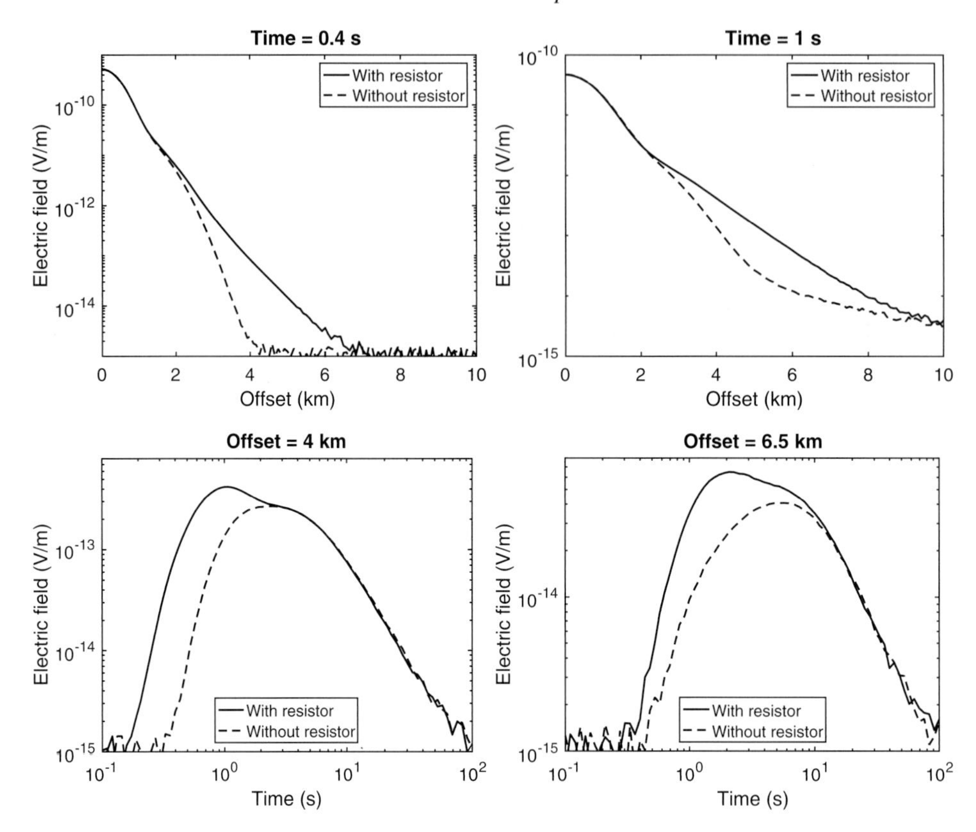

Figure 5.31 Electric field impulse response for the deep sea model without (solid lines) and with (dashed lines) as a function of offset (top) at 0.4 s (left) and 1 s (right) and as a function of time (bottom) at 4 km and 6.5 km offsets.

observation can be made in the right plot between 0.5 s and 7 s at 6.5 km offset. For both offsets there seems to be a decade of time interval available to detect the resistive layer with confidence. From these observations it can be concluded that the impulse response electric field gives almost an order of magnitude increase in the maximum amplitude difference between the presence and absence of a buried resistive layer than is obtained in the frequency domain at 0.5 Hz in the deep sea model. The detectability is present for a wide offset range and for a decade in time, the location of which is offset-dependent.

Shallow sea model

The electric field impulse responses are shown in Figure 5.32 for the sea-floor receivers (top) and streamer receivers (bottom) in the models with (left) and without (right) a resistive layer. It can be seen that the air wave arrives early and with

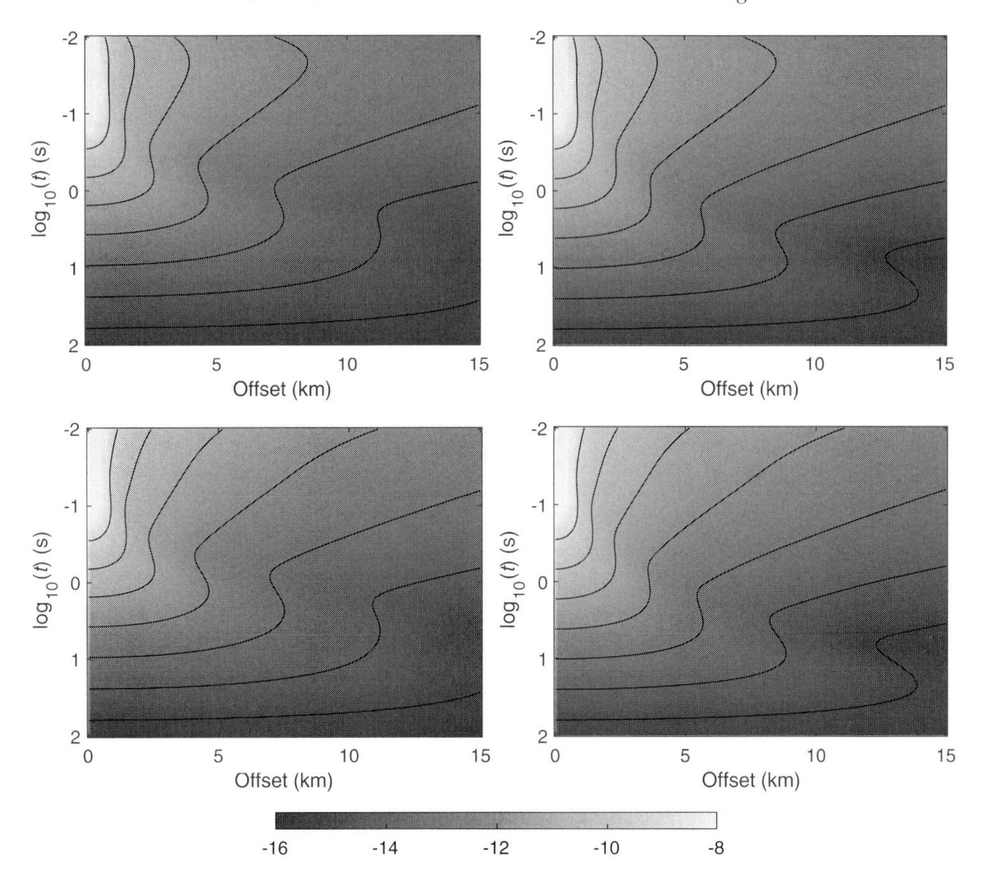

Figure 5.32 Electric field impulse response as a function of offset for the shallow sea model with (left) and without (right) a 40 m thick buried resistive layer 1000 m below the sea floor for sea-floor receivers (top) and streamer receivers (bottom). For the colour version, please refer to the plate section.

high amplitude at all offsets, while the subsurface response arrives late. This is comparable to the results of Figure 5.12 in the half-space model. In the left plot the presence of the resistive layer is clearly visible in the time–offset range where the subsurface response is present in the right plot for the model without the resistive layer. By comparing the plots in the bottom and top rows it can be seen that the air wave is stronger at earlier times in the streamer data than in the sea-floor data, because the receivers are closer to the sea surface for streamer receivers than for sea-floor receivers. The situation for the subsurface responses, however, does not seem so clear.

This is better visualised in Figure 5.33, showing the normalised amplitude of the difference between the two fields shown in Figure 5.32, but with the same noise

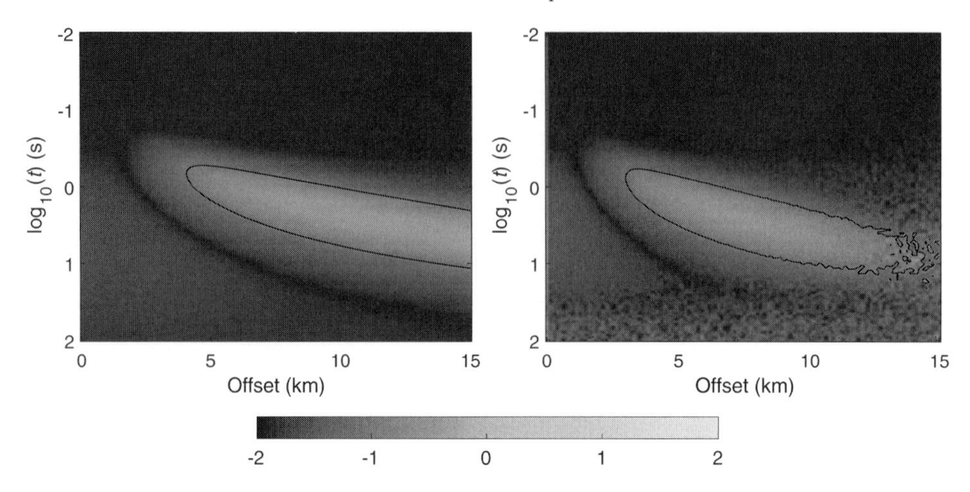

Figure 5.33 Electric field impulse response difference between models with and without the resistive layer, as shown in Figure 5.29, but using the noise model, as a function of offset for the shallow sea model for sea-floor data (left) and streamer data (right). The colour bar indicates the normalised difference in logarithmic scale. For the colour version, please refer to the plate section.

model as introduced in the deep sea example. The maximum difference at small offsets is a few per cent roughly between 1 s and 10 s and from zero to 2 km offset, which is a time–offset range where the field itself is still relatively strong. At larger offsets, from 3 km to 10 km, the difference reaches a maximum with increasing time. The time–offset range at which the difference is above 100% is quite large and slightly dipping to later times for larger offsets. The maximum difference is a factor 4.2, which is eight times bigger than in the frequency domain at the frequency of 0.5 Hz shown for the same model in Figure 5.21, in which the maximum amplitude difference was just over 50% for the sea-floor data. This maximum occurs at 7.8 km offset and $t = 3$ s. For streamer data the maximum difference is a factor 4.9, which is 12 times bigger than in the frequency domain at the frequency of 0.5 Hz shown for the same model in Figure 5.21, in which the maximum amplitude difference was just over 40%. This maximum difference occurs here at 7.6 km offset and at $t = 2.8$ s.

The electric field with added and multiplicative noise as a function of offset for two fixed time instants and as a function of time for two fixed offsets from the data with noise are shown in Figure 5.34 for sea-floor receivers in the top row and for streamer receivers in the bottom row as a function of offset in the left column and as a function of time in the right column. These line graphs are chosen such that they contain the maximum differences obtained. The fields corresponding to the model with the resistive layer are shown in solid lines and the fields for the model

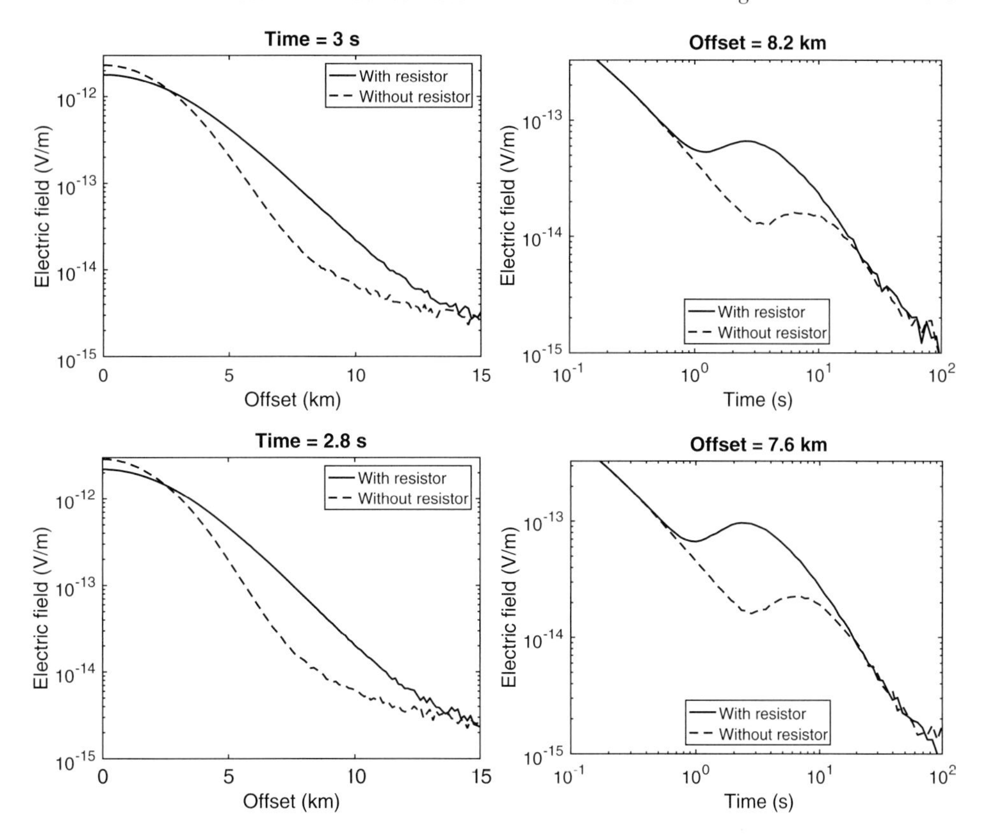

Figure 5.34 Electric field impulse response for the shallow sea model with (solid lines) and without (dashed lines) the buried resistive layer for the sea-floor data (top) and streamer data (bottom) as a function of offset at the time of the maximum difference (left) and as a function of time at the offset of the maximum difference (right).

without the resistive layer in dashed lines. The maximum difference in the sea-floor data occurs slightly later in time than for streamer data and also at a slightly larger offset. The plots in the left column show that the electric field for the model without the resistive layer at the given time instants is slightly larger than that of the model with the resistive layer up to 2 km offset, and smaller for larger offsets for the sea floor data, and this transition occurs at 2.5 km for the streamer data. Around 7.7 km offset, the maximum normalised difference of a factor of 4 is reached for the sea-floor data and a factor of 5 for the streamer data. The plots in the right column show the distinctive difference in arrival times of the resistive layer response. The curves for the models with and without the resistive layer are on top of each other for early and late times. For both acquisition geometries there seems to be more than a decade of time interval available to detect the resistive layer with confidence. Compared with the results from the deep sea model, we can see that the resistive layer can be detected at larger offsets and later times. The increase of the maximum difference between the presence and absence of the resistive layer from the frequency domain result at $f = 0.5$ Hz to the impulse response result is now an order of magnitude larger than compared to the same increase for the deep sea model.

For the shallow sea model, the presence of a resistive layer results in a slightly higher maximum amplitude difference of the impulse response for the streamer data than for the seafloor data. There is more than a decade in time where the presence of the resistive layer makes a significant difference in the electric field impulse response compared with the data from the model without the resistive layer. Compared with the deep sea model the range of offsets at which the presence of the resistive layer is clearly visible has increased and extends to well beyond 10 km. The streamer impulse response data give a better improvement of the maximum amplitude difference than the sea-floor data when a resistive layer is present at 0.5 Hz. In conclusion, the impulse response method seems more advantageous than the frequency domain method at a single frequency for the shallow sea model.

5.3.4 Anisotropic Model Results in the Time Domain

Time domain results are given as impulse responses in the same models as we use in the sections in which we look at single-frequency results.

Deep sea model

The electric field impulse responses are given in Figure 5.35 for the anisotropic models with (left) and without (right) the resistive layer present. The colour bar indicates the electric field amplitudes on a logarithmic scale for both plots. At small offsets and early times there is a clear difference in the impulse responses compared with the isotropic model results of Figure 5.29. This is because the vertical conductivity is a factor of 2 lower in the VTI model than in the isotropic model and the field that diffuses in the horizontal direction is attenuated less and travels faster. The left and right plots in Figure 5.35 are the same at small offsets and early times, but in the left plot, which corresponds to the model with the resistive layer, there is a clear change in amplitude slope in the onset of the signal from 3 km onwards compared with the right plot, which corresponds to the model without the resistive layer. The increase from 2 km to 3 km offset of a visible difference between the presence or absence of the resistive layer in the isotropic and VTI models can be understood from the factor $\lambda = \sqrt{2}$. In the model without the buried resistive layer the fastest-arriving event of the sea floor is the field diffusing horizontally just below the sea floor through the model, and its attenuation is controlled by σ_v.

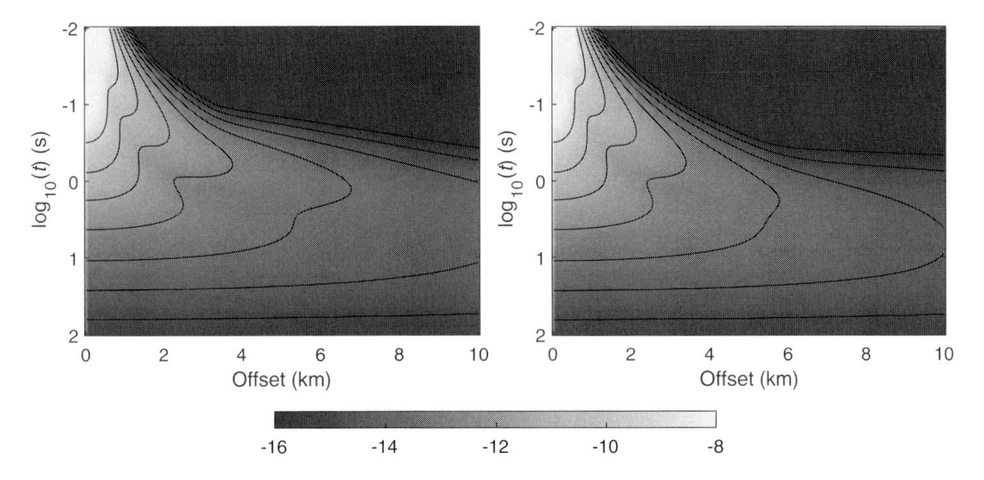

Figure 5.35 Electric field impulse response as a function of offset for the deep sea VTI model with $\lambda = \sqrt{2}$ for subsea resistivity values and with (left) and without (right) a 40 m thick buried resistive layer 1000 m below the sea floor. The colour bar indicates the amplitudes in logarithmic scale. For the colour version, please refer to the plate section.

In the model with the resistive layer there can be two fast-arriving events. One is the same as in the model without the resistive layer and the other is the resistive layer response. The latter has attenuation controlled mostly by the conductivity σ of the layer between the sea floor and the resistive layer and the conductivity σ_v of the resistive layer. Because λ is constant in the subsurface model there is not much difference in the influence on the attenuation along the horizontal parts of the travel paths of sea-floor and resistive layer responses. Hence, the difference in attenuation along the whole path is controlled by the vertical distance between the sea floor and the resistive layer in both the downward and upward directions. This amounts to an expected shift in offset by a factor of $2/\lambda$ for the resistive layer to become clearly visible in the data compared with its visibility in the isotropic model.

This is better visualised in Figure 5.36, where the normalised amplitude of the difference in the electric fields is shown as a function of offset and time on a logarithmic scale. The colour bar indicates the amplitude on a logarithmic scale, showing that the presence of the resistive layer causes 100% difference in green to two orders of magnitude in yellow, which is the saturation level in the colour bar. The left plot is the result of the normalised difference taken from the data shown in Figure 5.35 and shows very high values in time–offset ranges where the electric field in the earth's impulse response in the VTI model with and without buried resistive layer is extremely small. For this reason we use the same noise model and the corresponding normalised difference is shown in the right plot.

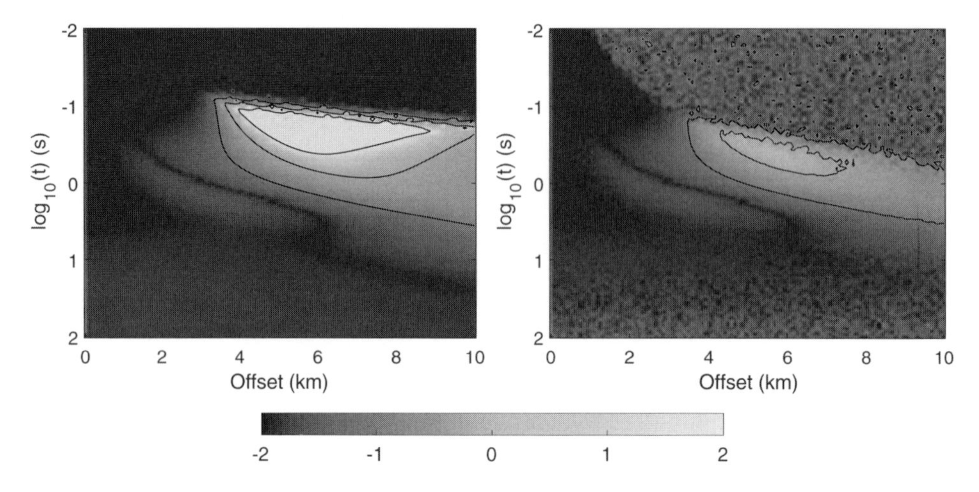

Figure 5.36 Electric field impulse response difference between the VTI models with and without resistive layer shown in Figure 5.35 as a function of offset for the deep sea model for noise-free (left) and noise (right) models. The colour bar indicates the normalised difference in logarithmic scale. For the colour version, please refer to the plate section.

The maximum difference is a factor of 40 in the time–offset range where the fields have an amplitude of at least 10^{-15} V/m. In the offset range of approximately 4 km to 8 km the difference is very large between roughly 0.3 s and 1.2 s. In the offset range from 6 km to 10 km the difference is still 100% to a factor of 3 in the time range from 1 s to 3 s. This offset–time range overlaps with the air wave that is present in both models, but still the presence of the resistive layer is clearly visible. Figure 5.37 shows the electric field with added and multiplicative noise as a function of offset for a fixed time instant (left), and as a function of time for a fixed offset (right), from the VTI model data from the models with and without the resistive layer. The fields corresponding to the model with the resistive layer are shown in solid lines and the fields for the model without the resistive layer in dashed lines. The left plot shows the field strength as a function of offset at $t = 0.4$ s and in the right plot as a function of time at 5.7 km offset. The left plot shows that the electric field for the model without the resistive layer at $t = 0.4$ s goes to zero very fast from 3.5 km offset onwards, whereas the electric field for the model with the resistive layer decays much more slowly and reaches the noise level around 9 km offset. At 5.7 km offset the maximum normalised difference of a factor of 40 is reached. In the right plot, the distinctive difference in arrival time of the resistive layer response can be observed. Higher amplitudes occur in the model with the resistive layer than in the model without this layer from just after 0.2 s to almost 2 s at 5.7 km offset. There seems to be a decade of time interval available to detect the

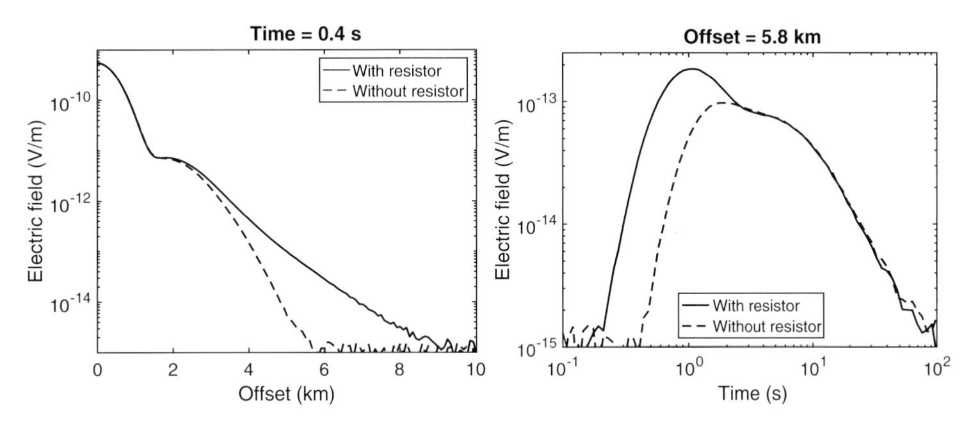

Figure 5.37 Electric field impulse response for the deep sea VTI model with (solid lines) and without (dashed lines) the buried resistive layer as a function of offset (left) at 0.4 s and as a function of time (right) at 5.7 km offset.

resistive layer with confidence. From these observations it can be concluded that the impulse response electric field gives a maximum amplitude difference between presence and absence of a buried resistive layer that is almost twice as strong as is obtained in the frequency domain at 0.5 Hz in the deep sea VTI model. The detectability is present for a wide offset range and for a decade in time. The time–offset range of the detectability is offset-dependent.

Shallow sea model

The electric field impulse responses for the shallow sea model with $\lambda = \sqrt{2}$ in all subsea layers are shown in Figure 5.38. The electric field at sea-floor receivers (top) and streamer receivers (bottom) are shown in the models with (left) and without (right) a resistive layer. Compared with the corresponding isotropic model results shown in Figure 5.32, it can be seen that up to $t = 0.5$ s there is no visible difference between the isotropic and VTI models. The sea-floor response in the model without the resistive layer arrives earlier at any given offset in the isotropic model than in the VTI model. This is visible by looking at 5.75 km offset in the right column plots and comparing the field strengths as a function of time in Figure 5.32 for the isotropic model and in Figure 5.38 for the VTI model. In the isotropic model the electric fields show a decreasing field strength with increasing time, to reach a local minimum of 7.4×10^{-14} V/m at $t = 2$ s, after which the field strengths increase again to a local maximum at $t = 3.5$ s, and then decrease again. In the VTI model the overall behaviour is similar, but when we look at the offset–time location with a similar local minimum value in field strength, we need to look at 6.75 km offset in the plots in the right column of Figure 5.38 and can see clear

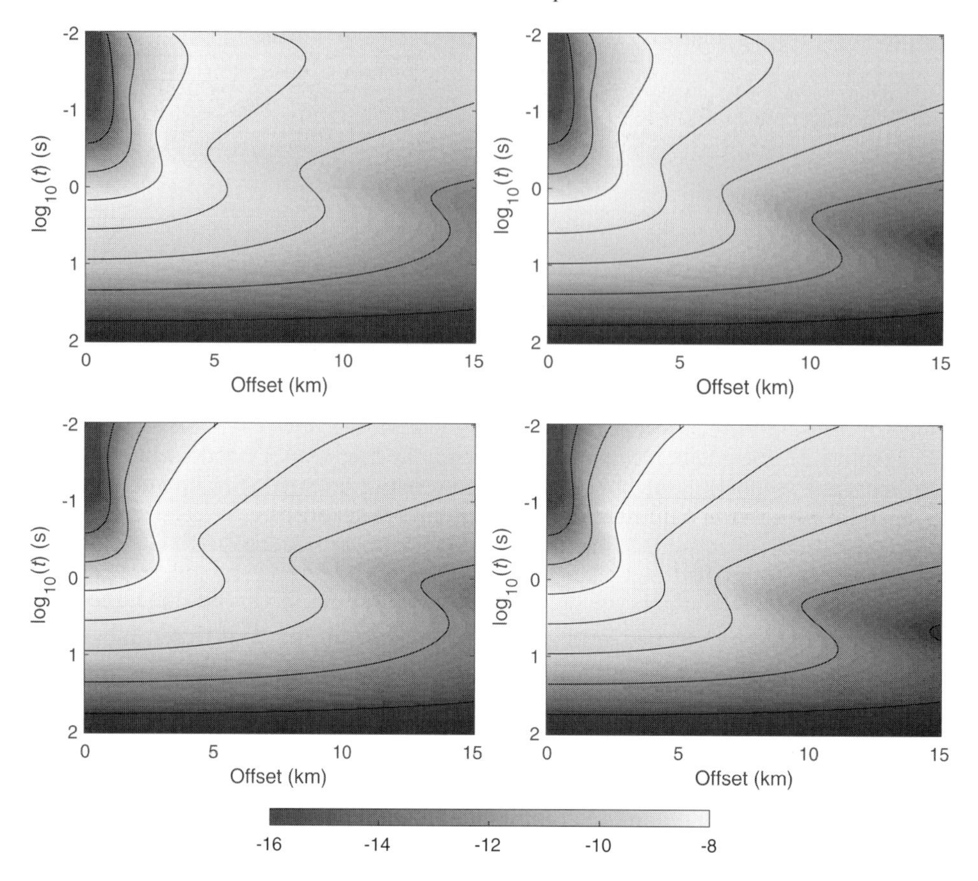

Figure 5.38 Electric field impulse response as a function of offset for the shallow sea model with (left) and without (right) a 40 m thick buried resistive layer 1000 m below the sea floor for sea-floor receivers (top) and streamer receivers (bottom). For the colour version, please refer to the plate section.

differences. In the VTI model the field strengths decrease to reach a local minimum amplitude of 7.25×10^{-14} V/m at $t = 1$ s after which they increase and reach a local maximum at $t = 3$ s. By comparing the plots in the right column with those in the left column of Figure 5.38, we can see that the presence of the resistive layer causes the local minimum field strength to arrive earlier in time and at larger offsets. The local minimum at the same field strength as in the isotropic model now occurs at 8.5 km offset and at $t = 0.6$ s.

The difference in the electric field in the presence or absence of the resistive layer is better visualised in Figure 5.39, which shows the normalised amplitude of the difference between the two fields shown in Figure 5.38, but with the same noise model as introduced in the deep sea example in Section 5.3.3. The maximum

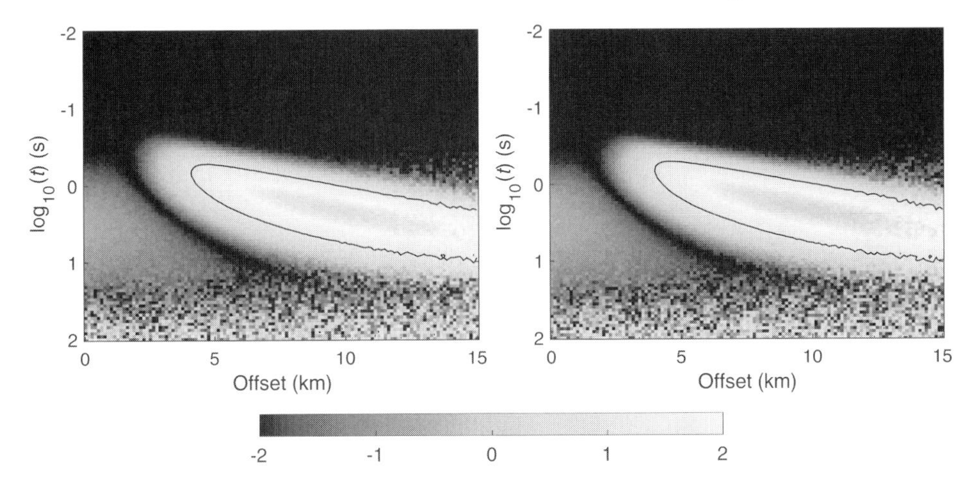

Figure 5.39 Electric field impulse response difference between models with and without the resistive layer, as shown in Figure 5.29, but using the noise model, as a function of offset for the shallow sea model for sea-floor data (left) and streamer data (right). The colour bar indicates the normalised difference in logarithmic scale. For the colour version, please refer to the plate section.

difference at small offsets is a few per cent roughly between 0.3 s and 10 s and from zero to 2 km offset, which is a time–offset range in which the field itself is still relatively strong. At larger offsets, from 4.5 km to 14 km, the difference reaches a maximum with increasing time. The shape of the time–offset range in which the difference is above 100% is very similar to that in the isotropic model, but is shifted to larger offsets and earlier times. The maximum difference is a factor of 5.5, which is seven times bigger than in the frequency domain at 0.5 Hz, shown for the same model in Figure 5.27, where the maximum amplitude difference was just over 80% for the sea-floor data. This maximum occurs at 11 km offset and $t = 2.5$ s. For streamer data the maximum difference is a factor 6.5, which is ten times bigger than in the frequency domain at 0.5 Hz, shown for the same model in Figure 5.27, in which the maximum amplitude difference was just over 60%. This maximum difference occurs here at 10 km offset and at $t = 2.3$ s. Comparing these results with those of the isotropic shallow sea model, we can observe that the offset where the maximum difference occurs in the VTI model is approximately $2/\lambda$ times the offset where the maximum difference occurs in the isotropic shallow sea model. The time where the maximum difference occurs in the VTI model is approximately $1/\sqrt{\lambda}$ times the time where it occurs in the isotropic model.

Figure 5.40 shows the electric field with added and multiplicative noise as a function of offset for a fixed time instant (left), and as a function of time for a fixed offset (right) for sea-floor receivers in the top row and for streamer receivers in

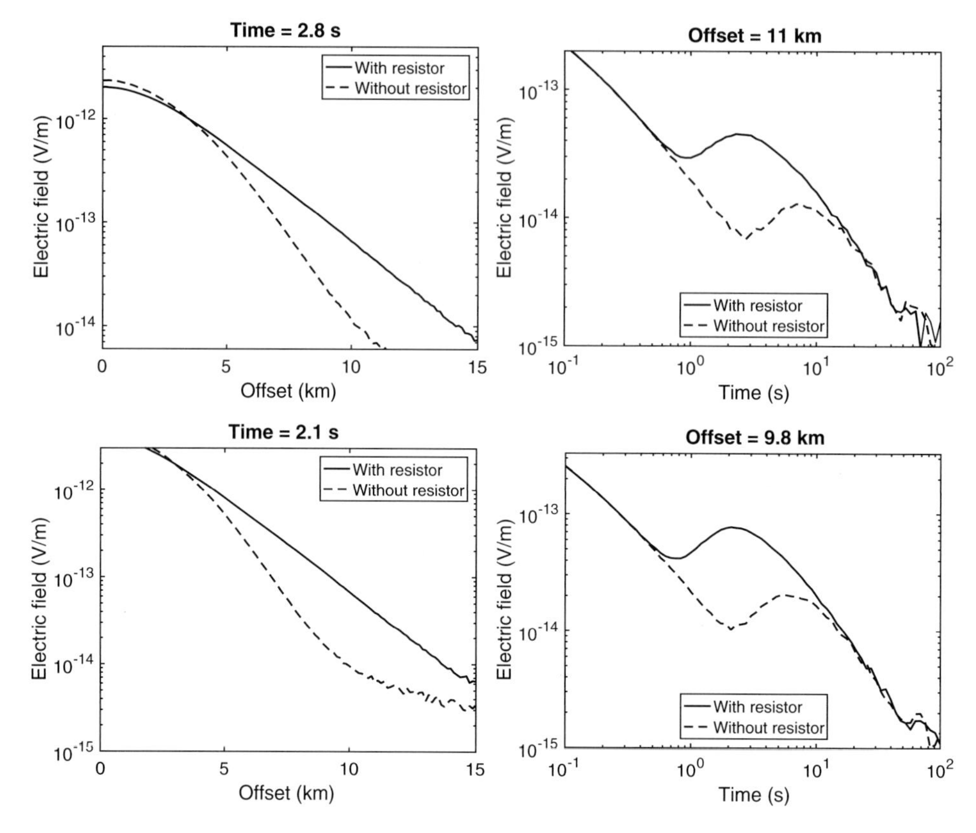

Figure 5.40 Electric field impulse response for the deep sea VTI model with (solid lines) and without (dashed lines) the buried resistive layer for the sea-floor data (top) and streamer data (bottom) as a function of offset at the time of the maximum difference (left) and as a function of time at the offset of the maximum difference (right).

the bottom row. These line graphs are chosen such that they contain the maximum differences obtained. The fields corresponding to the model with the resistive layer are shown in solid lines and the fields for the model without the resistive layer in dashed lines. The maximum difference in the sea-floor data occurs slightly later in time than for streamer data and also at a slightly larger offset. The plots in the left column show that the electric field for the model without the resistive layer at the given time instants is slightly larger than that of the model with the resistive layer up to 3.5 km offset, and smaller for larger offsets for the sea-floor data, and this transition occurs at 3.2 km for the streamer data. Around 11 km offset the maximum normalised difference of a factor of 5.5 is reached for the sea-floor data and a factor of 6.5 for the streamer data. The plots in the right column show the distinctive difference in arrival time of the resistive layer response. The curves for the models with and without the resistive layer are on top of each other for early and late times. For both acquisition geometries there seems to be more than a decade of time interval available to detect the resistive layer with confidence. Compared with the results from the deep sea VTI model, we can see that the resistive layer can be detected at larger offsets and later times. The increase of the maximum difference between the presence and absence of the resistive layer from the frequency domain result at $f = 0.5$ Hz to the impulse response result is now an order of magnitude larger than compared with the same increase for the deep sea model.

It can be concluded from the impulse response data for the shallow sea VTI model that the presence of a resistive layer is indicated by a slightly higher maximum amplitude difference for the streamer data than for the sea-floor data. There is more than a decade in time where the presence of the resistive layer makes a significant difference to the electric field impulse response compared with the data from the model without the resistive layer. Compared with the deep sea VTI model, the range of offsets at which the presence of the resistive layer is clearly visible is larger and extends to well beyond 10 km. The presence of a resistive layer is better determined by the streamer impulse response data than by the sea-floor frequency domain data at 0.5 Hz. The impulse response method seems more advantageous than the frequency domain method at a single frequency for the shallow sea VTI model.

5.4 The Electric Field in a Land CSEM Setting

For the land CSEM models the thicknesses and conductivities of the layers are the same as in the example in Chapter 1. The half-space is characterised by $\sigma = \sigma_v = 1/20$ S/m in the isotropic model, and in the VTI model we take $\sigma = 1/20$ S/m and $\sigma_v = 1/40$ S/m, hence $\lambda = \sqrt{2}$. In this case we have a buried resistive layer, it is 40 m thick and its top is 1 km below the surface. It is characterised by $\sigma = \sigma_v = 1/400$ S/m in the isotropic model and $\sigma = 1/400$ S/m and $\sigma_v = 1/800$ S/m, so also here $\lambda = \sqrt{2}$. A second land model places the same resistive layer at 3 km depth. Compared with the marine examples, the layer thickness is the same, but the resistivity contrast has been reduced from a factor of 70 to a factor of 20.

In the frequency domain we model all examples with a source operating with several oscillation periods. The periods of interest are greater than 31.6 μs and we specify the oscillation period rather than the frequency of operation. The source oscillation periods vary from 31.6 ms to 31.6 s. In the time domain we model all examples as impulse responses.

5.4.1 Isotropic Model Results in the Frequency Domain

Figure 5.41 shows the half-space response electric field amplitude as a function of offset and source oscillation period. We now consider land surveys for frequencies

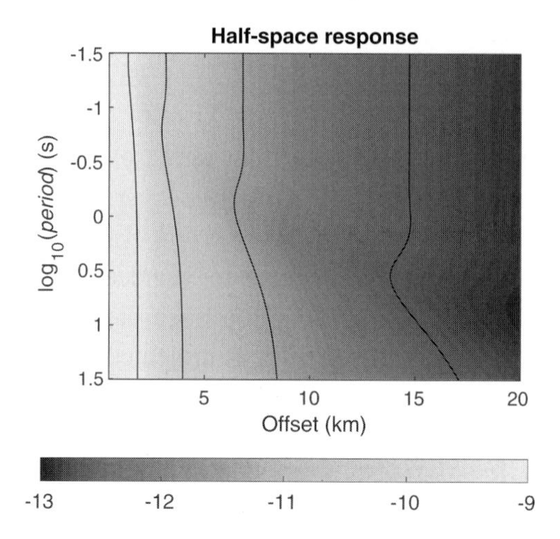

Figure 5.41 Electric field amplitude for the half-space land model as a function of offset and source oscillation period. The colour bar indicates the values of the logarithm of the amplitude. For the colour version, please refer to the plate section.

below 50 Hz, for which reason we show results as a function of source oscillation period instead of frequency, with the understanding that $f = 1/period$. Equation (4.130) gives the frequency domain expression for the electric field and for the in-line response $y = 0$ in that equation and $x = r$. We can see that the contribution of the air to the measurement is frequency-independent, and hence oscillation period-independent, and has an offset dependency of $1/x^3$. If the contribution of the air would dominate for all offsets and periods, the colour contours would be vertical in Figure 5.41. From equation 4.130 we can observe that when the oscillation period becomes very large, the contribution from the conductive half-space becomes a static field. Hence, we can expect that for small offsets at all periods, and for all offsets at small periods, the electric field is independent of the source oscillation period. The upper limit of the period to be considered small increases with increasing offset. Therefore a limited offset–period range exists where the subsurface response shows up as a frequency-dependent field in the measured electric field. The contribution from the conductive half-space is never larger than that from the air.

Figure 5.42 shows the amplitude of the electric fields in the model with the buried resistive layer buried at 1 km depth (left) and at 3 km depth (right). The presence of the resistive layer at 1 km depth is visible by comparing the left plot with Figure 5.41. By comparing the right plot with Figure 5.41 we can observe almost no differences in the amplitudes of the two responses at long periods and large offsets.

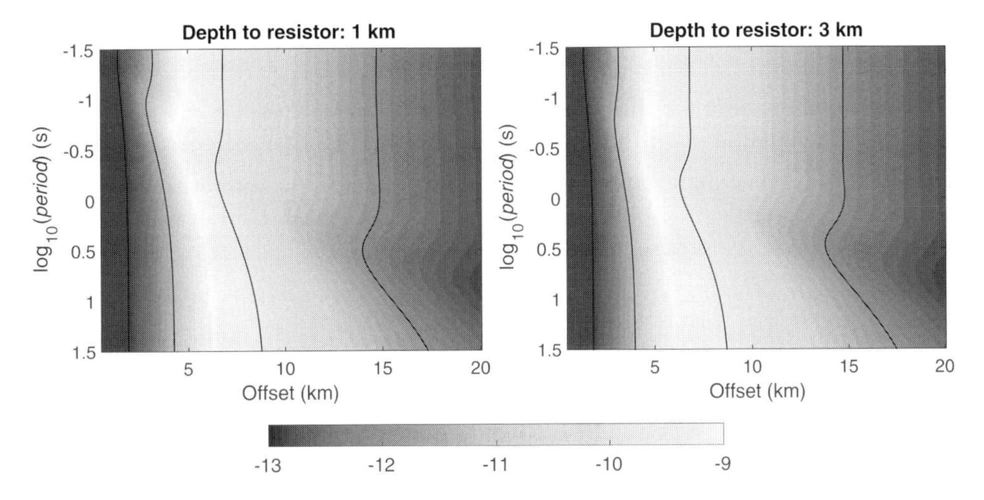

Figure 5.42 Electric field amplitudes for the land model with a buried resistive layer at 1 km depth (left) and at 3 km depth (right) as a function of offset and source oscillation period. The colour bar indicates the amplitudes on a logarithmic scale. For the colour version, please refer to the plate section.

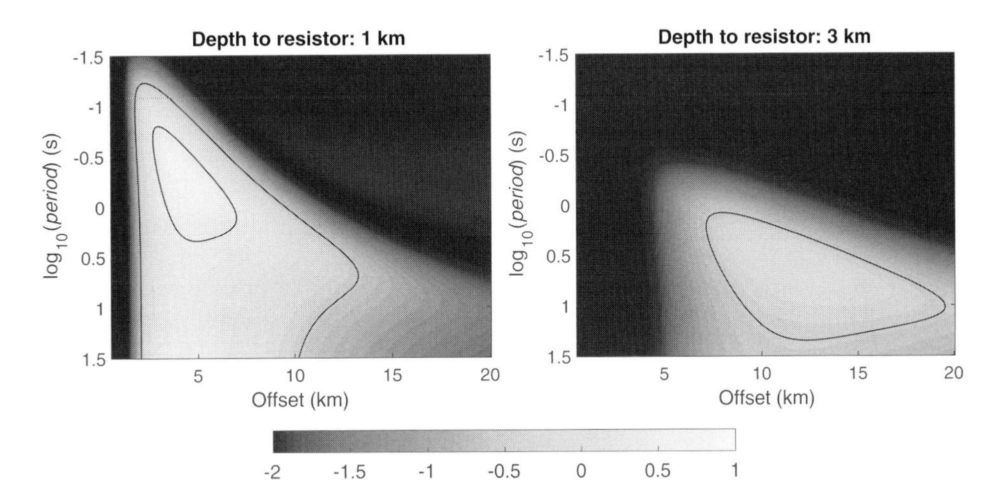

Figure 5.43 Normalised amplitude of the difference in the electric fields for the land model with a buried resistive layer at 1 km depth (left) and at 3 km depth (right) as a function of offset and source oscillation period. The colour bar indicates the amplitude of the normalised difference in the electric field on a logarithmic scale. For the colour version, please refer to the plate section.

These difference are visualised by plotting the normalised amplitude of the difference in the responses with and without the resistive layer. These are shown in Figure 5.43 for the resistive layer at 1 km depth in the left plot and at 3 km depth in the right plot. The colour scale gives the logarithmic values of the amplitudes of the normalised differences in the electric field. In the model with the resistive layer at 1 km depth, the maximum difference is a factor 0.42 and it occurs at 4.2 km offset and source oscillation period of 0.48 s. A small offset–oscillation period range with an almost triangular shape exists where the amplitude of the normalised difference is above 30%, marked by the inner contour line. As can be seen in the right plot, the triangular shape seems to exist but has been moved to longer oscillation periods of the source and to larger offsets. The maximum difference is around 15% for the reservoir at 3 km depth in the given range of offsets and oscillation periods. It occurs at 11 km offset and when the source oscillates with a period of 3.6 s. The field amplitudes as a function of offset at the optimal source oscillation period and as a function of oscillation period for the optimal offsets are shown in Figure 5.44. The top row shows the results for the resistive layer at 1 km depth and the bottom

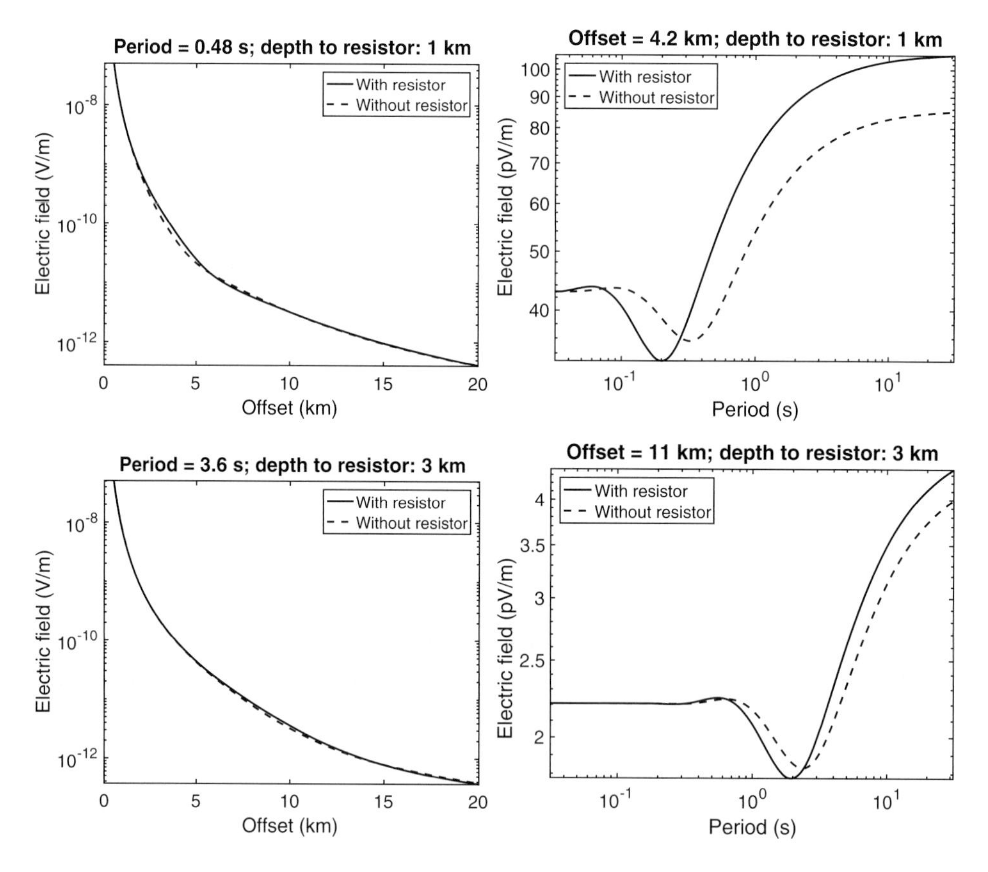

Figure 5.44 Electric field amplitudes for the land model with (solid lines) and without (dashed lines) a buried resistive layer at 1 km depth (top) and at 3 km depth (bottom) as a function of offset (left) and source oscillation period (right).

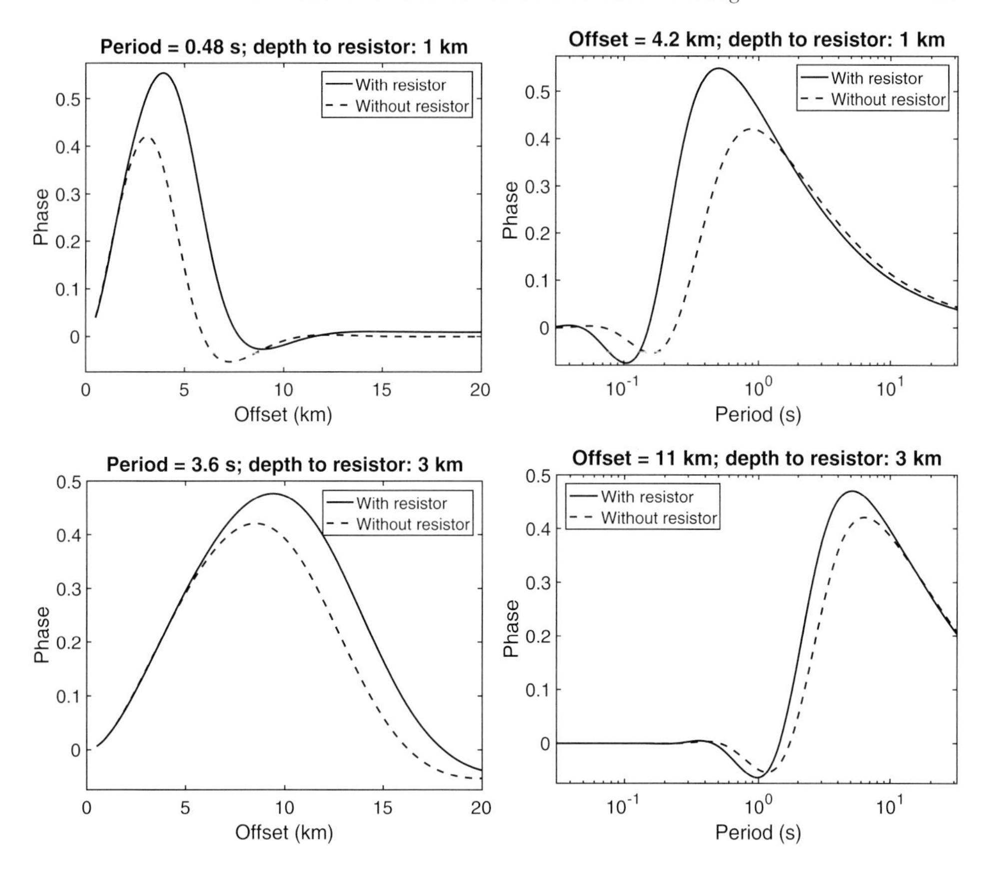

Figure 5.45 Electric field phases for the land model with (solid lines) and without (dashed lines) a buried resistive layer at 1 km depth (top) and at 3 km depth (bottom) as a function of offset (left) and source oscillation period (right).

row for the resistive layer at 3 km depth. It can be observed that the presence of the buried resistive layer is better visible in the plot as a function of oscillation period than as a function of offset.

The field phases as a function of offset and the optimum source oscillation period and as a function of oscillation period for the optimum offsets are shown in Figure 5.45. In the half-space response the contribution from the air is zero phase, because it is frequency-independent. As explained in relation to Figure 5.41, we expect that for a fixed source oscillation period the phase will go to zero at large offsets. For fixed offset the phase is expected to start at zero for small periods, because then the exponential attenuation is large and the static air wave will dominate. Then, if the oscillation period increases, the phase will reflect the subsurface response, while at very large oscillation periods the field becomes essentially a static field, with zero phase, in which the half-space response has the same strength as the air wave. This behaviour is visible in Figure 5.45, except at large offset for the deep resistive layer, because the offset is not large enough for an oscillation period of 3.6 s in a half-space resistivity of $\rho = 20\ \Omega$m. There is a large offset range at the optimal oscillation period available for detecting the presence of the resistive layer. There is a large range of oscillation periods at the optimal offset to detect the resistive layer. This occurs for both reservoir depth levels. Whether this leads to a practical method is largely determined by the actual complexity of the subsurface conductivity distribution and the present noise levels.

5.4.2 Anisotropic Model Results in the Frequency Domain

Figure 5.46 shows the amplitude of the electric field in the VTI half-space model. Arguments similar to those used with the isotropic half-space response shown in Figure 5.41 apply here as well. The visible frequency dependence has shifted to smaller periods, as can be expected with a vertical conductivity that is reduced by a factor of 2 compared with the isotropic half-space. Figure 5.47 shows the amplitude of the electric fields in the VTI model with the resistive layer buried at 1 km depth (left) and at 3 km depth (right). The same effect as seen in the VTI half-space response compared to the isotropic half-space response is visible in the effect of the presence of the resistive layer at 1 km depth. For the resistive layer at 1 km depth the detectability has increased compared with the isotropic case. By comparing the right plot to Figure 5.46 we can observe some weak differences in the

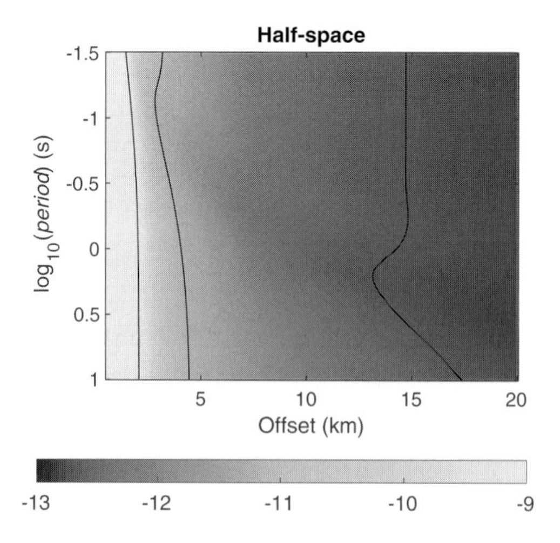

Figure 5.46 Electric field amplitude for the VTI half-space land model as a function of offset and source oscillation period. The colour bar indicates the values of the logarithm of the amplitude. For the colour version, please refer to the plate section.

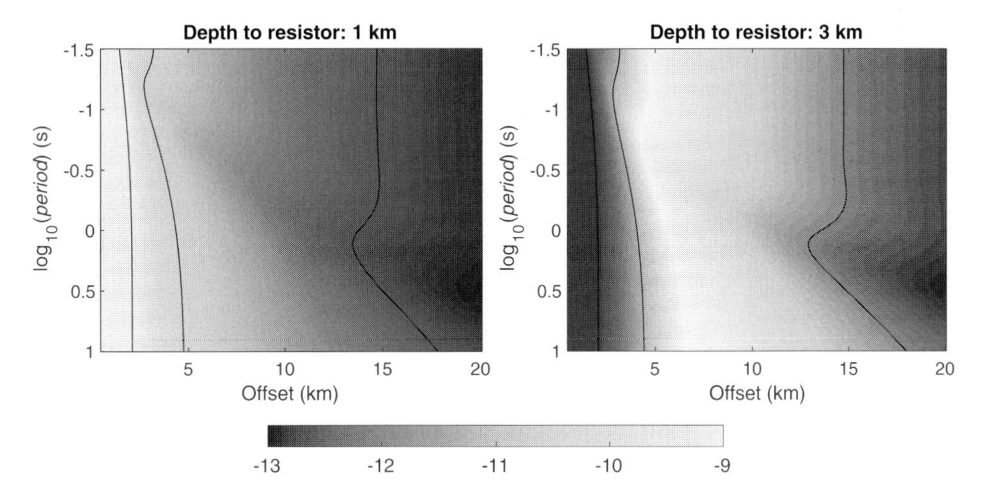

Figure 5.47 Electric field amplitudes for the land VTI model with the buried resistive layer at 1 km depth (left) and at 3 km depth (right) as a function of offset and source oscillation period. The colour bar indicates the amplitudes on a logarithmic scale. For the colour version, please refer to the plate section.

amplitudes of the two responses, which occur at large offsets and large oscillation times. These differences are visualised by plotting the normalised amplitude of the difference in the responses with and without the resistive layer. These are shown in Figure 5.48 for the resistive layer at 1 km depth in the left plot and at 3 km depth in the right plot. The colour scale gives the logarithmic values of the amplitudes of the normalised differences in the electric field. In the model with the resistive layer at 1 km depth the maximum difference is a factor of 0.66 and it occurs at 6 km offset, which is approximately a factor of of $2/\lambda$ times the offset of the maximum normalised difference in the isotropic model, and at a source oscillation period of 0.4 s. For the resistive layer at 3 km depth, the maximum difference is 0.23 and occurs at 16 km offset and oscillation period of 3 s.

The field amplitudes as a function of offset at the optimal source oscillation period and as a function of oscillation period for the optimal offsets are shown in Figure 5.49. The top row shows the results for the resistive layer at 1 km depth and the bottom row for the resistive layer at 3 km depth. Comparing the plots in Figures 5.44 and 5.49, we can observe that the location of the maximum difference in the VTI model is at an offset approximately $2/\lambda$ times the offset where it occurs in the isotropic model. The oscillation period where this maximum occurs in the VTI model is $1/\sqrt{\lambda}$ times the oscillation period where the maximum occurs in the isotropic model. This observation is similar to the one in Section 5.3.4, where we discuss the shallow sea impulse responses in a VTI model.

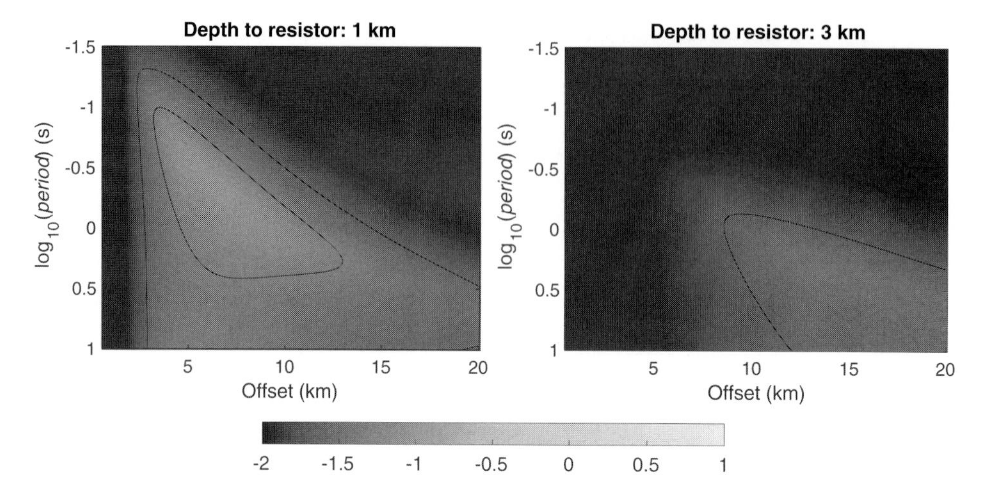

Figure 5.48 Normalised amplitude of the difference in the electric fields for the land VTI model with a buried resistive layer at 1 km depth (left) and at 3 km depth (right) as a function of offset and source oscillation period. The colour bar indicates the amplitude of the normalised difference in the electric field on a logarithmic scale. For the colour version, please refer to the plate section.

The field phases as a function of offset at the optimal source oscillation period as a function of oscillation period for the optimal offsets are shown in Figure 5.50. These can be compared with the phases in the isotropic model shown in Figure 5.45 to observe that the phase difference between the two models has increased in the VTI model compared with the isotropic model. The maximum difference in the phase occurs at larger than the optimal offset for the maximum amplitude difference and at smaller oscillation periods at the optimal offset for the maximum amplitude difference.

5.4.3 Isotropic Model Results in the Time Domain

Figure 5.51 shows the electric field impulse response of the half-space as a function of logarithmic time and offset. Here it is clear that the air wave is not part of the figure because it is frequency-independent: it is an impulse in time and the figure starts showing the field at 1 ms. Assuming a physical impulsive source can generate an impulse with a time bandwidth less than 1 ms, this is a physically reasonable plot. The electric field impulse responses are shown in Figure 5.52 for the 1 km deep (left) and 3 km deep (right) resistive layer. Comparing both plots with the half-space result shows two zones where the presence of the resistive layer strongly influences the measured electric field when the resistive layer is at 1 km depth. The

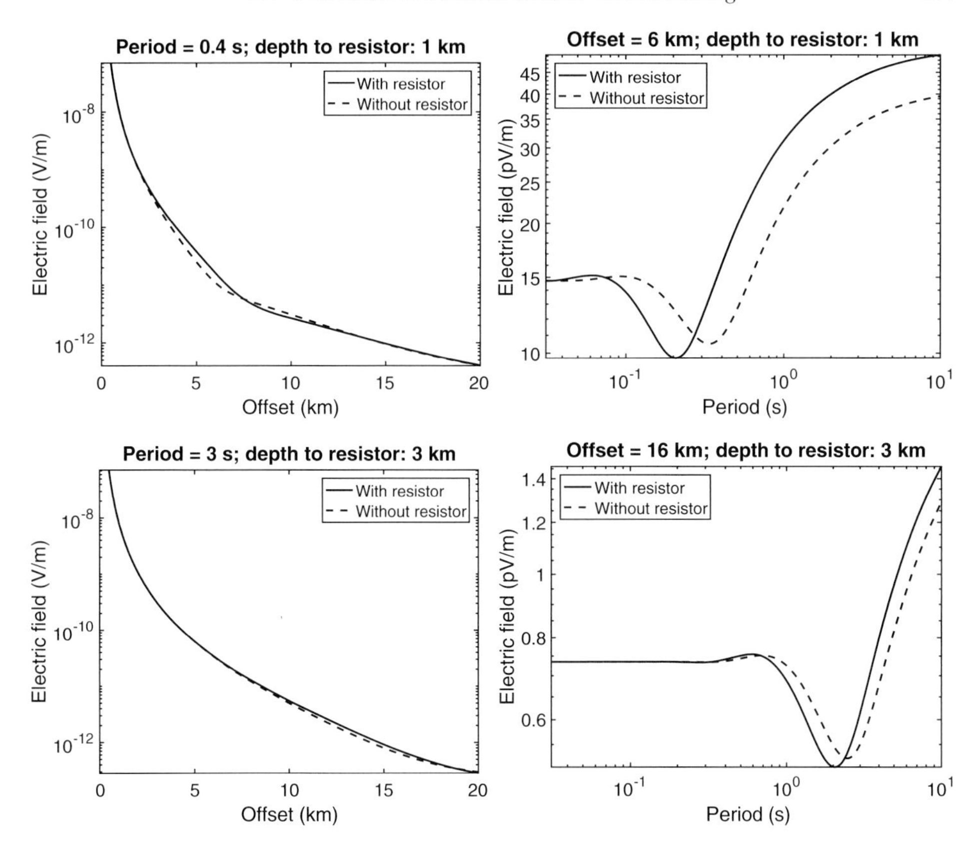

Figure 5.49 Electric field amplitudes for the land VTI model with (solid lines) and without (dashed lines) a buried resistive layer at 1 km depth (top) and at 3 km depth (bottom) as a function of offset (left) and source oscillation period (right).

first zone occurs at an offset–time interval where, in the absence of the resistive layer, the signal is very weak, which can be seen as an almost horizontal band lying between 10 ms and 0.1 s for all offsets, but is visible as an isolated event for offsets larger than 5 km. The field diffusing along but just below the surface is still increasing in that zone, as can be seen in Figure 5.51. This is the TE mode part of the resistive layer response. This can be understood as coming from two events that have the same diffusion times. The first is the TE mode part diffusing down, reflecting off the reservoir and diffusing up, then it is converted to an air wave and this wave gives a measurable response at any receiver. The reverse path also exists, which is the direct air wave coupling into the earth as a downwards-diffusing field at any receiver, reflecting off the resistive layer and diffusing up to the receiver, where it is measured. The second zone is the TM mode response, which follows more the half-space response. Due to the faster and less attenuating path through the resistive

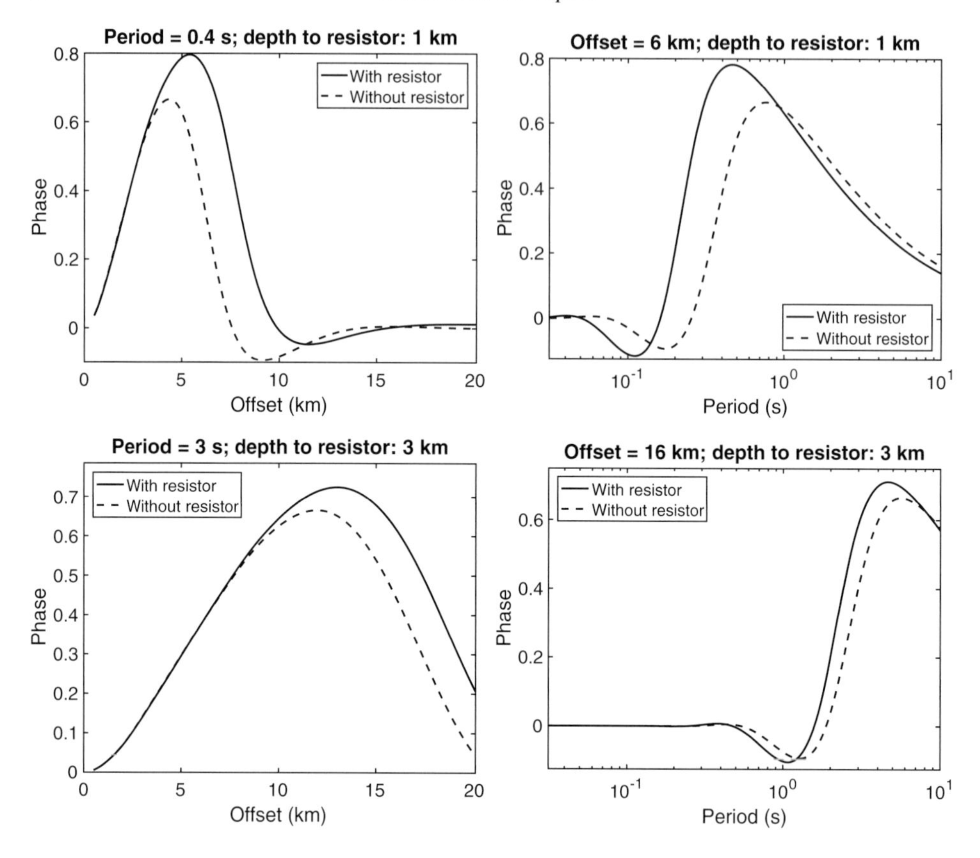

Figure 5.50 Electric field phases for the land VTI model with (solid lines) and without (dashed lines) a buried resistive layer at 1 km depth (top) and at 3 km depth (bottom) as a function of offset (left) and source oscillation period (right).

layer, the TM mode part of the response arrives earlier and with larger amplitudes than the half-space response at the receivers at large enough offsets. This can be seen by comparing the left plot of Figure 5.52 with Figure 5.51 between 10 ms and 1 s between 4 km and 15 km offset. When the resistive layer is at 3 km depth, the TE mode part of the response is visible but weak and the difference in the TM mode part of the response is much less clear.

This is better visualised in Figure 5.53 showing the normalised amplitude of the difference between the two fields shown in Figure 5.52, but with the same noise model as introduced in the deep sea example. The difference is more than two orders of magnitude in the zone that is saturated yellow in the left plot of Figure 5.53. The maximum difference is more than three orders of magnitude and occurs at $t = 14$ ms and at an offset of 4.4 km. For the example in which the resistive layer is buried at 3 km depth, the difference is still quite large with a maximum of

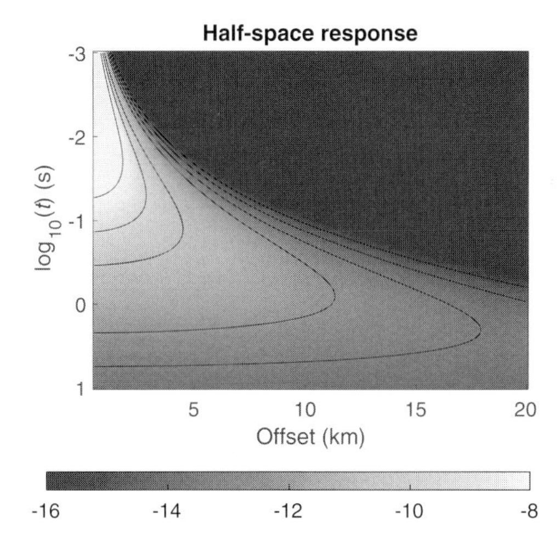

Figure 5.51 Electric field impulse response as a function of logarithmic time and offset for the isotropic half-space model. For the colour version, please refer to the plate section.

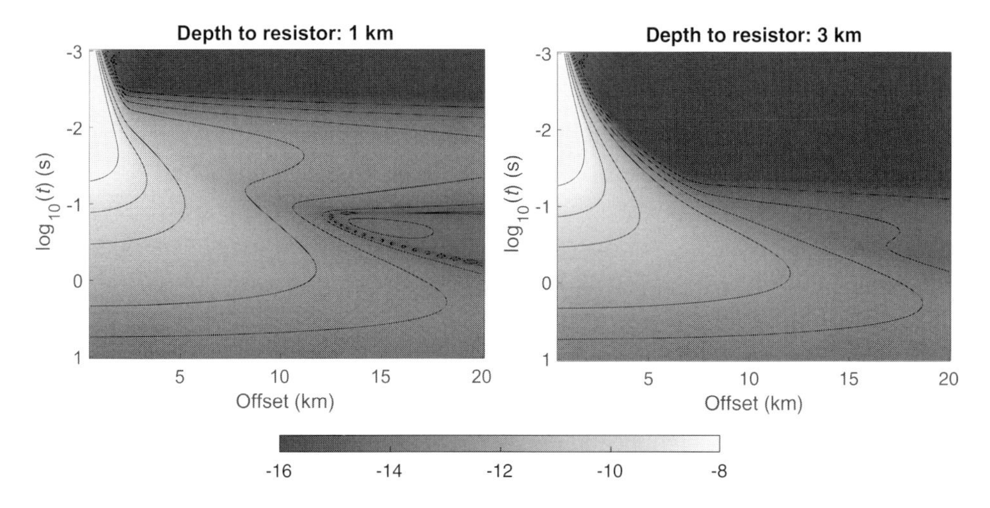

Figure 5.52 Electric field impulse response as a function of logarithmic time and offset for the land model with the 40 m thick buried resistive layer at 1 km depth (left) and 3 km depth (right). For the colour version, please refer to the plate section.

a factor of 6.3. This maximum occurs at $t = 0.17$ s and at an offset of 12 km. At the maximum difference between absence and presence of the resistive layer, the electric field strength in the presence of the resistive layer is one order of magnitude higher than the implemented noise floor. For this simple model, the 40 m thick

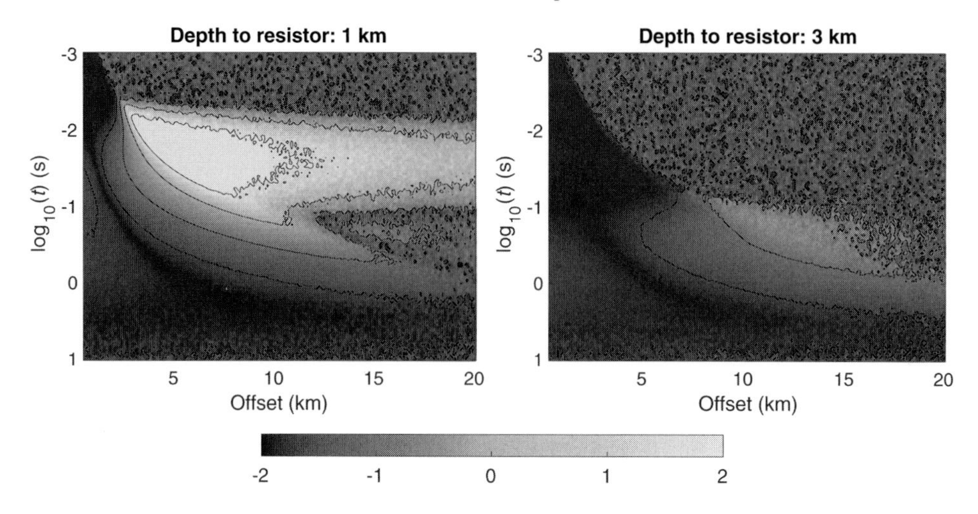

Figure 5.53 Electric field impulse response difference between models with and without the resistive layer as shown in Figure 5.52, but using the noise model, as a function of logarithmic time and offset for the 1 km deep (left) and 3 km deep (right) resistive layer. The colour bar indicates the normalised difference in logarithmic scale. For the colour version, please refer to the plate section.

resistive layer with a conductivity ratio of 20 to its surroundings seems feasible. In real situations this can be assumed only under the condition that the actual subsurface conductivity distribution allows the detection of the resistive layer and that noise levels are not higher than the measured electric fields on and off the target.

The electric field with added and multiplicative noise is shown in Figure 5.54 as a function of offset for optimum time instants and as a function of time for optimum offsets from the data with noise for the resistive layer buried at 1 km depth in the left plot and at 3 km depth in the right plot. These line graphs are chosen such that they contain the maximum differences obtained. The fields corresponding to the model with the resistive layer are shown in solid lines and the half-space response fields in dashed lines. The half-space response shown in the top-left plot goes to zero rapidly, whereas the response in the model with the resistive layer at 1 km depth has a much slower decay and the field strength stays well above the noise level up to 15 km offset. In the top-right plot it can be seen that the response of the model with the resistive layer at 1 km depth is much bigger than the half-space response from 4 ms to 0.1 s. For the resistive layer at 3 km depth the response grows to just above a factor of 6 between 8 km offset and 13 km offset at $t = 0.17$ s, as can be seen in the bottom-left plot. The noise is clearly visible in both responses shown in the bottom-right plot.

It can be concluded that when we look at impulse responses for the land model, the maximum amplitude difference indicating the presence of a resistive layer

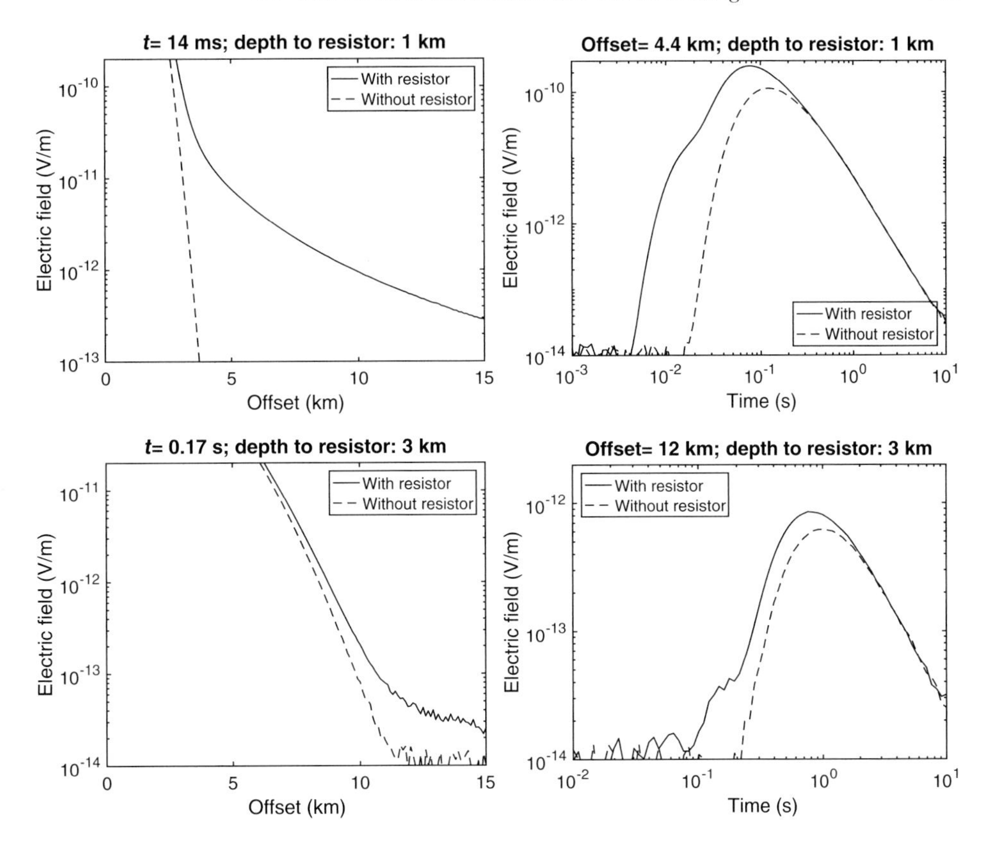

Figure 5.54 Electric field impulse response for the land model with (solid lines) and without (dashed lines) the resistive layer buried at 1 km (top) and 3 km (bottom) depth as a function of offset at the time of the maximum difference (left) and as a function of time at the offset of the maximum difference (right).

compared with absence of that layer is remarkably good for resistive layers at 1 km and at 3 km depths. The maximum difference for both resistive-layer models and the half-space model is much bigger than in the frequency domain. This is mostly caused by the fact that in the frequency domain the air wave is always present and strong, whereas in the time domain it occurs virtually at zero time, where the subsurface response is not yet present. On land, measuring impulse response functions is clearly the most favourable survey method.

5.4.4 Anisotropic Model Results in the Time Domain

Figure 5.55 shows the electric field impulse response of the half-space as a function of logarithmic time and offset. Here it is clear that the air wave is not part of the figure because it is frequency-independent: it is an impulse in time and the figure starts showing the field at 1 ms. The electric field impulse responses are shown in

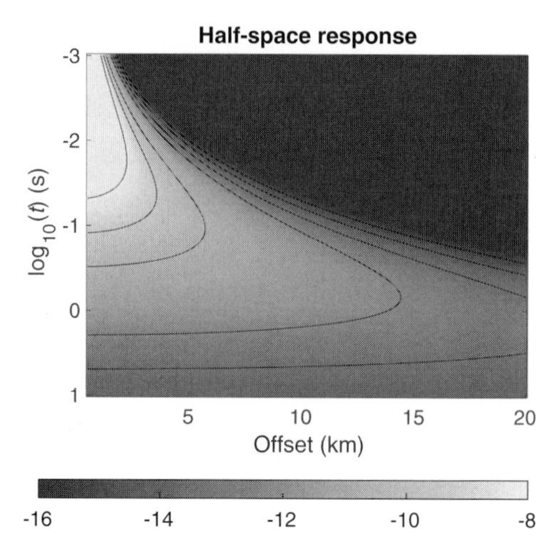

Figure 5.55 Electric field impulse response as a function of logarithmic time and offset for the VTI half-space model. For the colour version, please refer to the plate section.

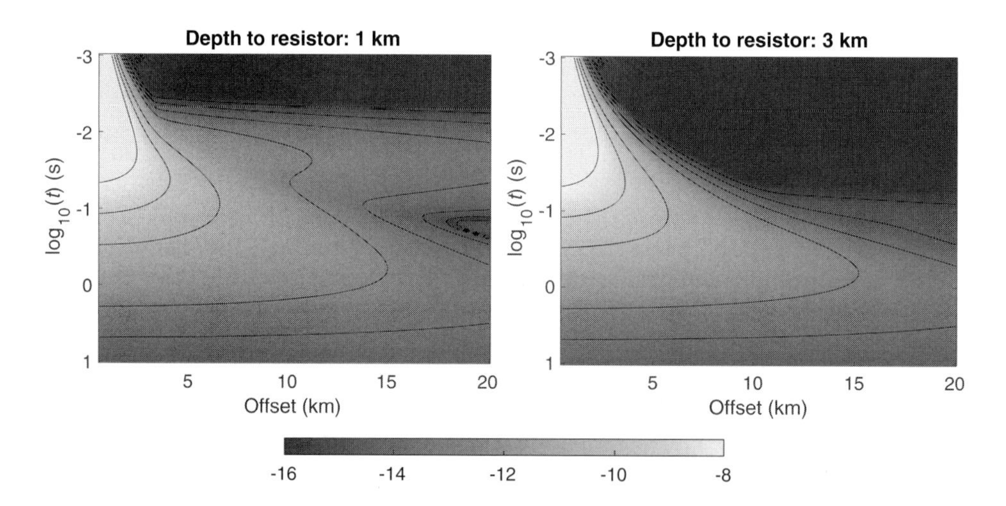

Figure 5.56 Electric field impulse response as a function of logarithmic time and offset for the land VTI model with the 40 m thick buried resistive layer at 1 km depth (left) and 3 km depth (right). For the colour version, please refer to the plate section.

Figure 5.56 for the 1 km deep (left) and 3 km deep (right) resistive layer. Comparing both plots with the half-space result shows two zones where the presence of the resistive layer strongly influences the measured electric field when the resistive layer is at 1 km depth. When it is at 3 km it is much less clear.

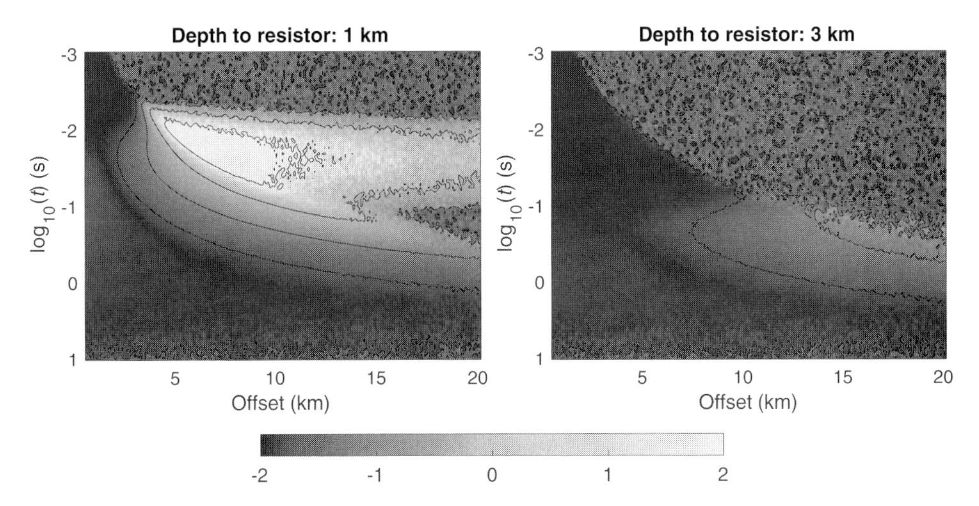

Figure 5.57 Electric field impulse response difference between the VTI models with and without the resistive layer as shown in Figure 5.52, but using the noise model, as a function of logarithmic time and offset for the 1 km deep (left) and 3 km deep (right) resistive layer. The colour bar indicates the normalised difference in logarithmic scale. For the colour version, please refer to the plate section.

This is better visualised in Figure 5.57, showing the normalised amplitude of the difference between the two fields shown in Figure 5.56, but with the same noise model as introduced in the deep sea example. The difference is more than two orders of magnitude in the zone that is saturated yellow in the left plot of Figure 5.57. The maximum difference is a factor 837 and occurs at $t = 16$ ms and at an offset of 6.2 km. For the example in which the resistive layer is buried at 3 km depth the difference is still quite large, with a maximum factor of 3.3. This maximum occurs at $t = 0.13$ s and at an offset of 16 km. The electric field strength is then just one order of magnitude above the implemented noise floor. Detecting a 40 m thick resistive layer with a conductivity ratio of 20 to its surroundings seems feasible with such a simple model. In a real situation this can be assumed only under the condition that the actual subsurface conductivity distribution allows for detecting the resistive layer and that noise levels are not higher than the measured electric fields on and off the target. We can see that, compared with the frequency domain results, the impulse responses present a much stronger difference between presence and absence of a resistive layer, whether at 1 km depth or 3 km depth. The impulse responses also allow for measurements at smaller offsets, which is more practical than frequency domain measurements.

The electric field with added and multiplicative noise is shown in Figure 5.58 as a function of offset for optimum time instants and as a function of time for optimum

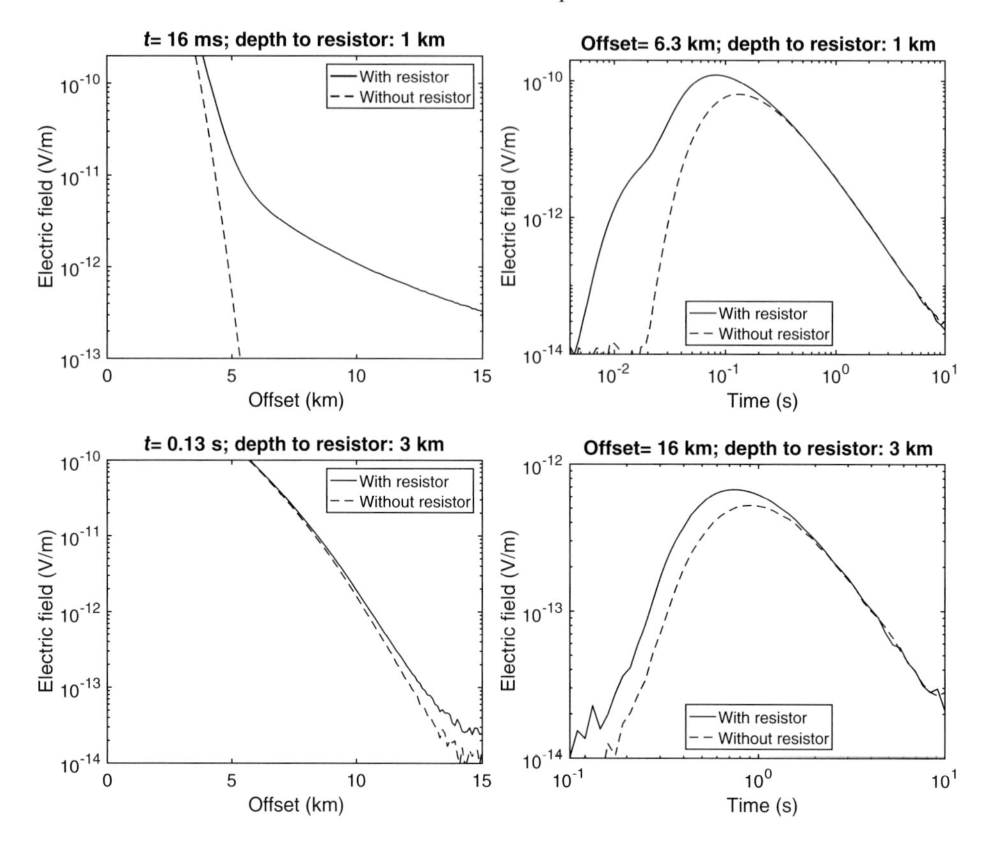

Figure 5.58 Electric field impulse response for the land VTI model with (solid lines) and without (dashed lines) the resistive layer buried at 1 km (top) and 3 km (bottom) depth as a function of offset at the time of the maximum difference (left) and as a function of time at the offset of the maximum difference (right).

offsets from the data with noise for the resistive layer buried at 1 km depth in the left plot and at 3 km depth in the right plot. These line graphs are chosen such that they contain the maximum differences obtained. The fields corresponding to the model with the resistive layer are shown in solid lines and the half-space response fields in dashed lines. The half-space response shown in the top-left plot goes to zero rapidly, whereas the response in the model with the resistive layer at 1 km depth has a much slower decay and the field strength stays well above the noise level up to 15 km offset. In the top-right plot it can be seen that the response of the model with the resistive layer at 1 km depth is much bigger than the half-space response from 5 ms to 0.1 s. For the resistive layer at 3 km depth, the response grows to a factor of 3 between 13 km offset and 15 km offset at $t = 0.13$ s, as can be seen in the bottom-left plot. It is just over half an order of magnitude bigger than the half-space response between 0.1 s and 1 s at 16 km offset. The maximum difference occurs when the signal is at noise level, and at higher signal levels the difference is a factor of 2.

We conclude that acquiring data to recover impulse responses gives better results on land with VTI media than using individual frequencies, as it does for isotropic media.

5.4.5 Land CSEM for Detecting a Buried Conductive Layer

Not all applications require the detection of the presence of a resistive layer in a conductive environment. Deep hot water saturated layers can be of interest for production of geothermal heat. As an example we show some results of impulse response electric fields in the time domain in the same measurement configuration as used in the previous land CSEM examples. The x-directed electric dipole current source is excited by a unit impulse in time and the x-component of the electric field is computed at the surface for several offsets as a function of time. The difference with the previous examples of Section 5.4.3 is that we use a 40 m thick conductive layer buried at 2 km below the surface in a resistive half-space. The half-space has a resistivity of $\rho = 30$ Ωm and the conductive layer has a resistivity of $\rho = 1$ Ωm. Now there is no fast and less attenuating path through the buried layer and we expect only a measurable difference in the TE mode part of the electric field. The impulse responses are shown in Figure 5.59, in which the half-space response is shown in the left plot and the response of the half-space containing the conductive layer in the right plot. The left plot of Figure 5.59 can be compared with Figure 5.51 and it can be seen that the responses show similar behaviour. In the right plot of Figure 5.59 we can see that the presence of the conductive layer is clearly visible in the small time interval between 0.03 s and 0.1 s for all offsets larger than 5 km. This is characteristic for the TE mode part of the electric field in the earth response where the presence of the conductive layer manifests itself as a new event that is absent in the half-space response. This type of event is also present in the response when a resistive layer is present in a conductive half-space, as shown in Figure 5.52. In that situation the resistivity contrast between the half-space and the layer was a factor of 20, whereas here the ratio is a factor of $1/30$, leading to a similar field in the TE mode part of the response of the buried layer, but with a distinct line with zero amplitude that is absent in the resistive layer response. In the right plot of Figure 5.59 we can see that the conductive layer buried at 2 km depth shows a very strong response. Figure 5.60 shows the normalised difference between the responses in the absence and presence of the buried conductive layer in the resistive half-space. The same noise model is used and the time–offset range where a measurable difference of more than two orders of magnitude is observed above the noise level is quite large. It has an onset at 14 ms at 5.4 km offset and this onset time increases exponentially to 24 ms at 20 km offset, as is evident by the linear increase in the logarithmic time scale of the plot. The end time has a more complicated offset dependency. At 5.5 km offset the two orders of magnitude difference is present up to 0.02 s; at 10 km offset it is present up to 0.1 s and up to 0.27 s at 20 km offset.

Figure 5.61 shows two line plots that are combined cross-sections of the left and right plots in Figure 5.59. In the left plot the electric field is shown as a function of offset at $t = 0.05$ s, which is the time instant at which the largest normalised difference occurs. In the right plot the electric field is shown as a function of time at 9.5 km offset, which is the offset at which the largest normalised difference occurs. In both plots the electric field in the presence of the buried conductive layer has negative and positive field strengths, which are indicated in the logarithmic field strength plots using solid lines for positive values and dotted lines for negative values. The half-space response is always positive and is indicated by dashed lines. In the left plot it can be seen that the offset dependency is polynomial from approximately 6 km offset onward. The behaviour is near-field and is proportional to $1/r^3$ where r is offset. In the right plot we can see that a new event arises much earlier than the half-space response, which can only occur when a large part of the path is travelled through the air. This is the TE mode part of the electric field in the earth response and at 9.5 km offset the half-space response takes over from about 0.75 s onward.

From these results we can conclude that the presence or absence of a conductive layer in a resistive half-space can be detected with confidence using an in-line

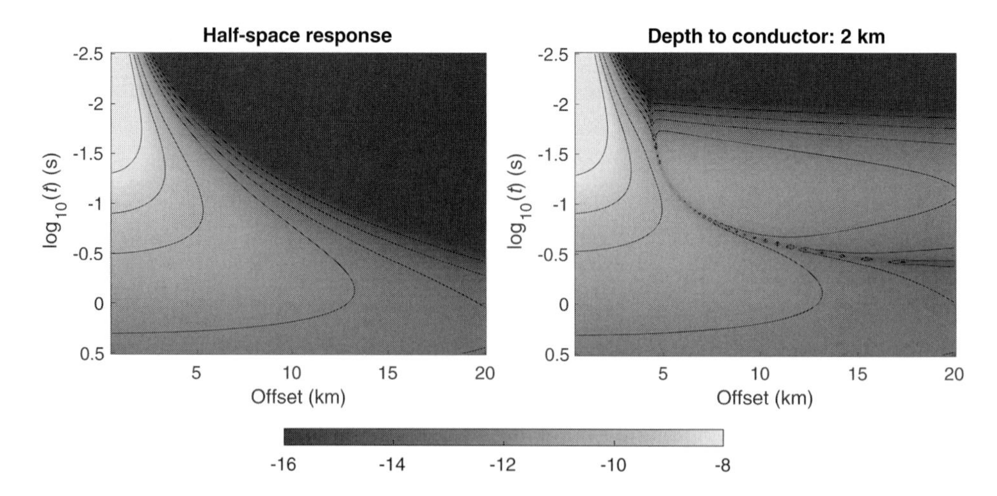

Figure 5.59 Electric field impulse response as a function of logarithmic time and offset for the isotropic 30 Ωm half-space model (left) and the half-space with 40 m thick 1 Ωm layer in the half-space (right). For the colour version, please refer to the plate section.

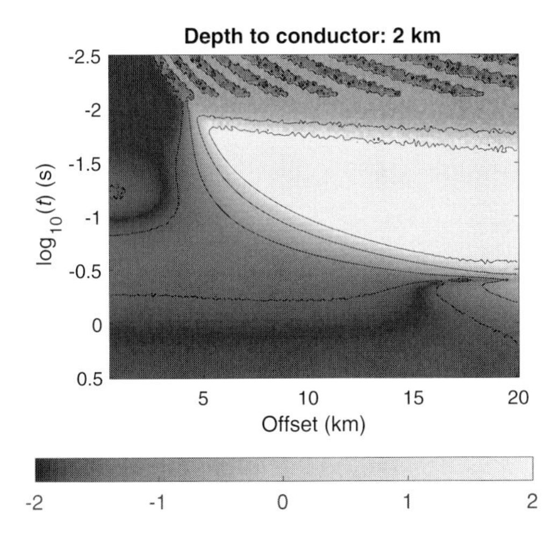

Figure 5.60 Electric field impulse response difference between the models with and without conductive layer as shown in Figure 5.59, but using the noise model, as a function of logarithmic time offset. The colour bar indicates the normalised difference in logarithmic scale. For the colour version, please refer to the plate section.

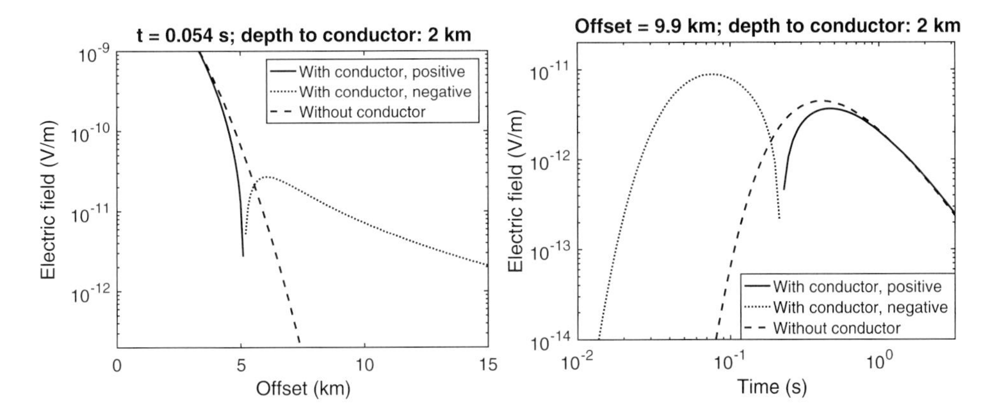

Figure 5.61 Electric field impulse response for the isotropic land model with (solid line: positive values and dotted line: negative values) and without (dashed lines) the conductive layer buried at 2 km depth as a function of offset at the time of the maximum difference (left) and as a function of time at the offset of the maximum difference (right).

horizontal electric current source and electric field receivers at the surface of a half-space. The conclusion is valid for the example of a 40 m thick layer that has a resistivity ratio of 30:1. The fact that the air wave couples into the earth and produces a downwards-diffusing field with relatively large field strength at offsets larger than at least 2.5 times the burial depth leads to a viable method for detecting such layers.

6

Source Control

There is no source control in passive geophysical methods. The receivers simply measure fields and waves. In earthquake seismology, for instance, the time, location and time function of an earthquake are unknown, and we may try to determine these parameters from the resulting seismograms. To do that, we need to know certain physical parameters of the Earth, in particular the seismic velocity structure. In active geophysical exploration we actively try to determine the physical parameters of the Earth and are free, within obvious limits, to choose the time, location and time function of the source.

In exploration seismology, it is widely accepted that the aim of source control is to generate a known signal with the widest possible bandwidth. Apart from the dynamite source, there is usually a problem in generating low frequencies with man-made seismic sources and, normally, we are not totally satisfied with the low-frequency content of the signal.

In controlled-source electromagnetic (CSEM) exploration there is no agreed definition of the aim of source control. In this book we argue that the goal of CSEM data acquisition should be to recover the Earth's impulse response or Green's function within the constraints imposed by the Earth. The reason for this is the same as for exploration seismology: we are trying to find out about the interior of the Earth and we need the greatest possible bandwidth. In general it is the high frequencies that are a problem, because of the intrinsic attenuation of electromagnetic propagation in conducting media. There are two main issues: the bandwidth, or frequency content of the source time function; and the time required to obtain adequate signal-to-noise ratio. This second issue is clearly to do with efficiency and cost.

This chapter focuses on the choice of time function of the current $I(t)$ of a current dipole source. As described in Chapter 4, the earth response is a convolution of the Green's function with the dipole moment time function, which is simply the source current time function multiplied by the distance Δx_s between the source electrodes.

This distance is fixed for a given source, so the source current time function $I(t)$ is the interesting part.

Various functions have been advocated for use as source current time functions in CSEM exploration: pseudo-random binary sequences, a step function, square waves, and special periodic functions. These are discussed here. The problem of recovery of the Green's function by deconvolution of the measured data for the measured source time function is considered in detail and the concept of *deconvolution gain* is introduced.

6.1 The Convolutional Model in CSEM

Let the source current be $I(t)$ and let the resultant voltage at the receiver be $V(t)$. The response of the earth can be regarded as a causal linear filter with impulse response $g(t)$ that depends on the position and direction of the injected current at the source and the position and orientation of the receiver electrodes. The concept is illustrated in Figure 6.1.

In the absence of noise, the three quantities are related by the convolution

$$V(t) = \int_0^t g(\tau)I(t-\tau)d\tau. \tag{6.1}$$

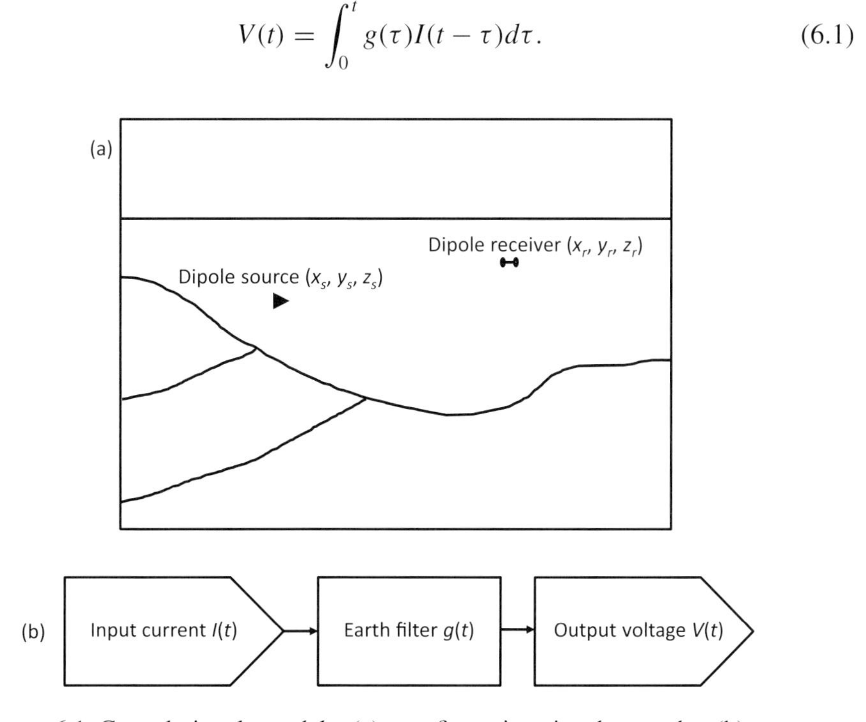

Figure 6.1 Convolutional model: (a) configuration in the earth; (b) one-dimensional linear filter representation.

The lower limit of the integral is zero because the earth is causal and cannot respond before there is an input. The upper limit is t also because of causality: the response at time t is for inputs no later than t. Since the flow of current in a conducting earth is a lossy process, the impulse response $g(t)$ must decay to zero as $t \to \infty$. Within the precision of the measurements, therefore,

$$g(t) = 0, \quad \text{for} \quad t > T_g, \tag{6.2}$$

where T_g is a time greater than that for which the response is too small to detect. That is, the earth impulse response $g(t)$ is *transient*: it has a beginning and an end. Sheriff (1973) defines 'transient' as 'A nonrepetitive pulse of short duration, such as a voltage pulse or a seismic pulse.' The word 'transient' is used here in Sheriff's sense, both as a noun and as an adjective.

Given that $g(t)$ is of finite duration, what is the best function for $I(t)$? This is the problem of source control. In this chapter we look at the properties of a variety of source time functions that have been used in CSEM: the pseudo-random binary sequence, or PRBS; the Heaviside function, or step function; the square wave function; and certain special periodic functions that are popular at the time of writing (2017).

If the source time function is also transient, it has a duration T_s, say, and the time to acquire the complete response is $T_s + T_g$. Equation 6.1 may then be written as

$$V(t) = \int_0^t g(\tau)I(t-\tau)d\tau, \quad \text{for} \quad 0 \le t \le (T_s + T_g). \tag{6.3}$$

6.2 Pseudo-Random Binary Sequence

A PRBS is normally a periodic sequence. We consider here the properties of a single period of a PRBS and refer to this as a PRBS. The generation of a PRBS may be performed using linear feedback shift registers, each of which contains a 1 or a 0, or it may be calculated algebraically using modulo-2 linear algebra (Golomb, 1955, 1982). The number of registers determines the maximum length of the sequence. If there are n registers, the maximum sequence length is $N = 2^n - 1$. The maximum sequence length or m-sequence is a series of bits containing 2^{n-1} ones and $2^{n-1} - 1$ zeros. The key to obtaining a maximum length PRBS is to set the initial values of the registers correctly. The number of correct settings increases with the number of registers. Determination of the correct settings can be obtained algebraically using modulo-2 linear algebra. A sequence of order 5 is illustrated in Figure 6.2(a).

There is no energy associated with the zeros. As a source time function for use in CSEM, the zeros in the sequence reduce the energy by almost a factor of 2. It is preferable to use -1 instead of 0, for then the power is constant and the energy is

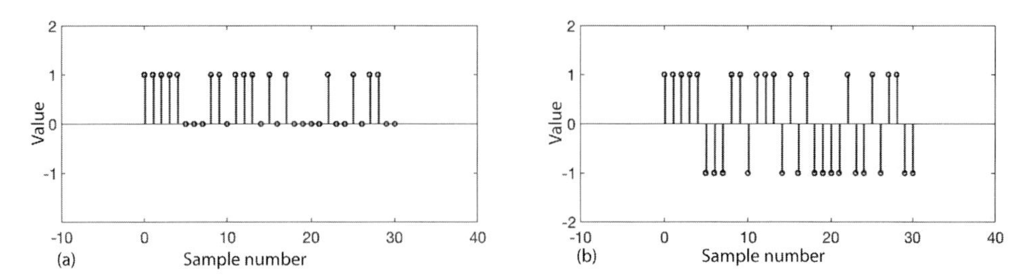

Figure 6.2 (a) One period of an order 5 PRBS: 1s and 0s; (b) the same as (a), but with 0s replaced by -1s.

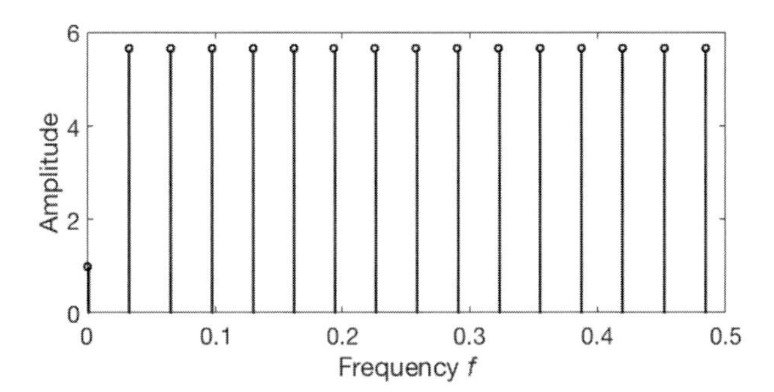

Figure 6.3 Amplitude spectrum of PRBS in Figure 6.2(b) in which $\Delta t = 1$.

almost double. This is illustrated in Figure 6.2(b). This sequence is the same as the first 31 samples of the PRBS of Figure 2.13.

A PRBS has important properties for signal processing. In particular, its amplitude spectrum is flat in the frequency range

$$\frac{1}{N\Delta t} \leq f \leq \frac{1}{2\Delta t}, \tag{6.4}$$

where Δt is the time sampling interval and $\frac{1}{2\Delta t}$ is the Nyquist frequency. The spectrum of the sequence shown in Figure 6.2(b) is obtained as

$$\hat{p}(f) = \Delta t \sum_{k=0}^{N-1} p_k e^{2\pi i f k \Delta t}, \quad \text{for} \quad f = 0, \Delta f, 2\Delta f, \ldots, (N-1)\Delta f, \tag{6.5}$$

where $p_k = p_0, p_1, p_2, \ldots, p_{N-1}$ is the PRBS and $\Delta f = \frac{1}{N\Delta t}$. The amplitude spectrum $|\hat{p}(f)|$ for the frequency interval $0 \leq f \leq \frac{1}{2\Delta t}$ is shown in Figure 6.3. The Nyquist frequency for $\Delta t = 1$ is 0.5.

6.3 Convolution and Deconvolution with a PRBS

Consider the convolution of p_k with the series $a_k = a_0, a_1, \ldots, a_{m-1}, a_m$.

$$y_t = \sum_{k=0}^{N-1} p_k a_{t-k}, \quad \text{for} \quad t = 0, 1, 2, \ldots, N+m-1, \tag{6.6}$$

and let a_k be the digital impulse $1, 0, 0, \ldots, 0$; that is, 1 followed by m 0s. For $m = 20$, the result y_t is shown in Figure 6.4.

We now show that this can be deconvolved for the known PRBS p_t. We use the frequency domain method described in Section 3.18. There are three steps. First, we transform to the frequency domain:

$$\hat{y}(f) = \hat{p}(f)\hat{a}(f). \tag{6.7}$$

Then we divide by the Fourier transform $\hat{p}(f)$ of the known source time function p_t. The sequence y_t is 51 samples and the sequence p_t is 31 samples. To perform this division in practice, we first add 20 zeros to p_t to make it the same length as y_t. We then divide the Fourier transform of y_t by the Fourier transform of p_t (including the additional 20 zeros):

$$\frac{\hat{y}(f)}{\hat{p}(f)} = \hat{a}(f). \tag{6.8}$$

Finally, we take the inverse Fourier transform of this to obtain the estimate $\text{Est}(a_t)$ of the sequence a_t:

$$\text{Est}(a_t) = \text{IFT}\left[\frac{\hat{y}(f)}{\hat{p}(f)}\right], \tag{6.9}$$

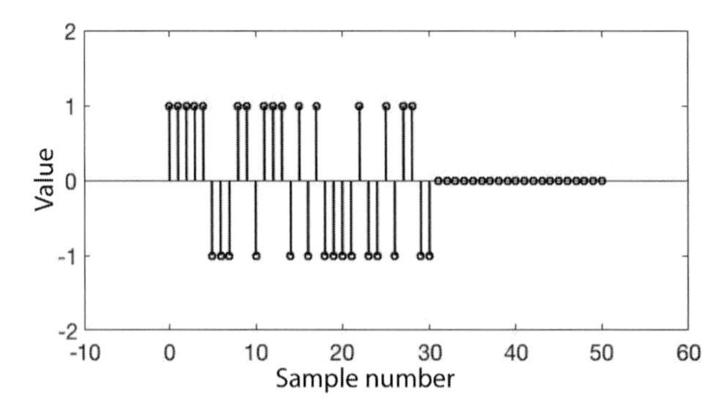

Figure 6.4 Signal y_t: the convolution of PRBS p_t with a digital impulse followed by 20 zeros.

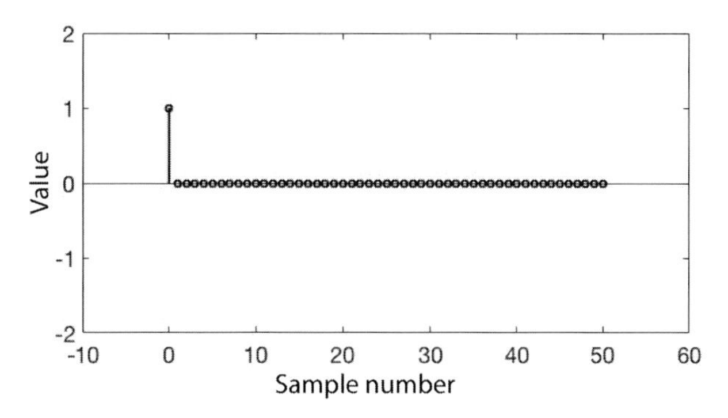

Figure 6.5 Recovered impulse: result of deconvolving y_t for p_t in the frequency domain.

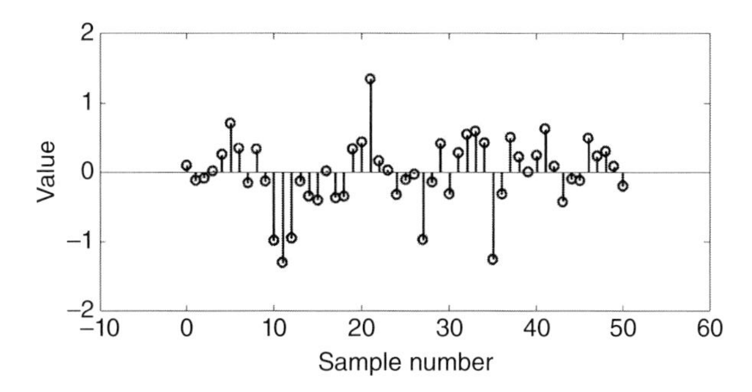

Figure 6.6 Example of Gaussian noise with standard deviation equal to 0.5.

where IFT[ ] is the inverse Fourier transform of the quantity in square brackets. The result is shown in Figure 6.5.

The 50 samples following the impulse are not precisely zero. They are actually very small positive and negative numbers of the order of 10^{-18}. This is noise introduced by the deconvolution process and caused by round-off errors in the computer. Within this very small error, we see that we have recovered our digital impulse plus 30 additional zeros, making it the same length as y_t.

6.4 Effect of Noise and Deconvolution Gain

Now we examine the effect of noise on the deconvolution process and introduce the concept of *deconvolution gain*. Real noise is normally not white and is often organised in some way. Nevertheless, for the purposes of this test, we use synthetic random noise generated by a computer. A short sequence of Gaussian random noise with standard deviation 0.5 is shown in Figure 6.6.

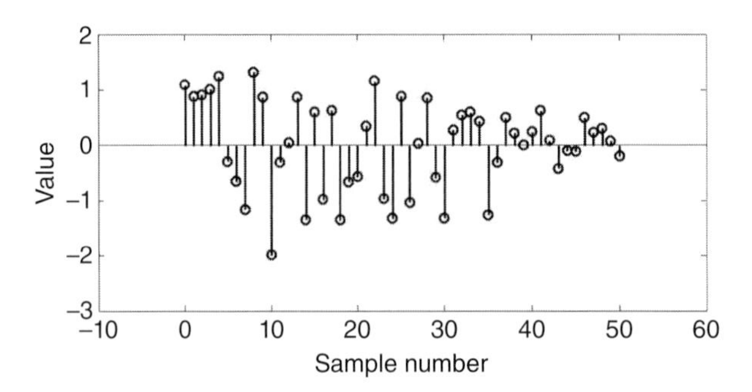

Figure 6.7 Signal $z_t = y_t + w_t$.

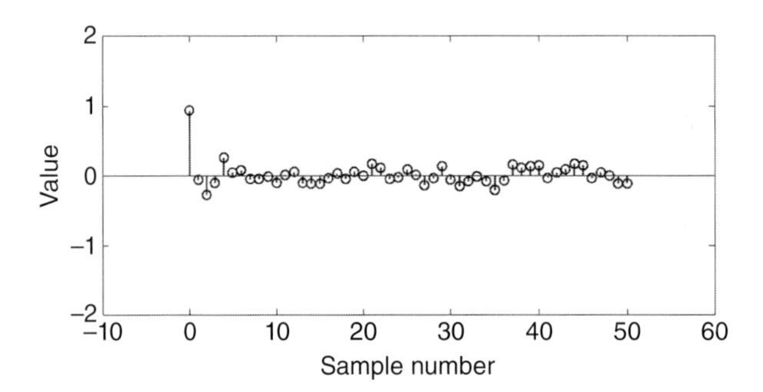

Figure 6.8 Deconvolution of z_t for p_t to recover impulse in the presence of noise.

This noise is added to the signal y_t and the result is the series

$$z_t = y_t + w_t, \tag{6.10}$$

shown in Figure 6.7, in which w_t is the noise of Figure 6.6. The last 20 samples of z_t are pure noise.

We now perform the deconvolution in the same way as for the noise-free case. We divide the Fourier transform of z_t by the Fourier transform of p_k (with 20 added zeros) to obtain the Fourier transform of the result of deconvolution:

$$\begin{aligned}\frac{\hat{z}(f)}{\hat{p}(f)} &= \frac{\hat{y}(f)}{\hat{p}(f)} + \frac{\hat{w}(f)}{\hat{p}(f)} \\ &= \hat{a}(f) + \frac{\hat{w}(f)}{\hat{p}(f)},\end{aligned} \tag{6.11}$$

and then take the inverse Fourier transform of this to obtain $\text{Est}(a_t)$, the estimate of the sequence a_k:

$$\text{Est}(a_t) = \text{IFT}\left[\frac{\hat{z}(f)}{\hat{p}(f)}\right]. \tag{6.12}$$

The result is shown in Figure 6.8.

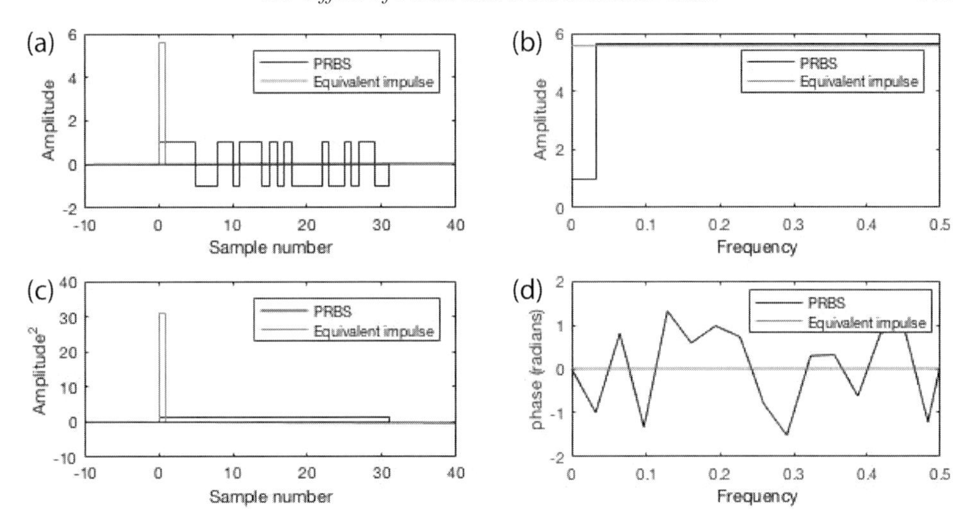

Figure 6.9 (a) PRBS and equivalent impulse; (b) amplitude spectra of PRBS and equivalent impulse; (c) energy distribution of PRBS and equivalent impulse; (d) phase spectra of PRBS and equivalent impulse.

We have recovered the impulse quite well (its amplitude is 0.93) and still have some noise. The noise is smaller than the original noise. The deconvolution process has increased the signal-to-noise ratio. We call this *deconvolution gain*.

We can calculate the theoretical deconvolution gain in a straightforward way by considering an impulse with the same energy as the PRBS. The deconvolution process has replaced the PRBS by an impulse of unit amplitude. The energy of the PRBS is the sum of the squares of the amplitudes. In this case this is 31. A digital impulse of $\sqrt{31}$ also has energy 31. We call this the *equivalent impulse*. The PRBS and its equivalent impulse are shown in Figure 6.9(a). The energy of the PRBS is spread uniformly over all 31 samples, whereas the same energy in the impulse is concentrated in one sample, as shown in Figure 6.9(c). The PRBS and its equivalent impulse have very similar flat amplitude spectra (Figure 6.9[b]); the PRBS has a low zero-frequency value of 1, compensated by a slightly higher amplitude at all other frequencies. The two signals have completely different phase spectra (Figure 6.9(d)); the impulse is zero-phase and the PRBS has pseudo-random phase.

The deconvolution process collapses all the energy of the PRBS into a single time sample. This may be visualised more easily by multiplying the amplitude of the signal in Figure 6.8 by $\sqrt{31}$. This is shown in Figure 6.10.

In this figure the noise is about the same magnitude as the noise in z_t, shown in Figure 6.7, but is different in detail, as a result of the deconvolution. The amplitude spectrum of the noise is essentially unchanged because the amplitude spectrum of the PRBS is a constant: the difference is in the phase.

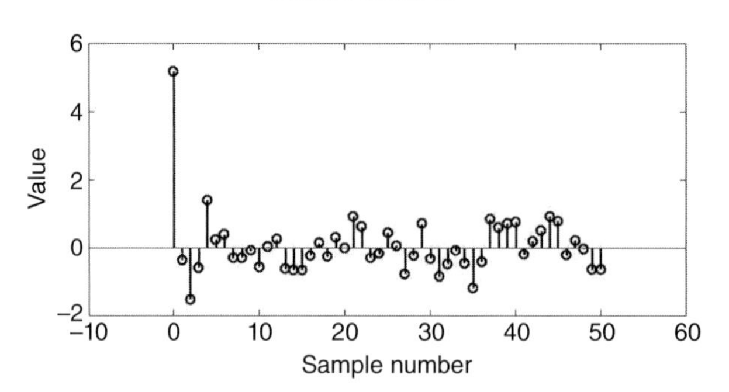

Figure 6.10 Same as Figure 6.8 but with amplitude scaled by $\sqrt{31}$.

It is important to notice where the deconvolution gain occurs. In the time domain the signal-to-noise ratio (SNR) before deconvolution is obtained from equation 6.10 as

$$\text{SNR} = \frac{|y_t|}{|w_t|}. \tag{6.13}$$

In the frequency domain it is obtained from equation 6.11:

$$\text{SNR} = \frac{|\hat{y}(f)|}{|\hat{w}(f)|}. \tag{6.14}$$

The division by $\hat{p}(f)$ in equation 6.11 does not change the signal-to-noise ratio. The increase in signal-to-noise ratio, or deconvolution gain, is obtained *only* after the transformation back to the time domain.

In summary, a single-period PRBS of length $N = 2^n - 1$ samples, where n is the order of the PRBS, is a very useful source time function for CSEM. It can be designed for any desired bandwidth and, by deconvolution, can increase the signal-to-noise ratio by a factor equal to $\sqrt{N}$.

Since the bandwidth of the earth impulse response is in principle unknown until we attempt to collect data, we do not know what parameters to use. In particular, we do not know the best values for Δt and N. There is no point in making Δt too small: it will result in energy being wasted in frequencies that are too high. It is often necessary to do experiments to determine the best value for Δt.

For a given Δt, the time duration of the PRBS is $T_s = N\Delta t$. Therefore deconvolution increases the signal-to-noise ratio as $\sqrt{T_s}$. This deconvolution gain is a powerful weapon in the fight with noise.

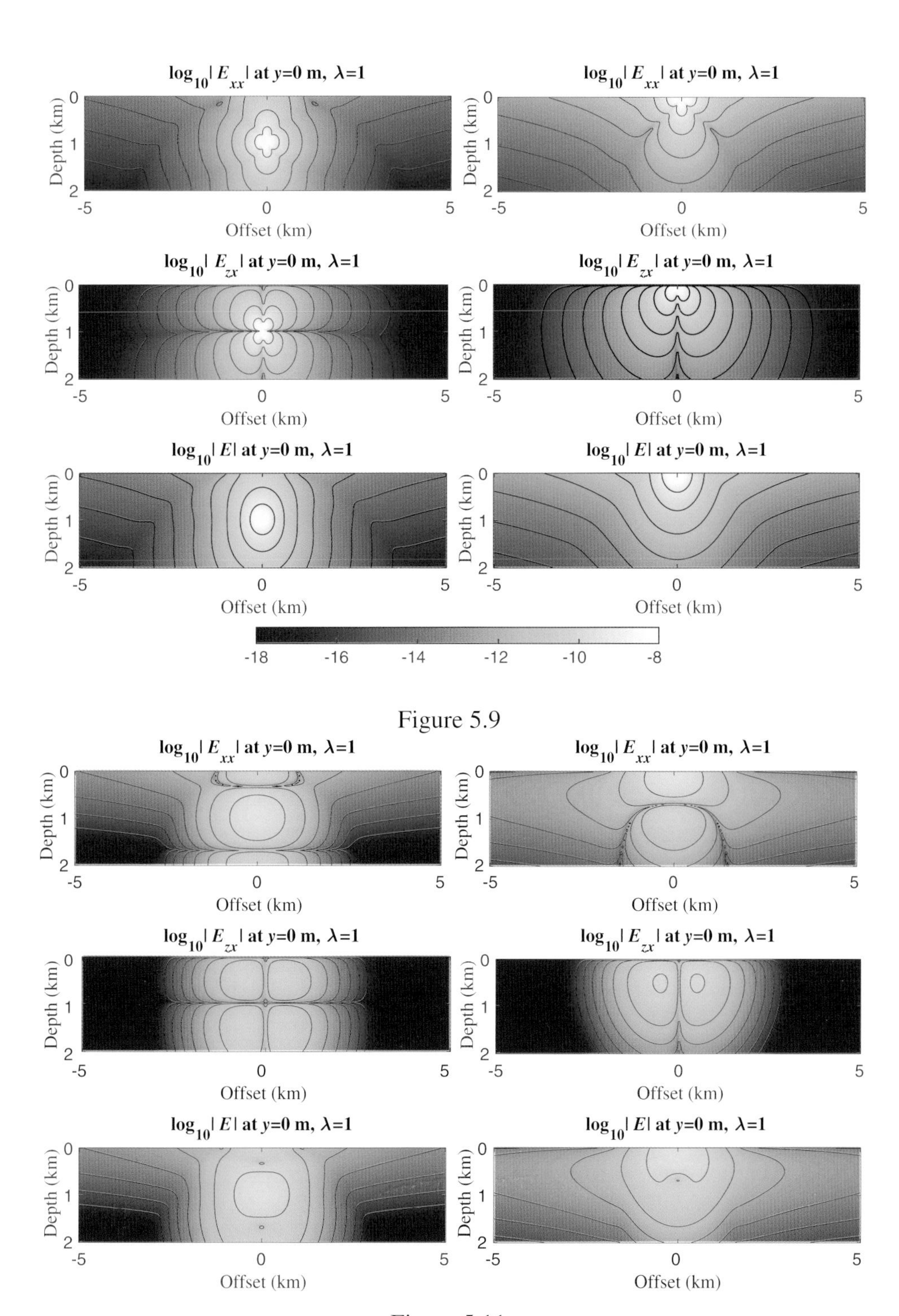

Figure 5.9

Figure 5.11

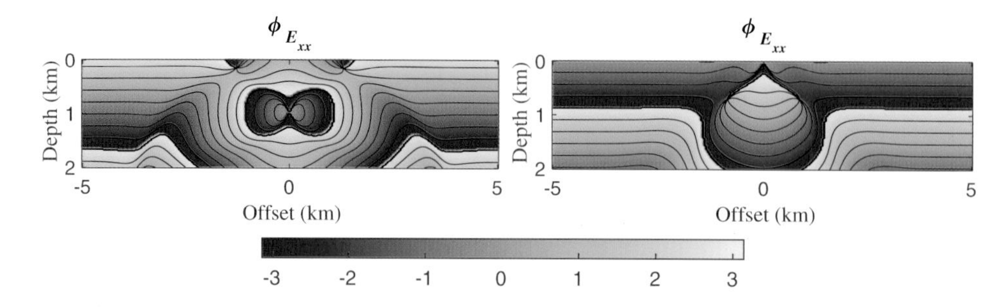

Figure 5.10

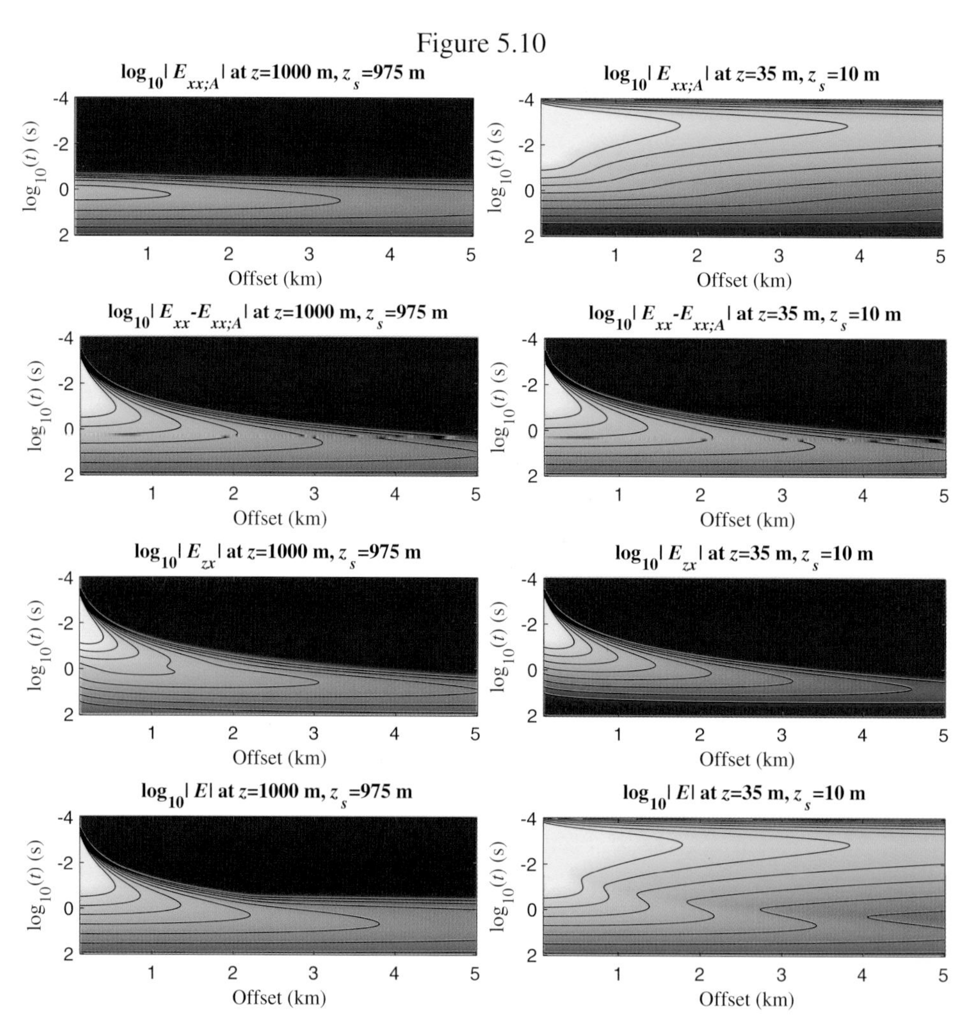

Figure 5.12

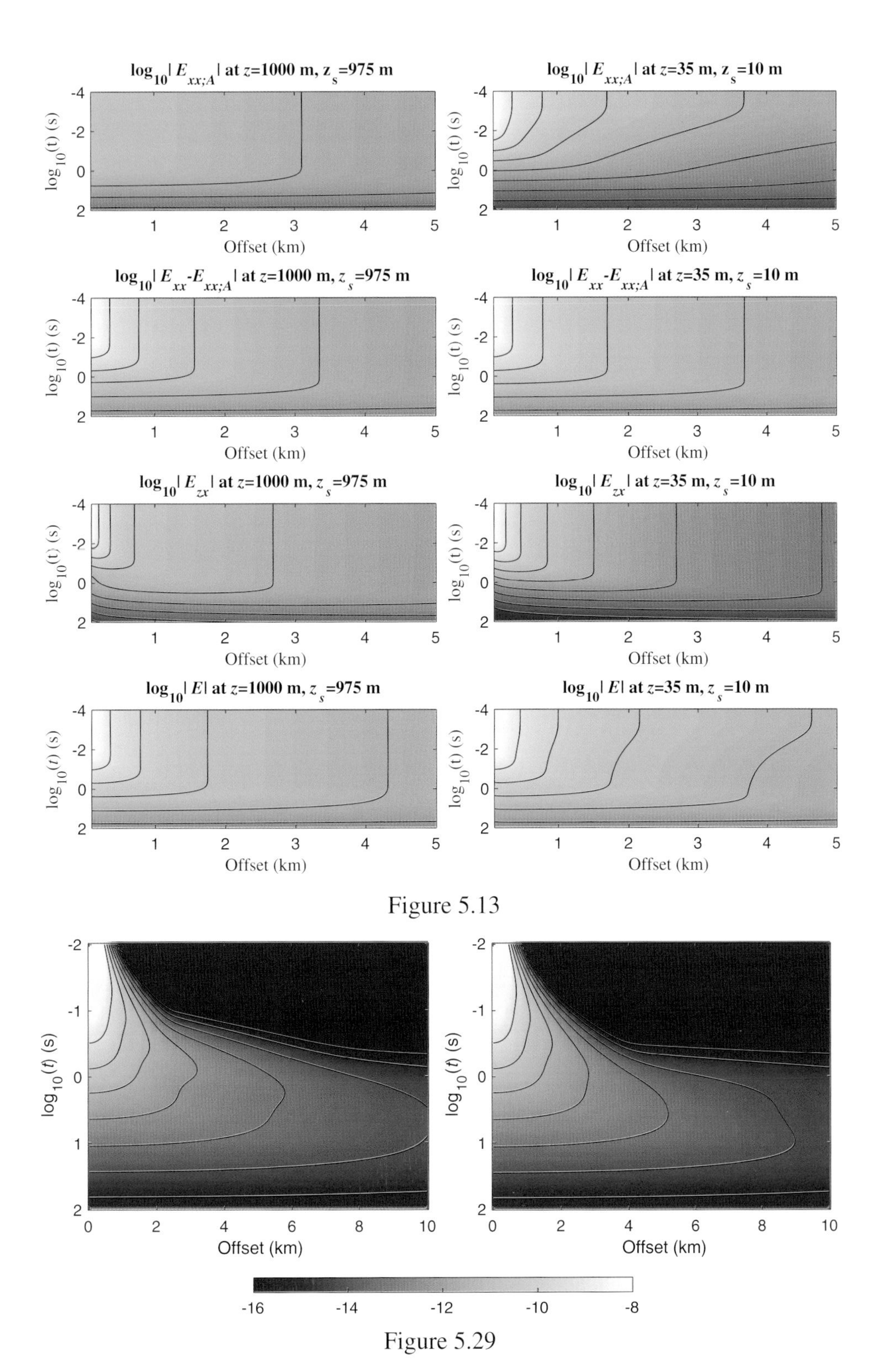

Figure 5.13

Figure 5.29

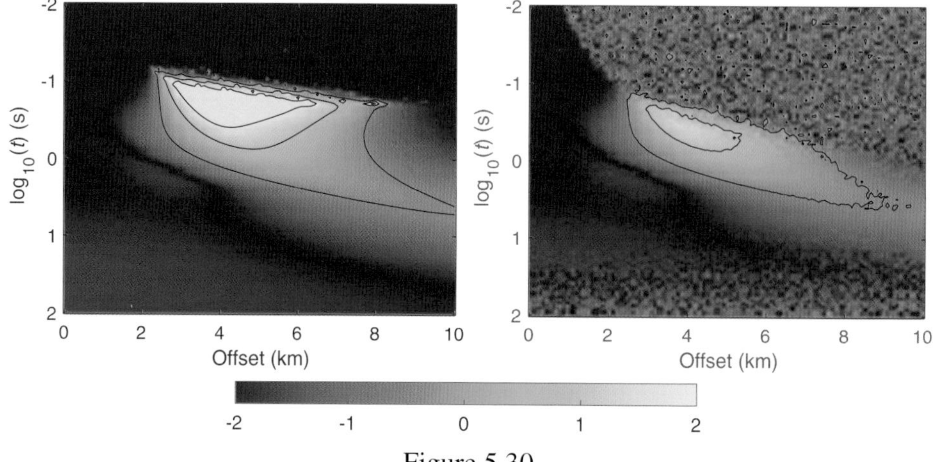

Figure 5.30

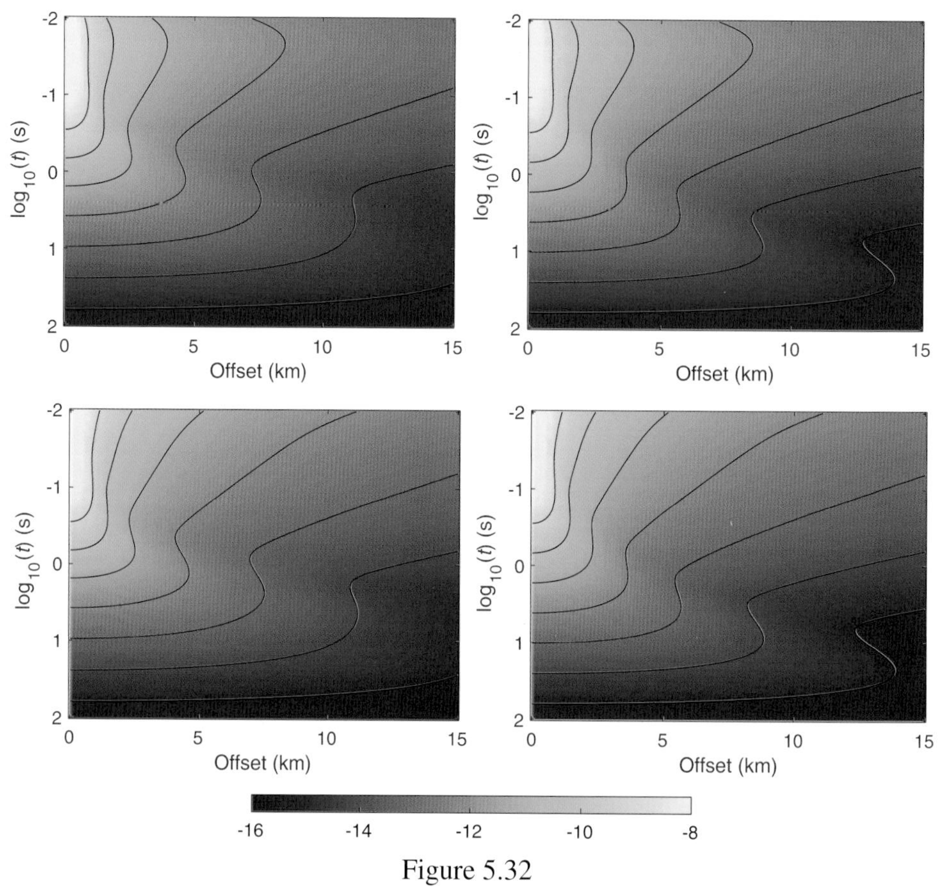

Figure 5.32

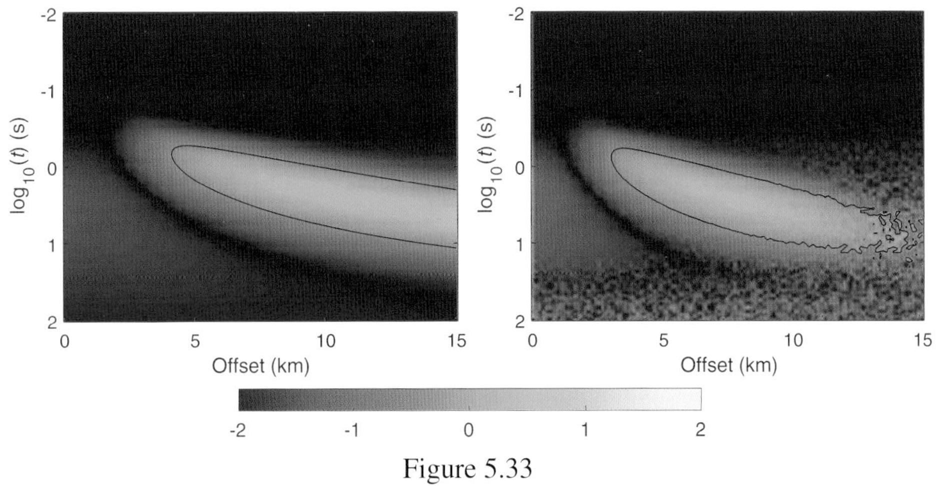

Figure 5.33

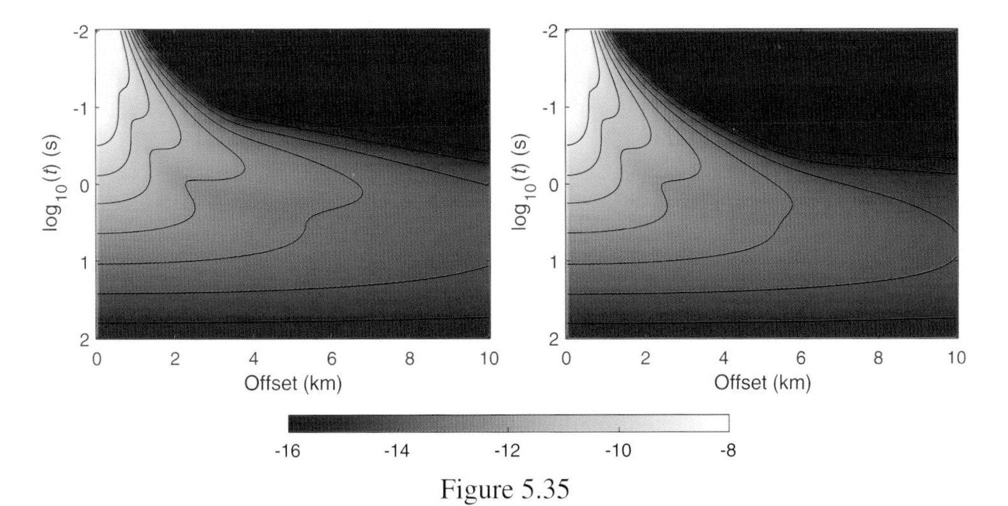

Figure 5.35

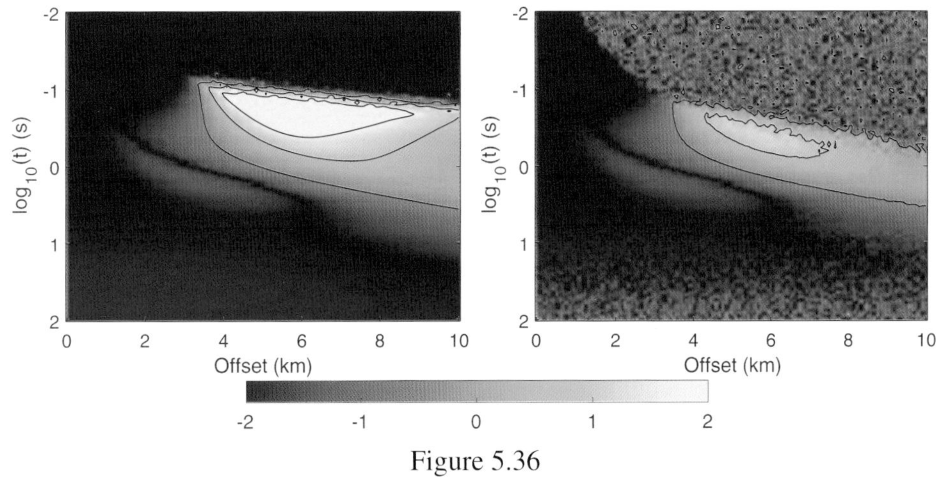

Figure 5.36

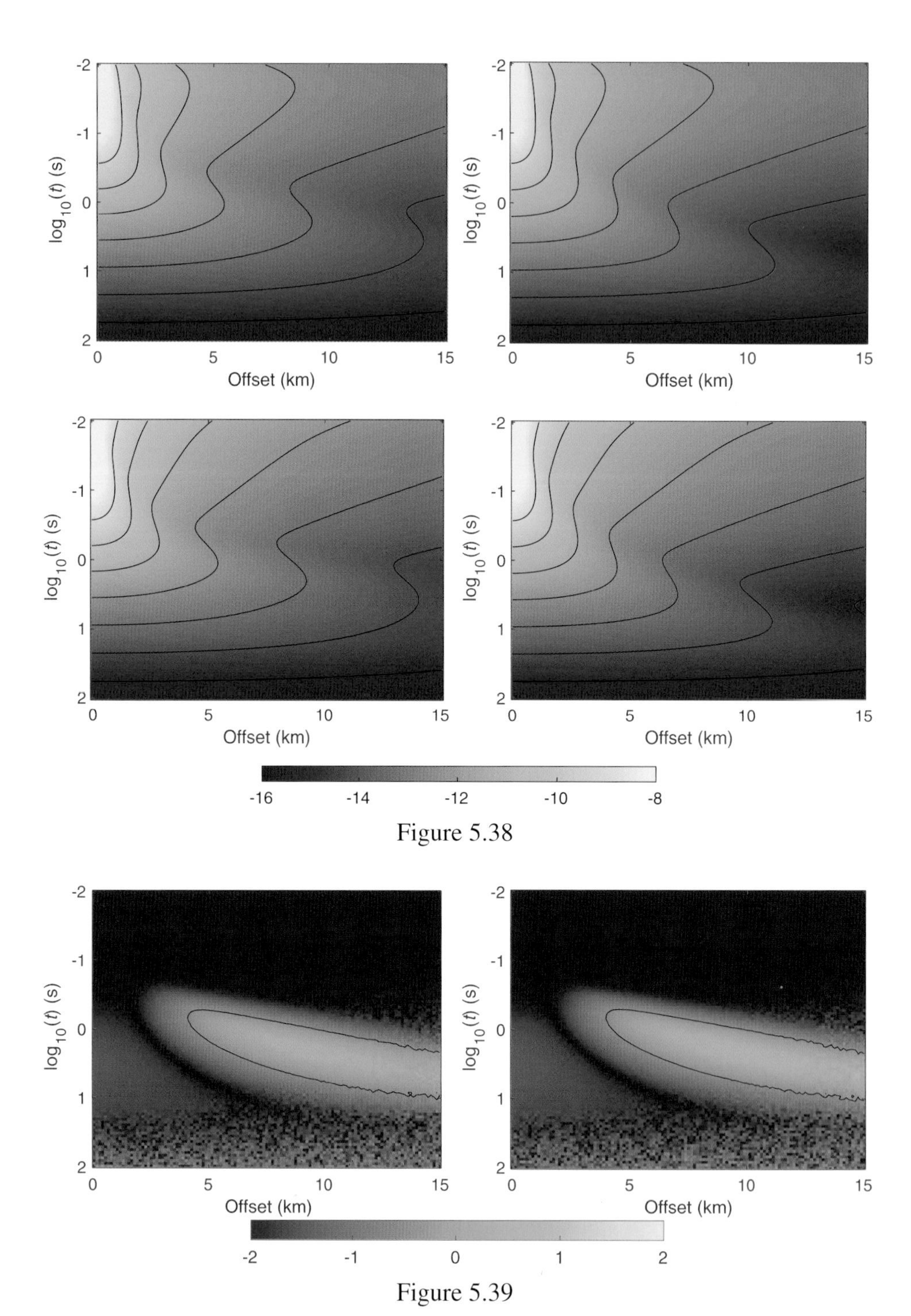

Figure 5.38

Figure 5.39

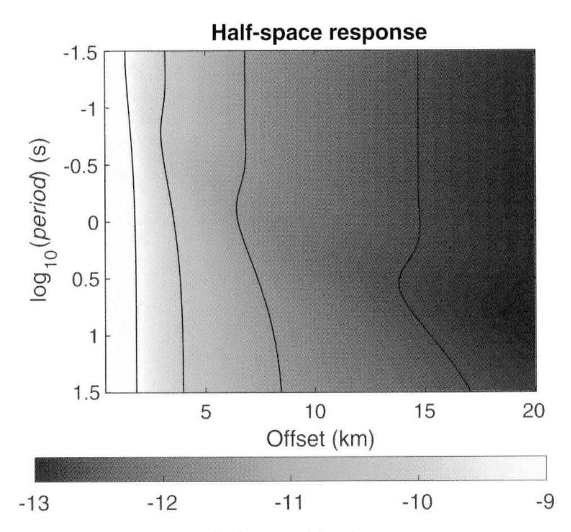

Figure 5.41

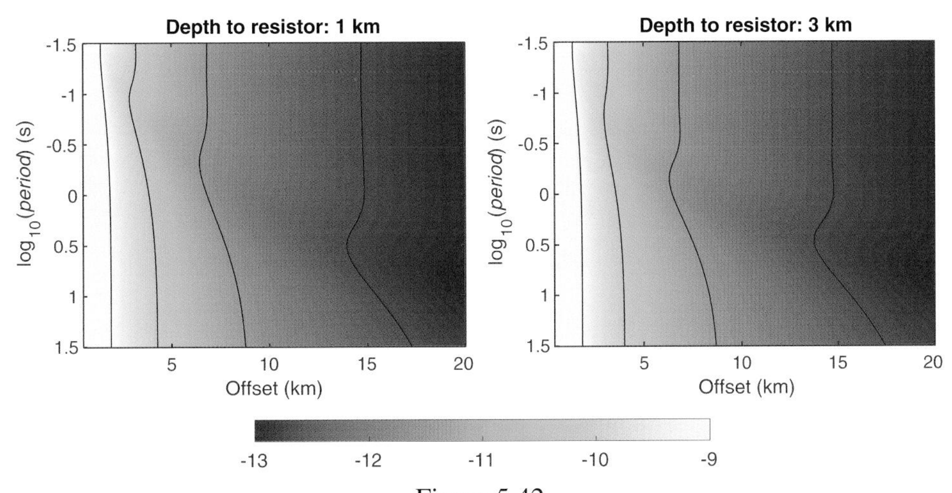

Figure 5.42

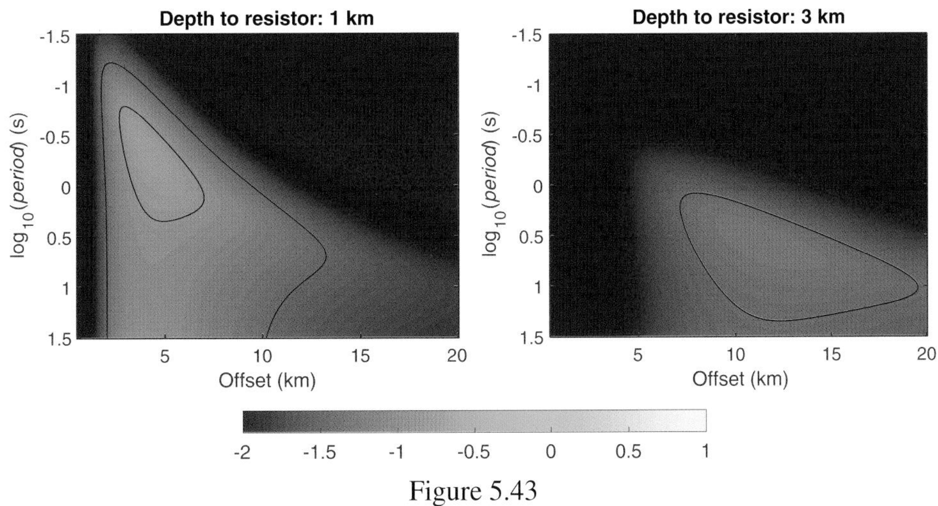

Figure 5.43

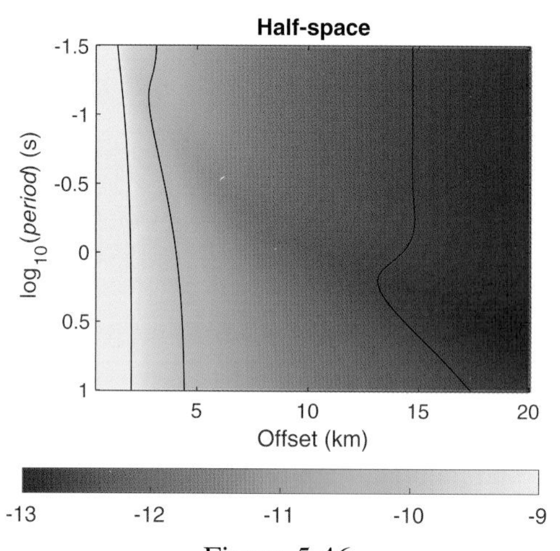

Figure 5.46

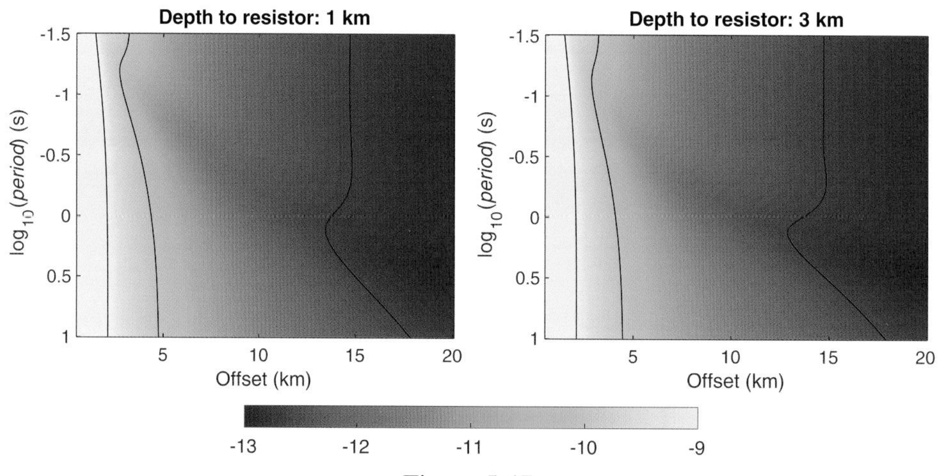

Figure 5.47

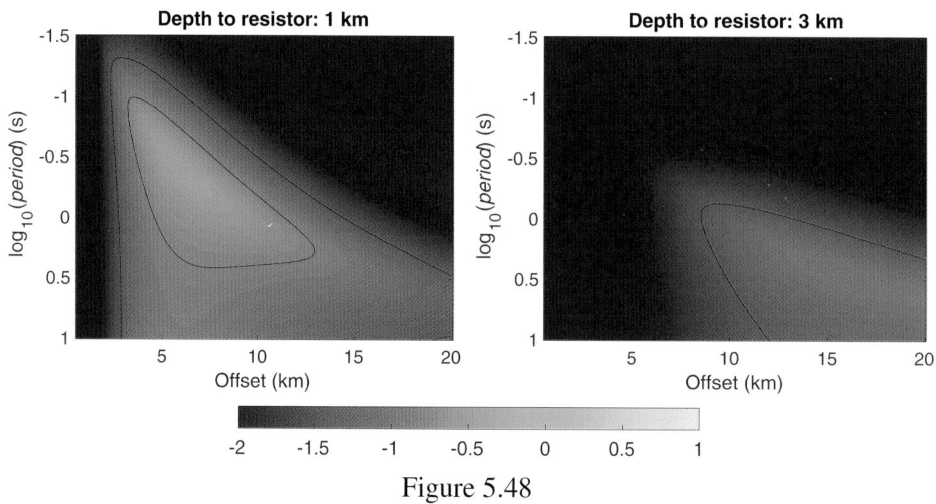

Figure 5.48

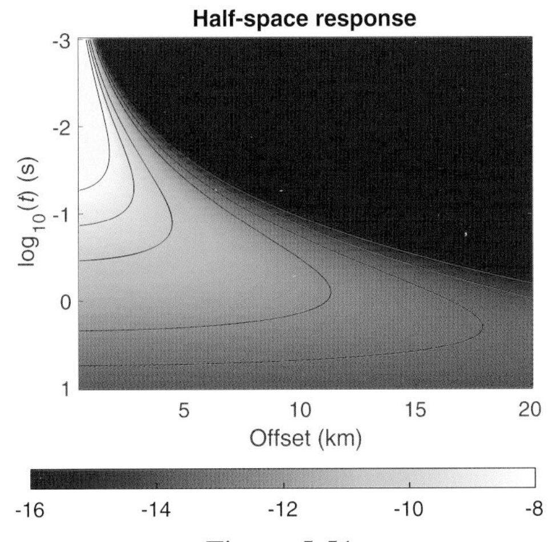

Figure 5.51

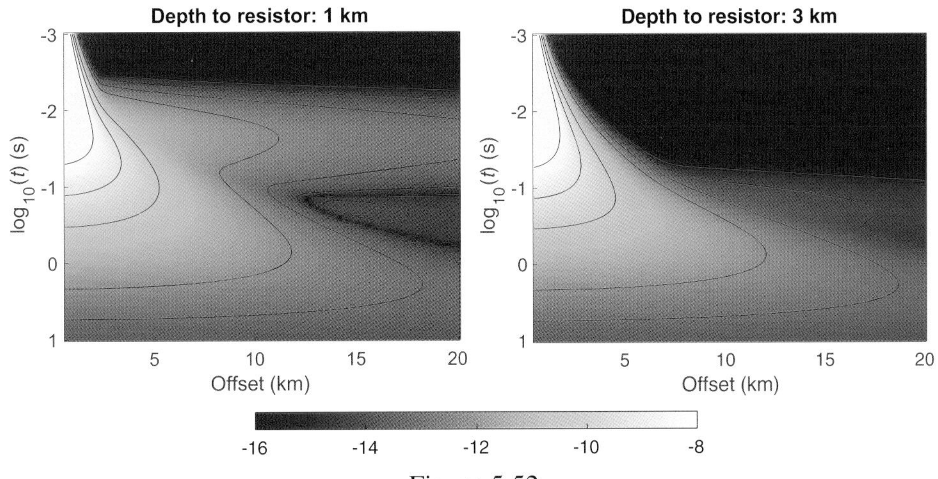

Figure 5.52

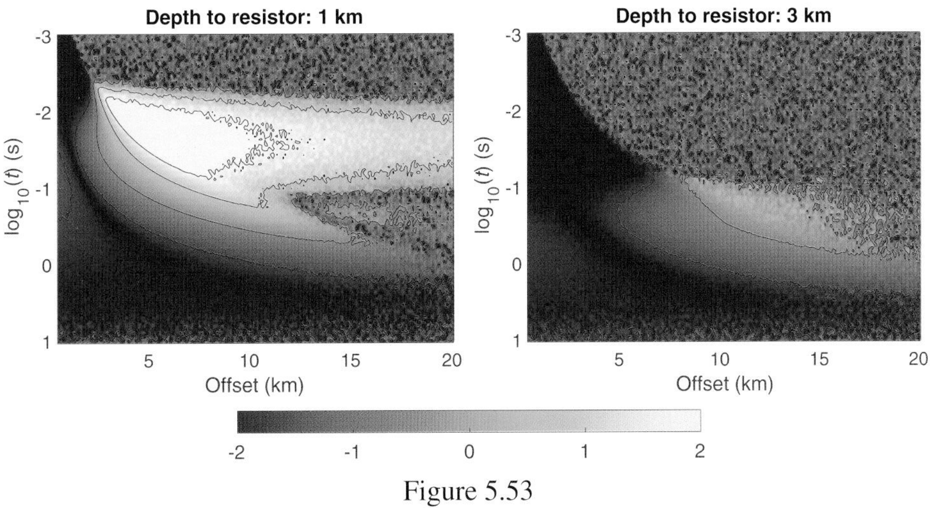

Figure 5.53

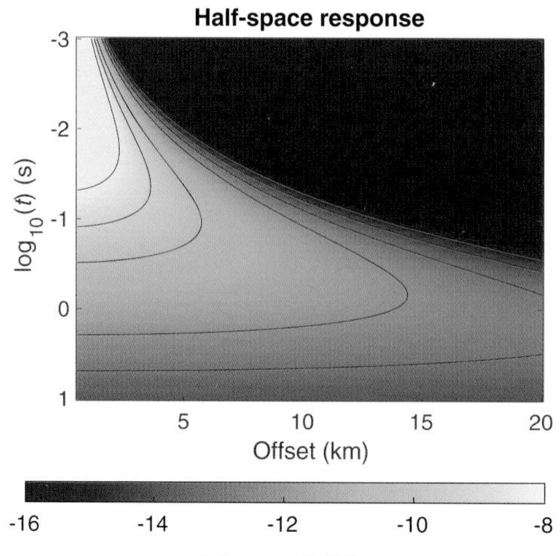

Figure 5.55

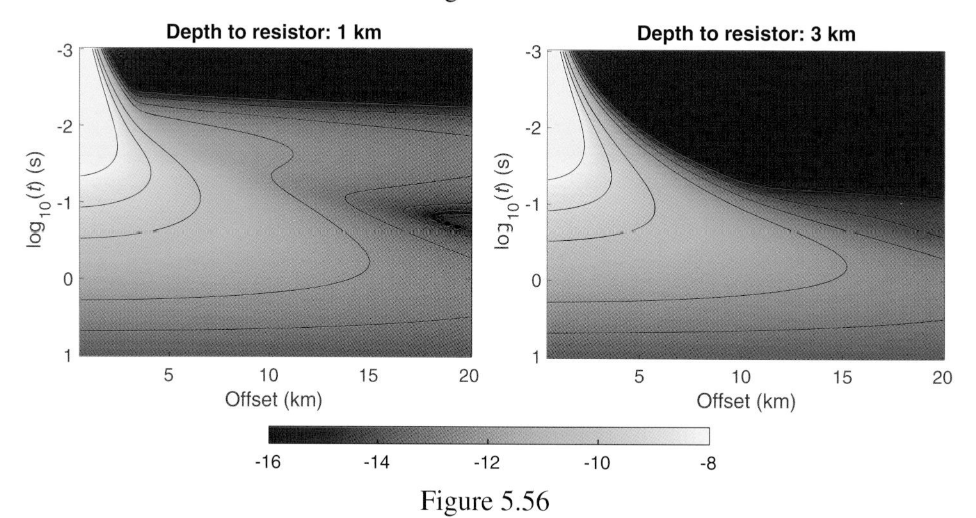

Figure 5.56

Figure 5.57

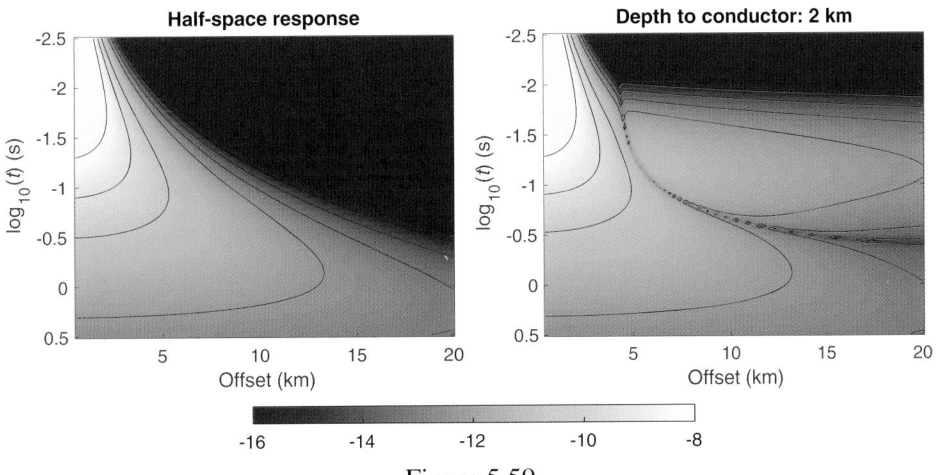

Figure 5.59

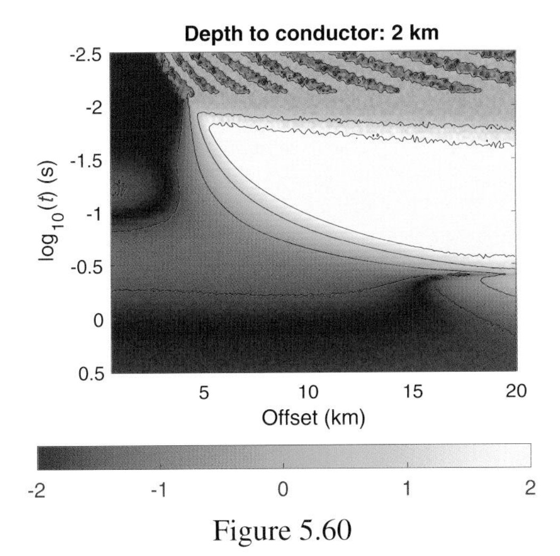

Figure 5.60

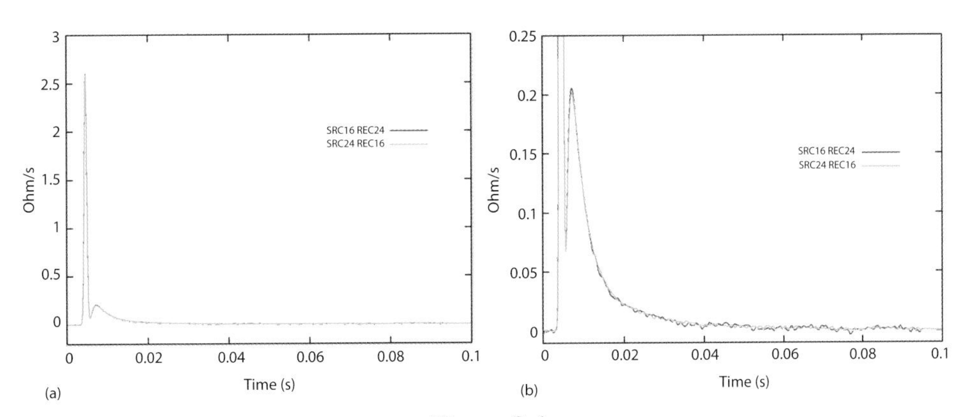

Figure 8.4

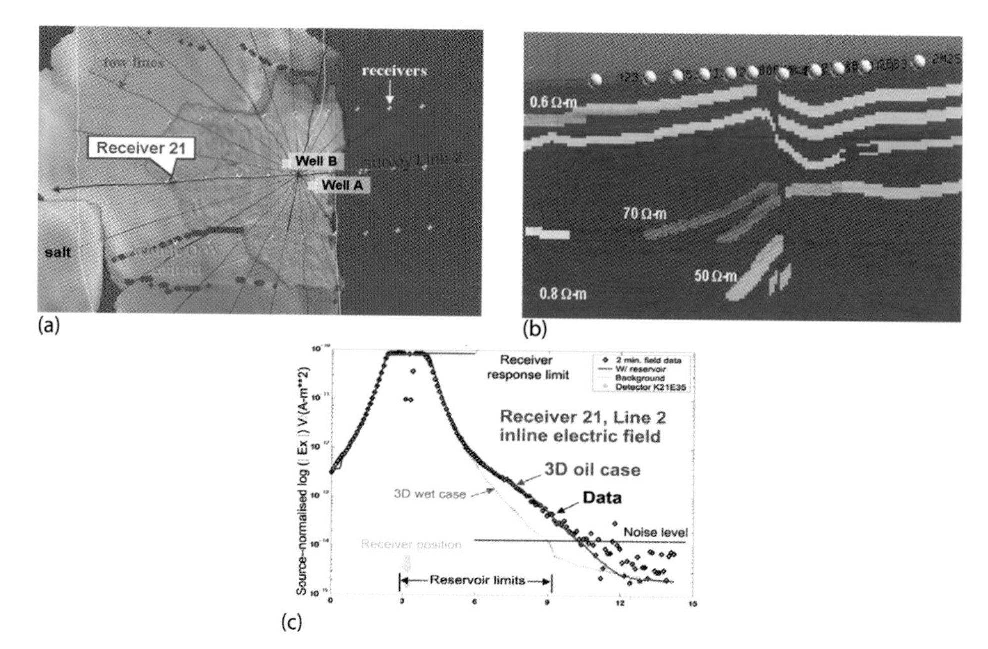

(a) (b) (c)

Figure 7.4

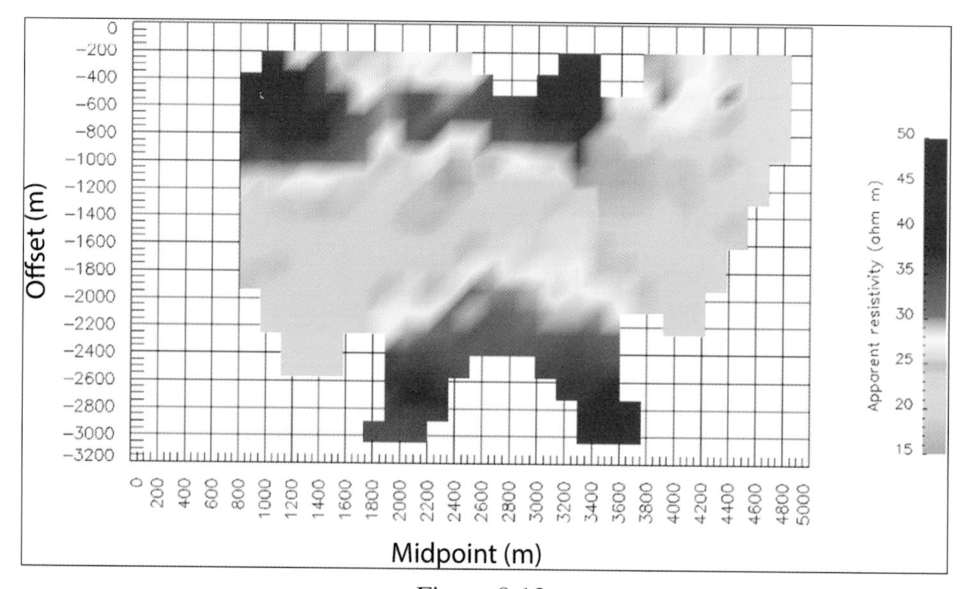

Figure 8.13

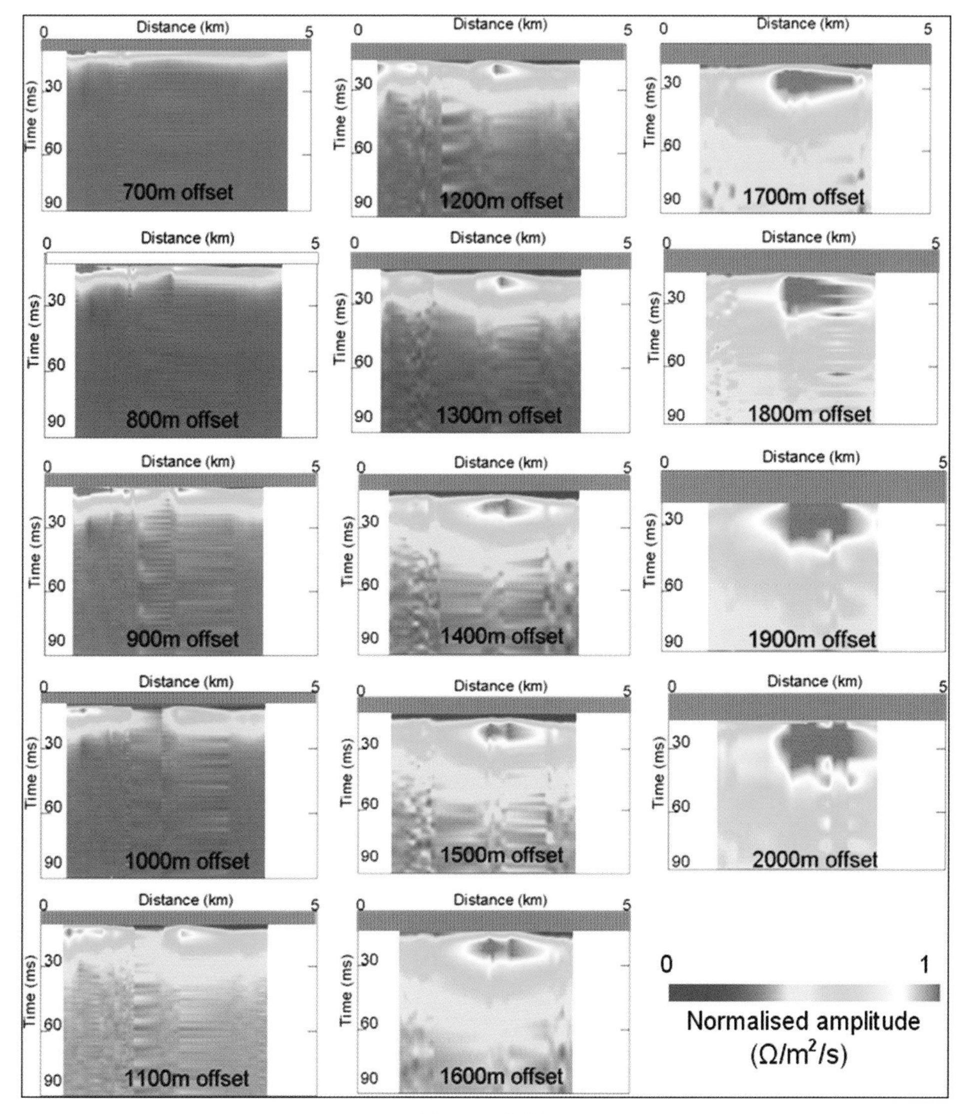

Figure 8.12

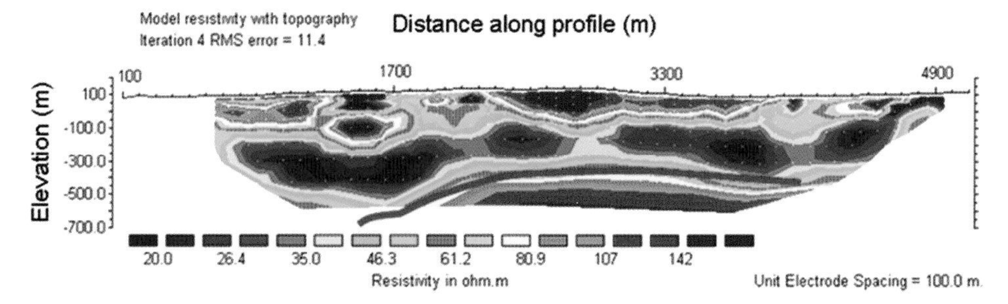

Figure 8.15

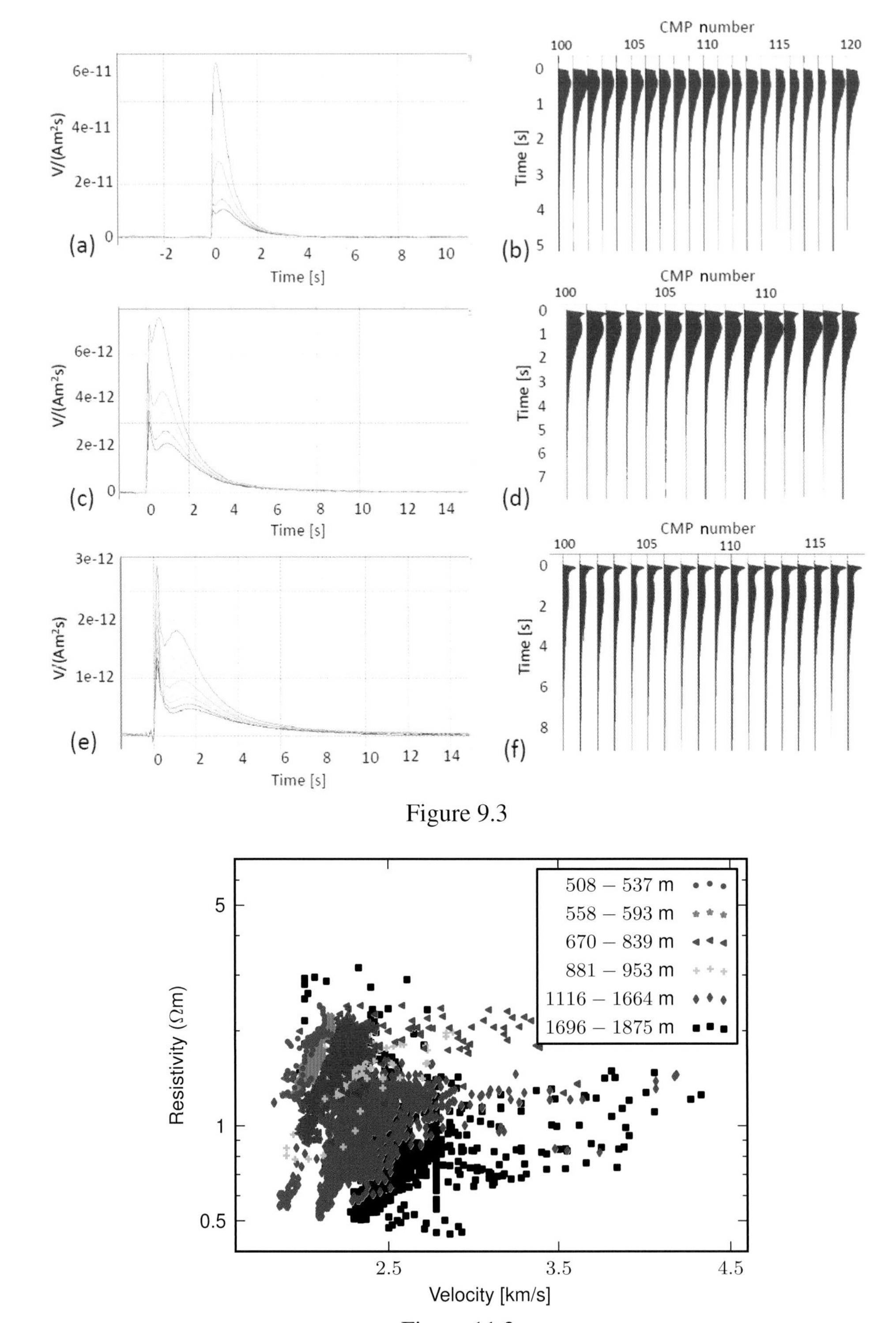

Figure 9.3

Figure 11.2

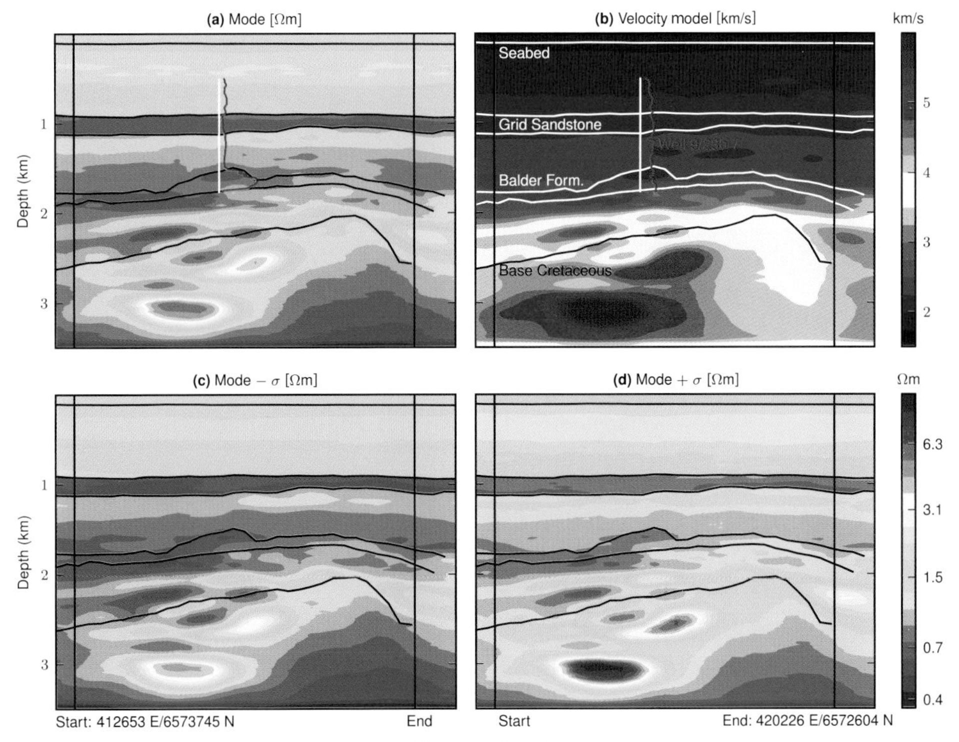

Figure 11.7

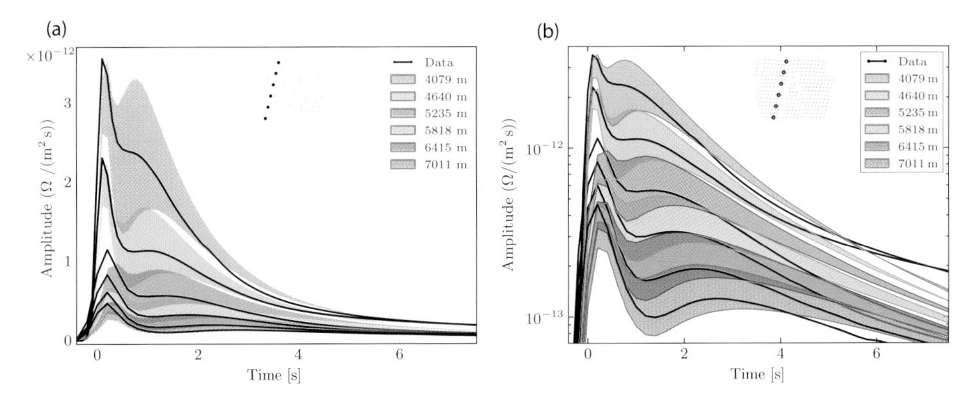

Figure 11.8

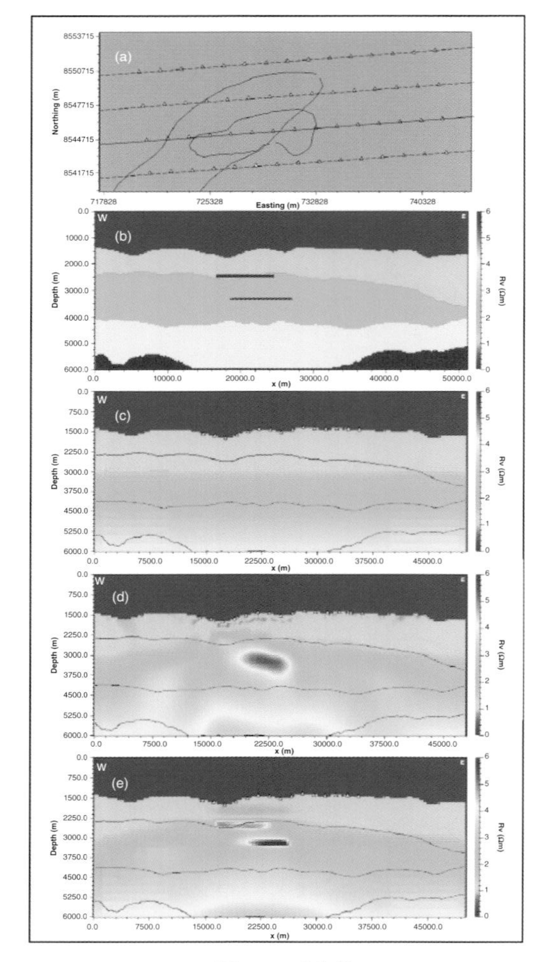

Figure 11.9

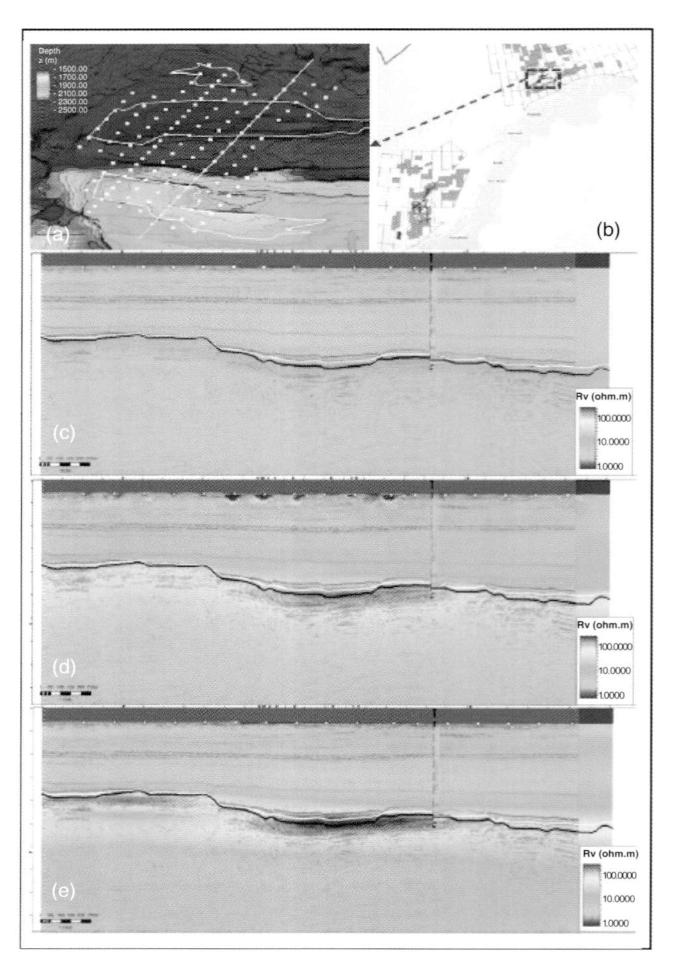

Figure 11.10

6.5 Heaviside Function, or Step Function

Kaufman and Keller (1983) define the transient field as the response to a step in current switched instantaneously into a magnetic or an electric dipole. For the magnetic dipole the current is switched on (Kaufman and Keller, 1983: 315) and for the electric dipole the current is switched off (Kaufman and Keller, 1983: 376). Edwards (1997) uses the switch-on case for the electric dipole source.

In Figure 6.11 we illustrate the step function source time function for the switch-on case. Convolution of the step function with a unit impulse yields the same step function. We can perform deconvolution by division in the frequency domain, as before, which gives the result shown in Figure 6.12. The zeros before and after the impulse at $t = 0$ are very small numbers of the order of 10^{-17}.

Since an impulse is the time derivative of a step, the impulse response can be obtained by differentiating the step response. Using a simple operator of the form

$$y_k = \frac{x_k - x_{k-1}}{\Delta t}, \tag{6.15}$$

the result is the same as shown in Figure 6.12, except the zeros are hard zeros.

We now attempt to recover the impulse in the presence of random noise with standard deviation 0.5. The step response of Figure 6.11 plus random noise is shown

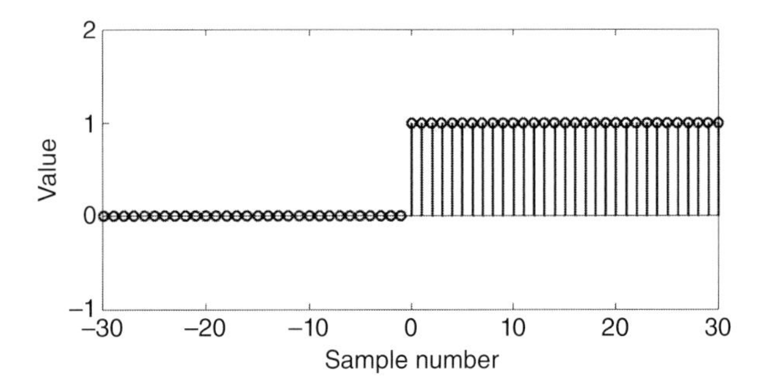

Figure 6.11 Switch-on step function.

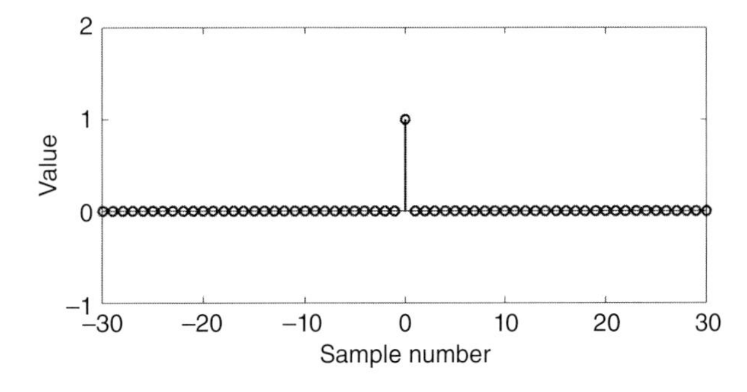

Figure 6.12 Step response deconvolved for switch-on step function input.

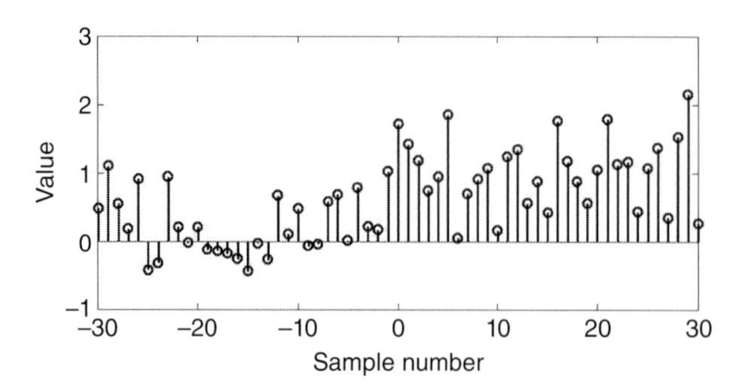

Figure 6.13 Step response plus noise with standard deviation = 0.5.

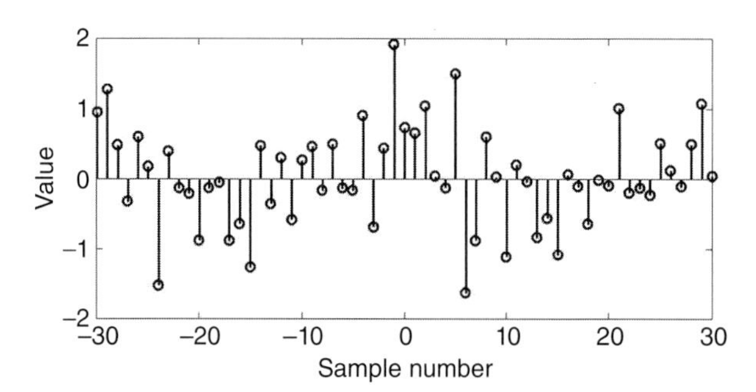

Figure 6.14 Recovered impulse by frequency-domain deconvolution.

in Figure 6.13. Deconvolution by straightforward division in the frequency domain yields the result shown in Figure 6.14. The result looks fairly random. The impulse at $t = 0$ is not even the largest value for this particular realisation of the noise.

Something is going wrong. The problem is that the complete step response can never be obtained, because it is infinitely long. The response is always truncated: it is a step function multiplied by a rectangle function. The multiplication by the rectangle function in the time domain is a convolution with the corresponding sinc function in the frequency domain. Since the sinc function is both positive and negative, the amplitude spectrum of the Fourier transform of the truncated step function can be zero at certain frequencies. In practice, therefore, there are some small numbers in the discrete amplitude spectrum. The amplitude spectrum of the step function of Figure 6.11 is shown in Figure 6.15. The theoretical amplitude spectrum of a step for positive frequencies is $1/\omega$, which, at $\omega = 0$ for example, is infinite. An untruncated step function has an infinite number of ones. Figure 6.15 shows an amplitude of 31 at zero frequency, because there are 31 ones in the truncated step function. The general shape of the theoretical response is apparent in the spectrum of Figure 6.15, but there are also many small values, caused by

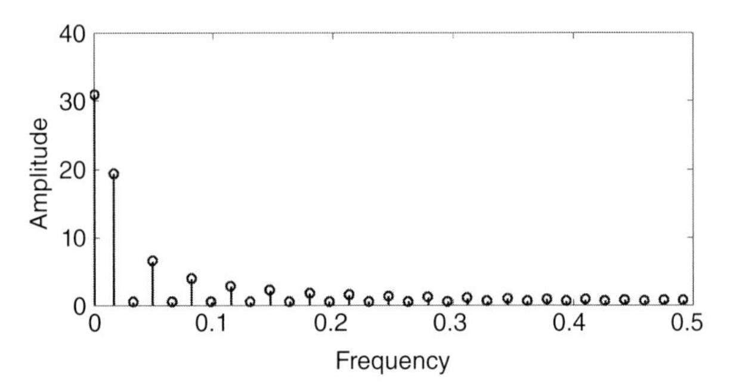

Figure 6.15 Amplitude spectrum of step of Figure 6.11.

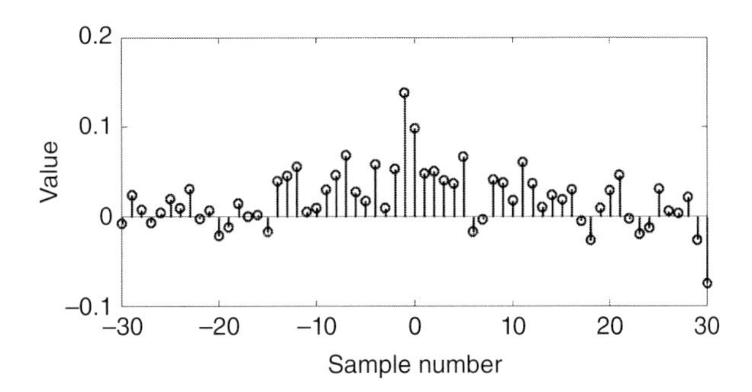

Figure 6.16 Recovered impulse by stabilised frequency-domain deconvolution using 1% white noise.

the convolution with the sinc function. Division by these small values increases the noise at those frequencies.

Deconvolution of the step function response by division in the frequency domain is not stable in the presence of noise. Division by zero can be avoided by adding a constant to the denominator. For example, the ratio $B(f) = C(f)/A(f)$ can be stabilised by multiplying the top and bottom by the complex conjugate of the denominator, thus making the denominator real, and adding a small positive real constant ϵ to the denominator:

$$B(f) = \frac{C(f)}{A(f)} = \frac{C(f)A^*(f)}{A(f)A^*(f)} \approx \frac{C(f)A^*(f)}{A(f)A^*(f)+\epsilon}. \tag{6.16}$$

This stops the noise blowing up to infinity and can give some improvement. Using $\epsilon = 1\%$ of the total energy, or 1% 'white noise' reduces all the amplitudes, but gives some improvement, as shown in Figure 6.16. The result is not good, however, and the value at $t = 0$ is still not the largest value. Stabilisation does not solve the basic problem of small values at certain frequencies in the discrete spectrum.

Another approach is to recover the impulse response by differentiating the step response, as was done for the noise-free case. The result of differentiating the signal

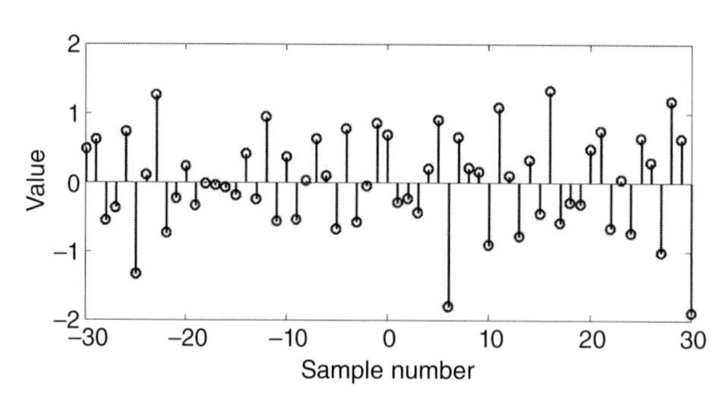

Figure 6.17 Result of differentiating the signal plus noise of Figure 6.13.

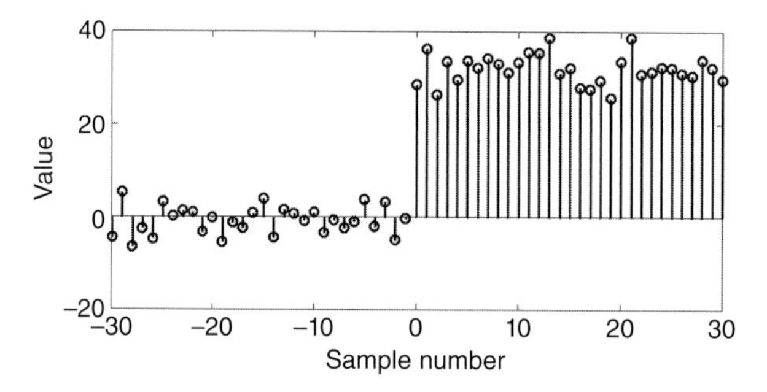

Figure 6.18 Stack of 31 experiments with switch-on step.

plus noise of Figure 6.13 is shown in Figure 6.17. It is not any better than the frequency-domain deconvolution. Since the process of differentiation is multiplication by $-i\omega$, the amplitude spectrum of the noise is multiplied by ω. This also occurs in the frequency-domain deconvolution of course. The deconvolution, or differentiation, magnifies the noise at high frequencies.

There is no deconvolution gain with the step function source. In the time domain it can be seen from Figures 6.11 and 6.12 that the only part of the source time function that affects the outcome is the step itself, which generates a single sample in the noise-free deconvolution. Comparing the performance of the step function with the PRBS, it is clear that the PRBS provides a far better use of the electrical energy.

The signal-to-noise ratio may be improved by repeating the step function experiment a number of times and adding the results. The signal component adds coherently and the random noise adds incoherently and tends to cancel. To make a comparison with the PRBS case, the numerical experiment for the switch-on step is performed 31 times and the results are added to give the result shown in Figure 6.18.

Frequency-domain deconvolution yields the result shown in Figure 6.19, in which the white noise stabilisation could be reduced to 0.0001%. The impulse

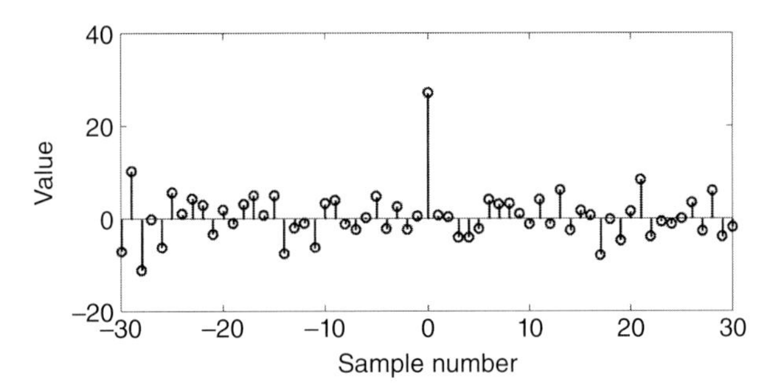

Figure 6.19 Frequency domain deconvolution of the time-series of Figure 6.18 with 0.0001% white noise.

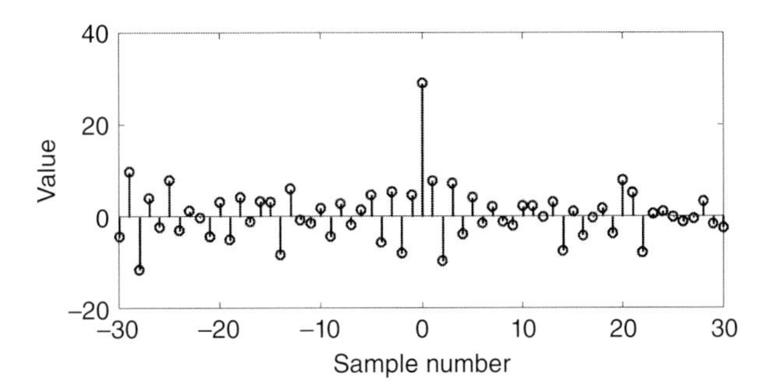

Figure 6.20 Differentiation of the time series of Figure 6.18.

can clearly be seen at $t = 0$ and the signal-to-noise ratio has improved exactly as expected.

Recovery of the impulse response by differentiation yields the result shown in Figure 6.20, in which both the signal and noise are similar to Figure 6.19.

In summary, a step function may be used to obtain the earth impulse response. There is no deconvolution gain.

For a given source current and source–receiver geometry, the signal-to-noise ratio can be increased only by repeating the experiment and summing the results. For random noise the signal-to-noise ratio increases as $\sqrt{N_E}$, where N_E is the number of experiments. If the duration of the earth impulse response is

$$T_g = M\Delta t, \tag{6.17}$$

where Δt is the time sampling interval, the minimum total recording time required is $N_E T_g = N_E M \Delta t$. This should be compared with the PRBS case, for which we have shown that, for random noise, the deconvolution gain for a PRBS of length N is $\sqrt{N}$. For the same noise conditions the PRBS deconvolution gain is the same as

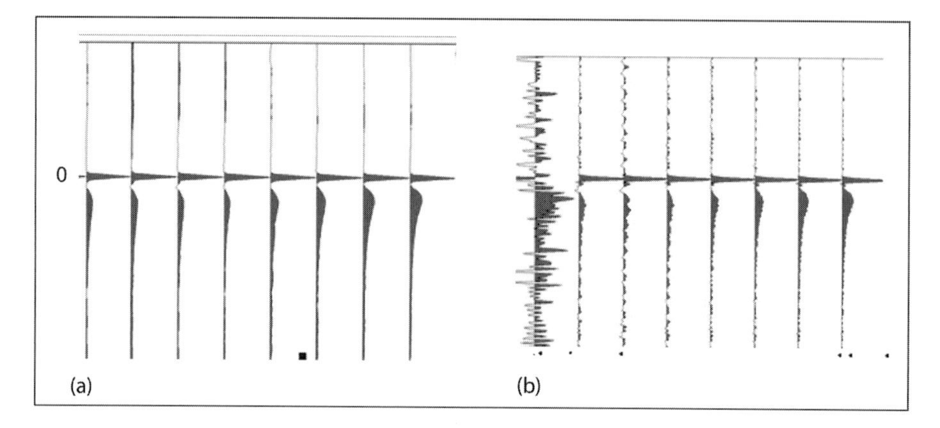

Figure 6.21 Recovered impulse responses on eight receivers with time increasing downwards and $t = 0$ in the centre: (a) deconvolution of the stack of 50 PRBS traces; (b) differentiation of the stack of 1550 step responses (redrawn from Wright et al., 2006).

the gain using repeated step functions if $N = N_E$. The time required for the PRBS case would be

$$T_s + T_g = N\Delta t + M\Delta t = (N + M)\Delta t, \tag{6.18}$$

where N is the number of PRBS samples. Comparing the times for the same noise conditions, for $M > 1$ and $N > 1$, we see that the step function time $NM\Delta t$ is greater than the PRBS time $(N + M)\Delta t$:

$$NM\Delta t > (N + M)\Delta t. \tag{6.19}$$

A step function source time function is inferior to a PRBS when noise is important for two reasons. First, the deconvolution of the step magnifies the high frequencies, whereas the deconvolution of the PRBS leaves the spectrum of the noise unchanged. Second, more time is needed to obtain data of a given signal-to-noise ratio. This has been shown theoretically previously, was demonstrated experimentally by Wright et al. (2006) and is shown in Figure 6.21. Figure 6.21(a) shows the result of deconvolving a stack of responses to 50 PRBSs of length 4095 samples; Figure 6.21(b) shows the result of differentiating a stack of 1550 responses to a step function. In both figures the sharp peak at $t = 0$ is the air wave; the lower amplitude response following it is the earth impulse response. The step function stack took longer to record than the PRBS stack, but the deconvolution result is worse. In noisy conditions, time is an important factor in the cost of data acquisition and the choice of source time function can therefore have a significant effect on survey costs.

6.6 Square Wave Function

Square wave functions are introduced in Section 2.6. If the half-period of the square wave is designed to be greater than the duration T_g of the earth impulse response, the response reaches a steady state before the next polarity reversal, which generates a response of opposite polarity. The polarity of the responses to negative polarity steps are reversed and added to the responses to positive polarity steps to increase the signal-to-noise ratio.

In a land survey in the South of France, Ziolkowski et al. (2007) used this source time function and summed 1500–3000 responses for each source position. Since the source transmitter does not generate a perfect step when the current is reversed, it is necessary to measure the source current and to deconvolve the received voltage for the measured current in order to obtain the earth impulse response (Ziolkowski et al., 2007). The results were sometimes as good as those shown in Figure 6.21(a), but sometimes no better than those in Figure 6.21(b).

This use of such long-period square waves is essentially a convenient way of generating the response to a series of step functions. The lowest non-zero frequency is $f_L = 1/2T$, where T is the period of the square wave. If f_L is lower than any other non-zero frequency of interest, the frequency bandwidth of the source is

$$\frac{1}{2T} \leq f \leq \frac{1}{2\Delta t}, \tag{6.20}$$

where Δt is the sampling interval, and it has an amplitude spectrum of approximately $1/(2\pi f)$ in that frequency range.

As mentioned in Chapter 2, conventional marine CSEM with ocean-bottom node receivers has used a continuous square wave source with a period T of 4 s. This has a line frequency spectrum with lines at the fundamental frequency $f = 1/T$ and at odd harmonics $3f$, $5f$, $7f$ and so on, with corresponding amplitudes 1, 1/3, 1/5, 1/7 and so on. Because the earth is a linear system, the only frequencies that contain information are these source frequencies. To recover the earth impulse response it is necessary to deconvolve the data for the source time function. This requires the source time function to have energy at *all* frequencies up to the Nyquist frequency. Since this is not the case for the square wave, the deconvolution cannot be performed and the earth impulse response cannot be recovered. This forces the data to be analysed in the frequency domain and only at the specific frequencies of the square wave.

An argument in favour of a square wave versus, for example, a PRBS, is that the square wave has more energy in the frequencies that are really needed and does not waste energy in frequencies that are not needed. This may be the case, but the argument ignores deconvolution gain.

Ziolkowski et al. (2011) describe a towed streamer marine experiment to compare the performance of a square wave with a PRBS over the Peon field in the Norwegian sector of the North Sea. A line was surveyed with a square wave source time function and then with a PRBS. The square wave was transient: 100 s long with a period of 10 s. The PRBS was a single period of order 10, with a bit rate of 10 s^{-1} and was 102.3 s in duration. Time between successive records was 120 s for both survey lines, allowing over 16 s for the duration T_g of the earth impulse response. The energy in the two source signals was nearly identical. The PRBS had 1023 samples, so the deconvolution gain was $\sqrt{1023} = 30$ dB. In the frequency domain the PRBS data had better resolution than the square wave data. And of course the PRBS data could be deconvolved to recover the earth impulse response, which was not possible with the square wave data.

The argument about the power of deconvolution gain is not very clear in Ziolkowski et al. (2011). We hope we have made it clearer here.

6.7 Special Periodic Functions

The limitations of square waves are well known and have been for a very long time. They have been used in CSEM simply for convenience. The main advantage is that the power is constant. The main disadvantage is the spectrum, with energy only in the fundamental frequency and odd harmonics and amplitudes decaying inversely with frequency. The zeros between the lines are not regarded as a disadvantage, because they are of importance only if there is an intention to transform back to the time domain. If the data processing and analysis are to be carried out in the frequency domain, zeros for certain spectral components are not a serious hindrance, provided there is a sufficient number of frequencies at which there is significant energy. The EM community has made an effort to generate periodic sequences that have power concentrated in a number of selected frequencies, preferably with equal power in each of the chosen frequencies. The idea is to have more usable frequencies than a square wave, but far fewer frequencies than a PRBS, with the advantage that the signal-to-noise ratio at the chosen frequencies is greater than with any other source time function. Lu and Srnka (2005, 2009), Mittet and Schaug-Pettersen (2008) and Mattsson et al. (2012) have proposed special periodic functions that are designed to be an improvement over square waves without using all the frequencies that are present in a PRBS. This section compares one of these functions with a PRBS.

Figure 6.22 compares the 'Optimized Repeated Sequence' (ORS) of Mattsson et al. (2012) with an order 8 PRBS. The ORS is 300 samples and the PRBS is 255 samples. The ORS has fewer frequencies with amplitudes that are mostly several

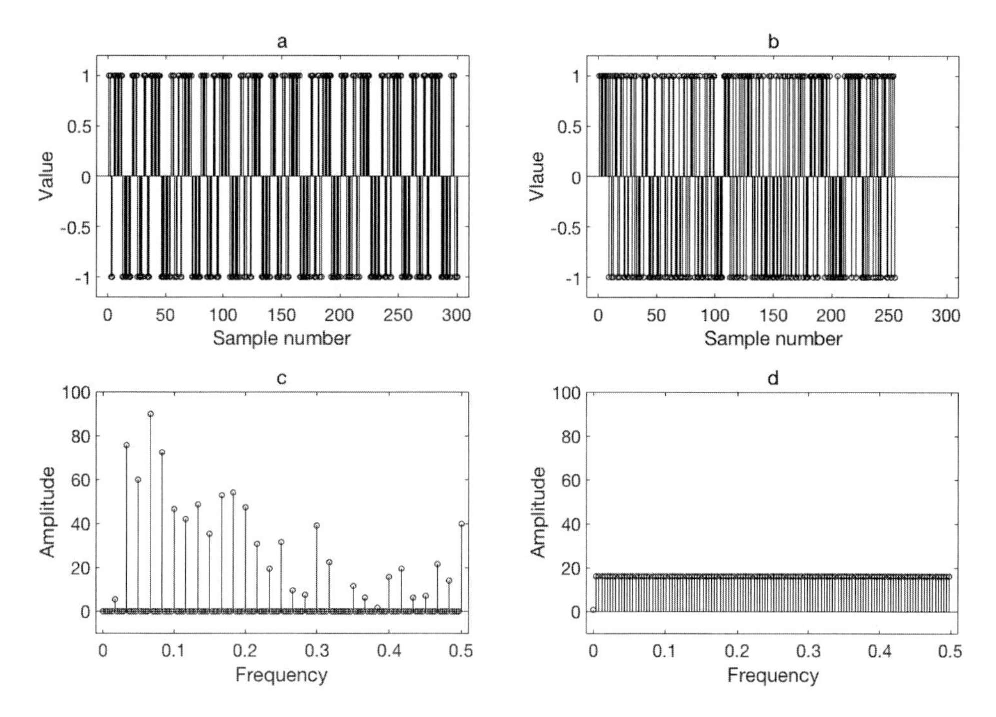

Figure 6.22 (a) 'Optimized Repeated Sequence' (ORS) (Mattsson et al., 2012); (b) order 8 PRBS; (c) amplitude spectrum of (a); (d) amplitude spectrum of (b).

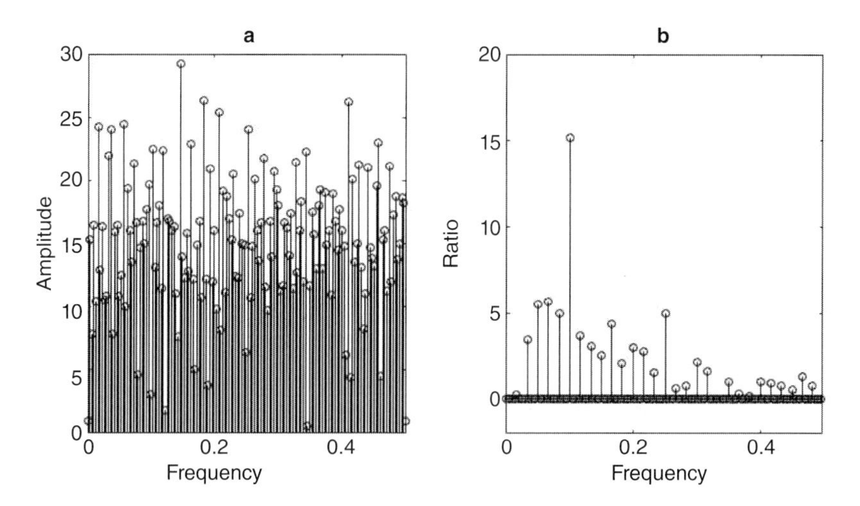

Figure 6.23 (a) Amplitude spectrum of PRBS after addition of 45 zeros; (b) spectral ratio of ORS/PRBS.

times greater than the PRBS amplitude, which is about 16, except at DC, where it is 1. To find the spectral ratio of these functions, 45 zeros are first added to the PRBS to make it the same length as the ORS. Then the amplitude spectrum of the ORS is divided by the amplitude spectrum of the PRBS with the additional zeros. Figure 6.23(a) shows the amplitude spectrum of the PRBS with the additional zeros. It is now not flat, but it does not go to zero at any frequency. Figure 6.23(b) shows the amplitude spectral ratio of the ORS to the PRBS. At a frequency of 0.1, with sampling interval $\Delta t = 1$, the amplitude ratio is nearly 16; at all other frequencies it is less than 6. Notice that there are many zeros in this ratio spectrum, originating from the zeros in the ORS amplitude spectrum.

These two functions are now convolved with the synthetic impulse response shown in Figure 6.24(a). Figure 6.24(b) shows the corresponding amplitude spectrum, which has a large DC component. The results of the convolution are shown in Figure 6.25, together with the corresponding amplitude spectra. It is clear that the convolution with the ORS gives higher amplitudes than the PRBS at certain frequencies, which is the main reason for designing the ORS. Similar results would be obtained with the other special periodic functions mentioned above. Notice that the amplitude spectrum of the ORS goes to zero at frequencies between the peaks.

As before, we now add normally distributed (Gaussian) noise, in this case with standard deviation 2.0154×10^{-11} V/m, as shown in Figure 6.26.

Because the PRBS contains energy over the whole frequency bandwidth, it can be used to deconvolve trace (b) in Figure 6.26 to recover an estimate of the impulse response. This is shown in Figure 6.27(a) and compared with the modelled impulse response. The noise in the recovered impulse response can be obtained by subtracting the modelled impulse response from the response recovered by deconvolution. This noise is shown in 6.27(b), where it is compared with the input noise. The noise after deconvolution has standard deviation 1.6257×10^{-12} V/m, which is much smaller than the noise before deconvolution. The deconvolution gain is 12.4, compared with the theoretical maximum of $\sqrt{2^8 - 1} = 15.97$.

The deconvolution option is not open to the ORS, nor is it open to any of the special periodic functions mentioned, because of the zeros in their amplitude spectra. Therefore there can be no deconvolution gain with any of these functions. The 12.4 PRBS deconvolution gain for the example considered here is greater than the amplitude of the spectral ratio of ORS to PRBS for all frequencies but one.

The PRBS deconvolution provides the option of further processing and inversion in the time domain or the frequency domain. The time domain option is not open to data obtained with these special periodic functions.

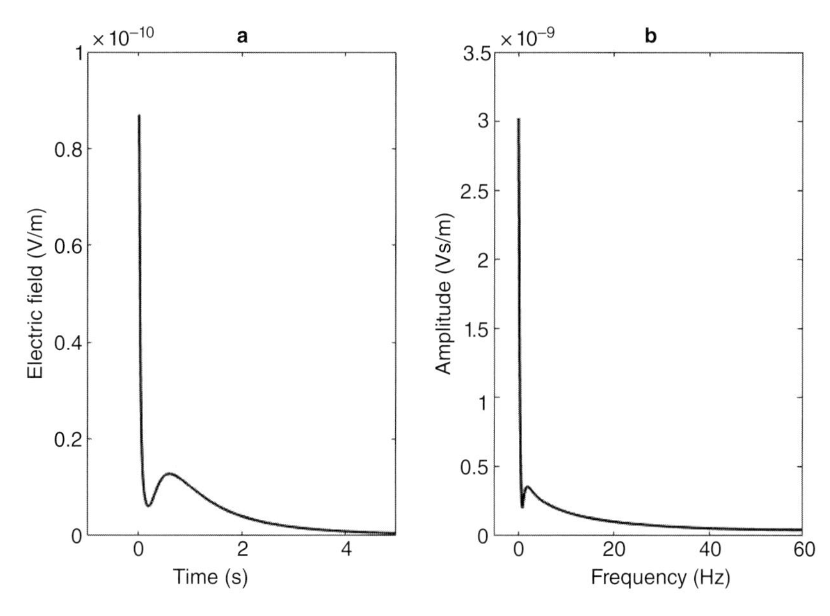

Figure 6.24 (a) 1D impulse response; (b) amplitude spectrum of (a). Configuration is horizontal 1 A-m electric dipole at 10 m depth in water layer 380 m deep, resistivity 0.27 ohm-m, overlying half-space, resistivity 3 ohm-m and point horizontal dipole receiver at depth 100 m and horizontal offset 2000 m.

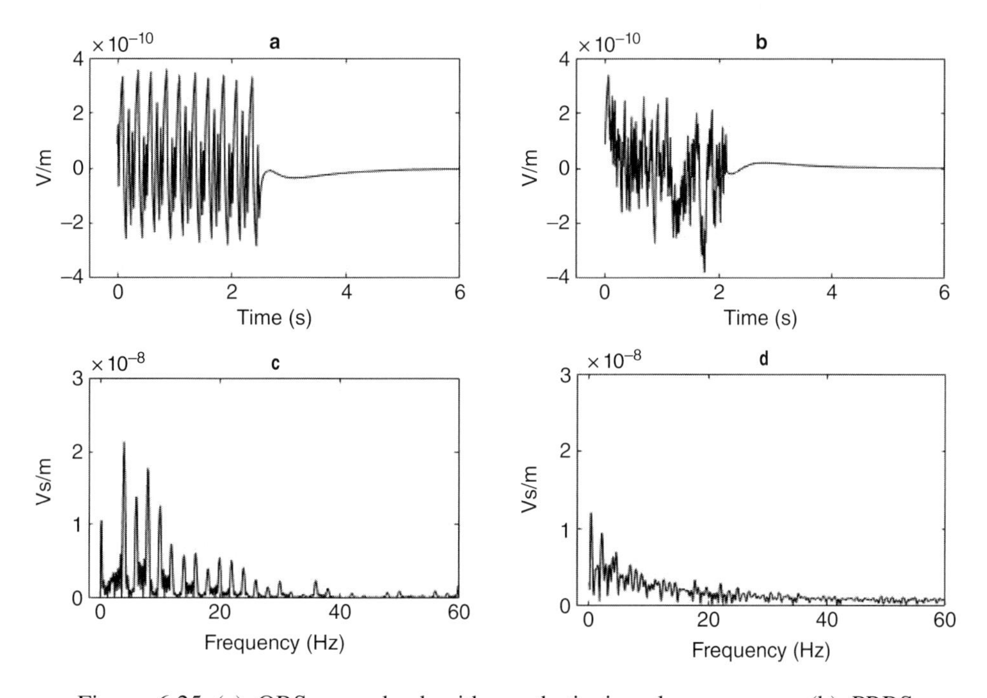

Figure 6.25 (a) ORS convolved with synthetic impulse response; (b) PRBS convolved with synthetic impulse response; (c) amplitude spectrum of (a); (d) amplitude spectrum of (b).

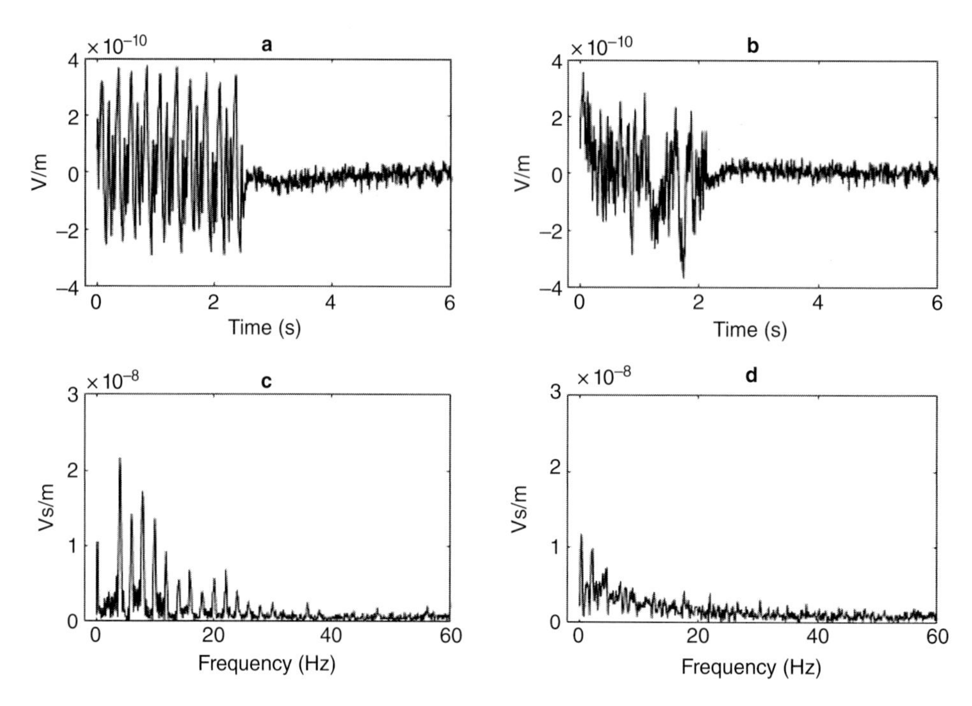

Figure 6.26 The same as Figure 6.25, but with the addition of white Gaussian noise: (a) ORS convolved with synthetic impulse response plus noise; (b) PRBS convolved with synthetic impulse response plus same noise; (c) amplitude spectrum of (a); (d) amplitude spectrum of (b).

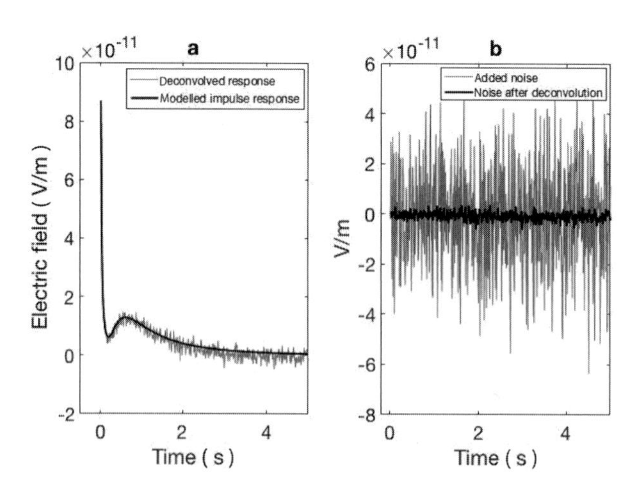

Figure 6.27 (a) 1D impulse response of Figure 6.24(a) plus result of deconvolving signal of Figure 6.26(b) with PRBS of Figure 6.22(b); (b) the noise added in Figure 6.26 compared with the noise level after deconvolution.

7
Deep Water CSEM

This chapter describes the elements of the data acquisition and processing of marine controlled-source electromagnetic (CSEM) data using a deep-towed source and stationary autonomous sea-floor receiver nodes. This CSEM configuration was proposed by academics as a way to overcome the attenuation of electromagnetic signals in sea water. It presents technical problems that are outlined here, together with proposed solutions. The chapter begins with a short introduction to the method and the attenuating effect of sea water. It discusses acoustic positioning of source and receiver, the deep-towed source and its orientation, the receiver nodes, the response at the receiver, determination of receiver orientation and the consequent ability of the configuration to detect subsurface resistive layers. Source–receiver synchronisation and data processing are briefly described and an important test of the method is included.

7.1 Introduction

As described in Section 1.6, marine CSEM was developed by the academic electromagnetic community, who wanted to determine the conductivity of the oceanic crust in the deep ocean and recognised that this could not be done with measurements of the natural electromagnetic field made on the sea floor. The attenuation of magnetotelluric signals by conducting sea water reduces the detectable signal bandwidth to frequencies below a few cycles per hour. These frequencies are too low to resolve the conductivity of the oceanic lithosphere. CSEM was the solution to this problem. Early CSEM surveys used a square wave source time function with a period of about 4 s. The fundamental frequency was thus 0.25 Hz. The skin depth (equation 2.1) of sea water with resistivity 0.3 ohm-m for a frequency of 0.25 Hz is 551 m. We arbitrarily define *deep water* as water deeper than 500 m and *shallow water* as water shallower than 500 m. The deep water CSEM configuration now in use is briefly described in Section 2.3. It consists of a deep-towed neutrally buoyant source and autonomous receiver nodes placed on the sea floor.

Norwegian scientists at Statoil recognised that this system could be used to find hydrocarbons in deep water. At the turn of this millennium, deep water plays in the sediments on the continental edge became important because oil company geologists had already identified that this environment had potential for enormous hydrocarbon reserves and deep water drilling had become feasible. The barrier to exploration was the high cost of drilling in deep water, which was much greater than in shallow water on the continental shelf.

All geophysical methods suffer from noise. Obtaining a good signal-to-noise ratio is usually a difficult problem. For electromagnetic methods on land, including magnetotellurics, for example, noise can be the biggest problem in analysis of the data. For acquisition of CSEM data, the deep ocean is the least noisy environment.

The marine CSEM system developed by the academic community was first used for hydrocarbon detection in an experiment in 2000, funded by Statoil, and conducted offshore Angola over a known deep water oil field. The method and results were published in two papers in 2002 (Eidesmo et al., 2002; Ellingsrud et al., 2002). These papers demonstrated that hydrocarbons could be detected. It was clear that this new geophysical exploration method could be used to reduce the risk of drilling 'dry' wells in deep water exploration for hydrocarbons.

7.2 Attenuation of Electromagnetic Signals in Sea Water

Attenuation of electromagnetic signals is described briefly in Section 2.2 and more extensively in Chapter 4. The conductivity of sea water increases with both temperature and salinity, as shown in Table 7.1. In modelling electromagnetic responses in sea water, a value for conductivity of about 3 S m^{-1} is often used. In surveys, conductivity is often measured with CDT (conductivity, depth and temperature) meters that also measure depth and temperature. Clearly, sea water conductivity does not have a uniform value throughout the whole water column.

Equation 2.1 gives the *skin depth* for a plane wave propagating in an isotropic medium like water. The skin depth decreases with increasing frequency and increasing conductivity. The first deep water commercial surveys used a square wave source time function with a fundamental frequency of 0.25 Hz. For water of conductivity 3 S m^{-1}, the skin depth at 0.25 Hz is about 580 m. To minimise the attenuation of the electromagnetic signal in the water, the dipole source is towed close to the sea floor – typically about 50 m above the sea floor – and receivers are placed directly on the sea floor.

Table 7.1 *Electrical conductivity σ of sea water at atmospheric pressure*

Temperature (°C)	Salinity (g kg^{-1})				
	20	25	30	35	40
	σ (S m^{-1})				
0	1.745	2.137	2.523	2.906	3.285
5	2.015	2.466	2.909	3.346	3.778
10	2.300	2.811	3.313	3.808	4.297
15	2.595	3.170	3.735	4.290	4.827
20	2.901	3.542	4.171	4.788	5.397
25	3.217	3.926	4.621	5.302	5.974

Source: Reproduced from Kaye and Laby (1971).

7.3 Acoustic Positioning

The source and receivers are several kilometres from the vessel and their positions need to be determined. This is performed with an ultra-short baseline (USBL) acoustic positioning system, operating at frequencies around 10 kHz, which provides the range and direction from the vessel to the device.

Each device is equipped with an acoustic transponder, which replies with its own signal when interrogated with a coded acoustic signal from the vessel. The time T_d from the emission of the coded signal from the vessel to the time of reception at the vessel of the acoustic response from the device is measured and the range r_{vd} from the vessel to the device is given by

$$r_{vd} = \frac{T_d c_w}{2}, \tag{7.1}$$

where c_w is the velocity of sound in water. The precision of the range estimate depends on both the precision of the measurement T_d and the precision of the knowledge of c_w. It is likely that c_w has the greater uncertainty.

At the vessel there are three receiving hydrophones at known positions on three orthogonal axes. The acoustic response signal from the remote device arrives as a plane wave. The direction of propagation, defined by azimuth and dip, can be calculated from the arrival times at the three hydrophones. If the signal is simply a pulse with one principal frequency, the phase difference is used and phase ambiguity is eliminated by putting the hydrophones less than half a wavelength apart, often less than 10 cm. It is for this reason that the system is known as *ultra* short baseline. The precision with which the direction can be determined is inversely proportional to the spacing between the receiver hydrophones. For a single measurement, the direction may not be very precise.

The vessel transponder system is fixed to the vessel, which is moving. Motion of the vessel may be divided into three types of linear motion and three types of rotational motion. Linear motion along the longitudinal axis is known as *surging*; along the transverse axis it is known as *swaying*; along the vertical axis it is known as *heaving*. Rotational motion around the longitudinal axis is known as *rolling*; around the transverse axis it is known as *pitching*; around the vertical axis it is known as *yawing*. Motion compensation of acoustic positioning estimates is made with motion reference units (MRUs) that use gyroscopes and accelerometers to make the required measurements. Modern MRUs use micro electromechanical system-based (MEMS-based) gyroscopes and accelerometers, and are small and lightweight, with low power consumption and are capable of data rates of 200 Hz.

7.4 Deep-Towed Current Dipole Source

The elements of the neutrally buoyant deep-towed source are illustrated in Figure 7.1. The principle of the deep-towed source design owes much to work done at the Scripps Institution of Oceanography (Chave and Cox, 1982) and at the University of Cambridge (Sinha et al., 1990). The source strength is the *dipole moment*. It is a vector with magnitude equal to the product of the current and the dipole length. The electrical objective is to maximise the current. A direct current (DC) is ideal, because its polarity can be reversed at will to produce a variety of signals, as discussed in Chapter 6. The objective is therefore to deliver a large DC current, say 1000 A, to a device towed near the sea floor in deep water. The towing cable is a few kilometres long.

The current dipole circuit is shown in Figure 2.1. Consider a DC current I applied from the vessel to a deep-towed current dipole source below the vessel, near the sea floor. The wires to the electrodes are each several kilometres long; let each wire have resistance R. The voltage drop down each wire is then $V = IR$ and the power loss down each wire is

$$P = VI = I^2R. \tag{7.2}$$

If $I = 1000$ A and $R = 0.5$ ohm, the total power loss down the two wires would be 1 MW, which is enormous. One kilogram of diesel fuel contains 48 MJ. A 40% efficient diesel generator would use about 200 kg of diesel fuel per hour to deliver this power.

Since the power loss is proportional to the square of the current in the cables between the vessel and the source, it is very important to reduce this current. This issue has been well understood ever since electrical power was first transmitted. The solution (Sinha et al., 1990) is to deliver high-voltage, low-current AC (alternating

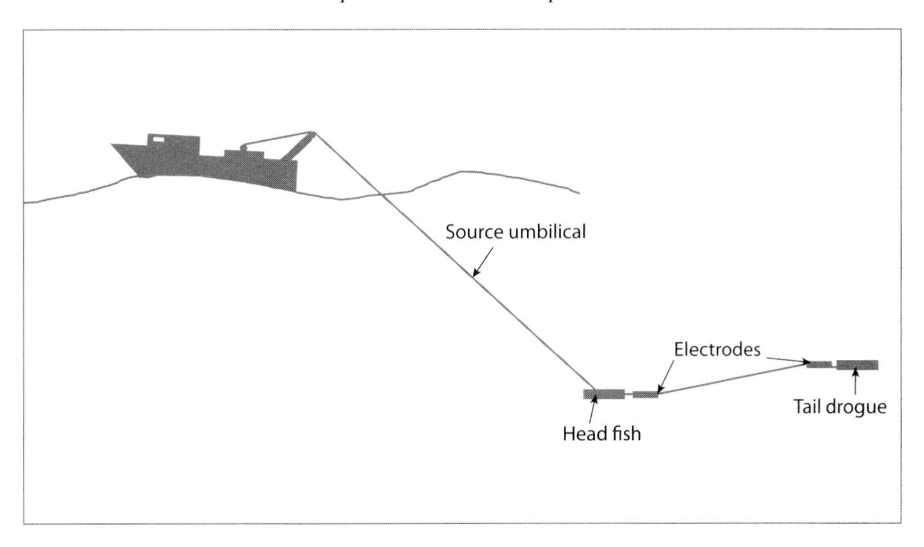

Figure 7.1 Neutrally buoyant deep-towed source. The source is towed from the vessel via an umbilical cable; it consists of five principal components: a head fish, two electrodes, a connecting cable, and a tail drogue. The electrodes are typically about 300 m apart.

current) power from the vessel to the source via the umbilical, and to transform this at the source to low-voltage, high-current AC. This is then converted to DC with a rectifier and smoothed to form steady DC. The DC current is fed into a waveform generator to perform current waveform generation, including polarity reversals. The head fish on the source contains this equipment. It also contains the positioning transducer. There is a similar positioning transducer on the tail drogue.

The power transmission requires a high voltage at the vessel. This imposes strict safety requirements. In the original design of Sinha et al. (1990), the ship's three-phase mains supply was 440 V, 50 Hz, unstabilised. This was converted to 2000 V RMS, 256 Hz, stabilised and tied to a frequency standard derived from the ship's master clock system.

The source must be towed close to the sea floor. Typical towing speeds for seismic reflection are 4–5 knots. It is very difficult to tow anything deep at such high speeds. Typical tow speeds for the deep-towed current dipole source are around 1.0–2.0 knots. At such speeds it is hard to control the behaviour of the source, which is typically 300 m between electrodes. The motion of the vessel, particularly the heave and pitch, are transmitted to the source via the umbilical. Deep ocean currents can cause the source to move from side to side, such that the dipole does not follow the ship's track, but tows at an angle to the track. And since the source is not perfectly neutrally buoyant, its profile is not perfectly horizontal: it may be tilted with the tail drogue above or below the head fish.

An excellent description of the problems of control of the behaviour of the source dipole and methods of solution is provided in the US patent description by Summerfield et al. (2012). In addition to current delivery, the source is capable of being steered up and down and from side to side. The head fish and tail drogue can have independent thrusters and fins to achieve the desired behaviour. The head fish has an echo sounder to determine its height above the sea floor. This information is fed back to the vessel along the umbilical and an operator on the surface survey vessel 'pays out' or 'reels in' to make the head fish go deeper or shallower. This operation can be automated.

In summary, the deep-towed source is a complicated device. Considerable effort is devoted to controlling it in the water and measuring the positions of the head fish and tail drogue, which are close to the two source electrodes. The result, from a geophysical point of view, is that the position and orientation of the source dipole vector can be determined, but with some uncertainty. The uncertainty in source position and orientation is probably the greatest source of error in the system.

7.5 Ocean-Bottom Receiver Node

The ocean-bottom receiver node was described briefly in Section 2.2. It has evolved from ocean-bottom magnetotelluric (MT) receivers. A Scripps MT receiver is shown in Figure 7.2. This instrument has its own batteries for power and measures two orthogonal horizontal components of the electric field E_1 and E_2 and two orthogonal horizontal components of the magnetic field B_1 and B_2. More recent instruments also measure the vertical components of the electric and magnetic fields E_z and B_z. All six components are recorded continuously, together with a clock signal to enable the received signals to be synchronised with the source and processed when the device is brought to the surface.

Communication with the instrument is via sound waves. The bandwidth that can be transmitted is small, so very little quality control of the data is possible. Each CSEM instrument is equipped with an acoustic transponder for positioning. When it is time to collect the data, the concrete anchor is released acoustically on command from the vessel.

7.6 In-line and Broadside Responses

The direction in line with the source is known as the in-line direction. The direction perpendicular to the source in the horizontal plane is known as the broadside direction. Figure 2.2 shows the field of a dipole source in a full space. If the source is horizontal and the section of Figure 2.2 is vertical, the field is characterised by

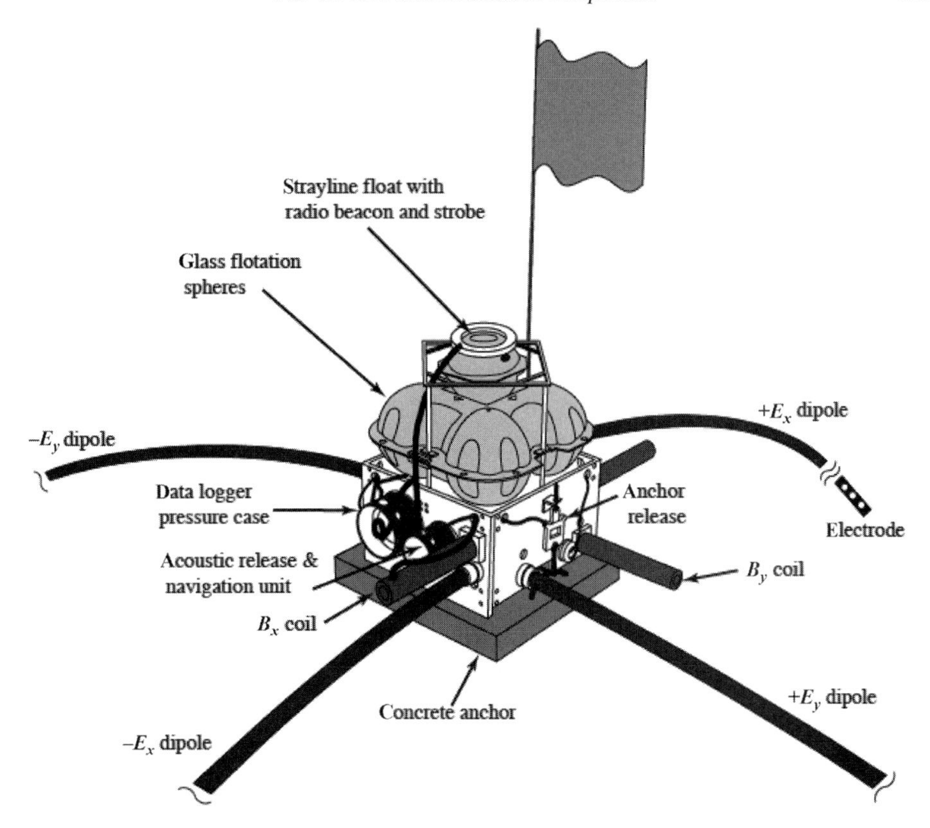

Figure 7.2 Diagram of the Scripps Institution of Oceanography broadband MT instrument (redrawn from Constable et al., 1998).

vertical current loops. If the section is horizontal – that is, plan view – the field is characterised by horizontal current loops. Purely horizontal current loops are known as the transverse electric or TE mode, as defined in Chapter 4. Purely vertical loops are the transverse magnetic or TM mode.

The real Earth is not a whole space and there are typically resistive layers below the sea floor. The vertical current loops cut through the layered sequence and are affected galvanically by variations in the resistivity of the layers. The horizontal current loops are coupled inductively to the strata and are largely unaffected by any more-resistive horizontal layer. The in-line response is dominated by the TM mode and is sensitive to subsea resistive layers, while the broadside response is dominated by the TE mode and is largely unaffected by subsea resistors.

It follows that differences between in-line and broadside responses are indicative of subsea resistive layers and the potential for hydrocarbons. This insight, perhaps obvious once it is pointed out, is one of the key ideas of Ellingsrud and Eidesmo of Statoil, who confirmed it experimentally in the laboratory before conducting their famous offshore Angola experiment. It is critical to know the orientation of the ocean-bottom receivers.

7.7 Receiver Orientation

Determination of ocean-floor receiver orientation has been studied in the frequency domain by Constable and Cox (1996) and Mittet et al. (2007). Mittet et al. (2007) explain that using compasses to determine the receiver orientation is not practical because the compass would have to be recalibrated each time the battery is changed before deployment of the receiver, which is not straightforward on a moving vessel. Therefore, the receiver orientation is determined from the received data.

The procedure is to steer the vessel on a path that crosses over the receiver. The dipole source is aligned in the direction of the ship's track, or it may have a deviation, known as the yaw, which may be known from acoustic positioning, as described above. The orientation of the two perpendicular electric receivers E_1 and E_2 relative to the ship's track, the 'x-direction', is shown in Figure 7.3.

The electric fields at time t in the x- and y-directions are

$$E_x(t,\theta) = E_1(t)\cos(\theta) - E_2(t)\sin(\theta),$$
$$E_y(t,\theta) = E_1(t)\sin(\theta) + E_2(t)\cos(\theta). \qquad (7.3)$$

From symmetry, referring again to Figure 2.2, if the x-direction is the dipole in-line direction and the geological structure is symmetrical about the x-axis, the electric field is a maximum in the x-direction and the source field in the y-direction

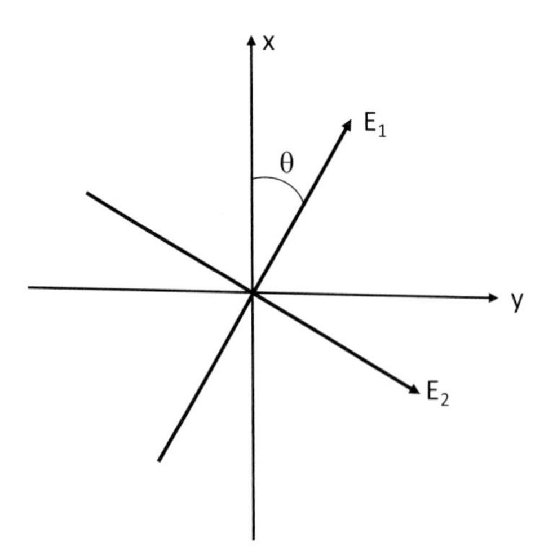

Figure 7.3 Orientation of receiver with respect to ship's track. The two perpendicular electric field measurements are E_1 and E_2.

should be zero. Noise is always present of course, so it is sensible to estimate the energy $En_x(\theta)$ in the x-direction and $En_y(\theta)$ in the y-direction in a time window of length T_w, as

$$\begin{aligned} En_x(\phi) &= \sum_{t_1}^{t_1+T_w} E_x^2(t, \phi), \\ En_y(\phi) &= \sum_{t_1}^{t_1+T_w} E_y^2(t, \phi), \end{aligned} \tag{7.4}$$

in which T_w is of the order of one minute. This should be calculated for $-\pi/2 \leq \phi \leq \pi/2$. For $\phi = \theta$, $En_x(\phi)$ should be a maximum and $En_y(\phi)$ should be a minimum. The vessel is moving, so independent estimates of the value of θ can be made by using non-overlapping time windows beginning at times t_1, t_2, t_3, etc., and averaging the results. This can also average errors introduced by variations in the yaw of the source dipole.

This procedure can give a value for θ that is incorrect by a value of π. It can be corrected by using the known source and receiver positions and the polarity of the data.

The magnetic field measurements provide additional data for determination of the receiver orientation. The magnetic field is perpendicular to the electric field, so the magnetic field strength should be a minimum in the x-direction and a maximum in the y-direction.

Once the receiver orientation has been determined, the recorded data can be rotated into the survey coordinate directions.

This method relies on symmetry. The required symmetry may be absent: the source dipole is likely to be displaced to one side or other of the ship's track because of deep ocean currents; the source dipole may be tilted; the receiver may not be directly beneath the ship's track; and there may be no symmetry in the geological structure. Without the required symmetry, the problem is harder. More information must be introduced. Key and Lockwood (2010) tackle this problem. The extra information is synthetic data that simulates the measured data. A model of the water and subsea conductivities is introduced and the problem is to minimise the difference between the synthetic and measured data by varying the receiver rotation angle and also the tilt, which can be in two directions. That is, determination of the orientation of the receiver, defined by three angles, is treated as part of the inversion of the data. In the modelling it is essential to know the position and orientation of the source. Myer et al. (2012) used compasses in the receiver nodes and applied the method of Key and Lockwood to estimate receiver orientation. They estimated the 2σ (two standard deviations) orientation error to be 6.6°.

7.8 Acquisition Geometries

The layout of receivers and ship's tracks in the survey area is determined by the subsurface target. Very often marine CSEM data are acquired after other geophysical data, particularly seismic data, have been obtained. So there is often a target that is to be investigated, possibly with a view to making a decision on whether to drill.

In general, the receivers are likely to be laid out on a rectangular grid with regular spacing Δx in the x-direction and Δy in the y-direction. Synthetic data are generated at the receivers for various geometries using the best available models of subsea resistivities. Noise is added to the synthetic data and the resulting data are then processed, inverted and interpreted to see if the subsea resistivity variations can be resolved. Resolution increases with denser spatial sampling. A factor of 2 increase in spatial sampling density results in a factor of 4 increase in the capital cost of receiver equipment, and probably a factor of 4 in data processing costs.

The maximum depth at which any target can be detected is normally less than 3 km below the sea floor. Maximum source–receiver offsets are normally 2–4 times the maximum required target depth. These parameters and the ship track spacing determine the costs of acquisition, processing and inversion. The object of this modelling exercise is to find the least costly set of parameters that will provide adequate information to make an informed decision, or rank the prospect against competing prospects.

7.9 Source–Receiver Synchronisation and Data Processing

The receivers on the sea floor have clocks that are synchronised with GPS before they are deployed from the vessel. When they are recovered, the clocks are checked for drift. Myer et al. (2012) found drift rates of ± 1.5 ms/day for 50% of their instruments and also found that 79% of the instruments were within the instrument clock specification of ± 4.3 ms/day. The instruments stay on the sea floor for about two weeks, so the clock drift, which is not necessarily linear, can become an important source of error for frequencies above 1 Hz. More accurate small low-power atomic clocks have been developed in the last ten years and source–receiver synchronisation is precise, with receiver instruments equipped with modern clocks.

The received data are divided into non-overlapping time windows of the order of one minute in duration. The exact length of the window depends on the source time function, which is usually a continuous periodic signal, as mentioned above. For a 0.25 Hz square wave, for example, a one-minute window would be exactly 15 cycles. Since the source clock and the receiver clock are independently synchronised to GPS time, with the source synchronised continuously during transmission, the time windows can be chosen correctly. At an average speed of 0.75 m/s (1.5 knots), the vessel moves 45 m in one minute.

Conventionally, in this system all processes on the data are performed in the frequency domain. The first step in data processing, therefore, is to take a Fourier transform of the data in each window. Then each frequency component in the data can be processed independently, taking into account the difference in amplitude of the different frequency components. That is, the received amplitude is divided by the transmitted amplitude, thus normalising the data to a 1 A-m response for each frequency. For a pure square wave source time function, for example, the ratios of the amplitude of the fundamental frequency to the amplitudes of the odd harmonics is $1{:}\frac{1}{3}{:}\frac{1}{5}{:}\frac{1}{7}{:}\ldots$ The true amplitude at any frequency is obtained from the Fourier transform of the synchronous time window of the source dipole transmitter output current. The output current is measured on some commercial sources (Myer et al., 2012: E285).

The data are rotated into survey coordinates. An important display is amplitude versus offset for a given frequency at a given receiver. The amplitude at a given frequency is displayed against the source–receiver distance, or offset. The range of frequencies is usually limited. For the special periodic source time function used by Myer et al. (2012), for example, the four strongest frequencies were 0.25, 0.75, 1.75 and 3.25 Hz. At one receiver the amplitude versus offset curves for these frequencies decayed to the background noise level at about 9, 6, 4 and 3 km, respectively.

A test of the marine CSEM method with a continuous square wave source time function was provided by Srnka et al. (2006), as described in Section 7.10.

7.10 Amplitude *versus* Offset Example

A demonstration of the amplitude versus offset principle was provided by Srnka et al. (2006). Figure 7.4(a) shows a plan of the survey area over a known field in water about 1000 m deep, showing the oil–water contact determined from seismic and well data, the CSEM receiver positions, including receiver 21, wells A and B and survey line 2. Figure 7.4(b) shows the section through the model of the field along survey line 2, the positions of the CSEM receivers on the sea floor and estimated values of the resistivities of subsurface bodies. Figure 7.4(c) shows the amplitude versus offset plot of the data at receiver 21 compared with two different modelled cases: the 3D oil case and the 3D wet case.

The source time function was a 0.25 Hz square wave. The data processing time window was two minutes, or 30 cycles of the square wave. In-line amplitudes at the fundamental source frequency of 0.25 Hz were obtained by the data process-

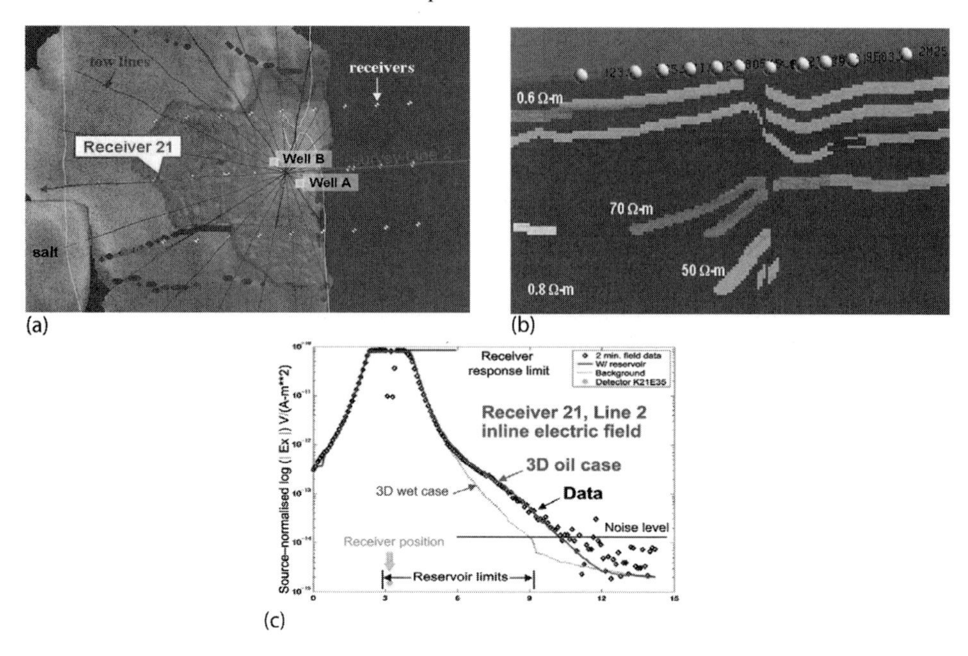

Figure 7.4 (a) Plan of the survey area; (b) vertical section through 3D resistivity model along survey line 2; (c) amplitude versus offset data at receiver 21. Horizontal scale is kilometres (redrawn from Srnka et al., 2006). For the colour version, please refer to the plate section.

ing described in Section 7.9. There is a considerable computational effort behind the modelled cases that is worth mentioning. ExxonMobil had found that simple methods for data interpretation were usually inadequate. For prospect evaluation and delineation, 3D model-based interpretation iterated with seismic depth control was required to produce reliable results.

First, a 3D seismic velocity model was obtained from seismic data and well logs. Wells A and B were logged. 'The (resistivity) model was developed by iterative 3D modelling guided by the seismic data (background), log data from wells A and B, and the full frequency–offset spectrum of the CSEM data for multiple survey lines and receivers. The reservoir intervals are the four highly resistive dipping units' (Srnka et al., 2006: 973). To obtain resistivities from seismic data, particularly velocities, requires rock physics as a link. One possible approach is outlined by Chen and Dickens (2009).

In the 3D wet case the reservoirs are assumed to contain brine, not hydrocarbons, and their resistivities are low. In the 3D oil case the reservoirs are assumed to contain oil and gas and have the high resistivities shown in Figure 7.4(b). The amplitudes are considerably higher for the 3D oil case than for the 3D wet case. The data match the 3D oil case, not the 3D wet case.

This example shows that the method is capable of detecting the presence of subsea hydrocarbons over a known field with seismic data and well control. Without the benefit of the careful 3D modelling, based on the seismic data, well logs and rock physics, it might be hard to interpret the amplitude versus offset data.

If there had been a large range of source tow lines parallel to survey line 2, the broadside amplitude-versus-offset curve at receiver 21 could have been displayed as well. This should have looked similar to the 3D wet case curve. The difference between the in-line and broadside curves would then have been indicative of subsurface resistive units and the offsets at which these differences occurred would have been indicative of the depth of these units.

Ultimately, of course, the objective is to derive the subsurface resistivities from the data.

8

Land CSEM with a Transient Source Signal

This chapter discusses the acquisition and processing of land controlled-source electromagnetic (CSEM) data using a transient source signal. After a short introduction, the data acquisition setup for acquisition of 2D and 3D data is described, followed by a short section on the recovery of the complete Earth impulse response from the data, and the subsequent elimination of the air wave. Acquisition parameters are determined by the target depth, the noise and the known analytic response of a half-space to an impulsive dipole source. A section is devoted to the attenuation of cultural noise. We show how an approximate subsurface resistivity model can be obtained directly from the impulse response data and how, after integration of the impulse responses, well-known DC resistivity inversion can be applied to the resulting step response data to recover a more detailed subsurface model.

8.1 Introduction

Land CSEM exploration has major advantages over deep ocean marine CSEM using remote autonomous receiver nodes. First, the precise positions and orientations of the source and receivers are known from surveying. Second, it is possible to have real-time quality control of the data and to make adjustments to data acquisition parameters if necessary. On land, electrical noise can be very troublesome. There are two main kinds: natural magnetotelluric signals, which are noise for active CSEM data, and man-made noise, which is known as *cultural noise*. An additional problem is the air wave, a solution for which is found by using an appropriate wide-band source time function, for instance a pseudo-random binary sequence (PRBS), as discussed in Chapter 6, and deconvolving the received data for the source time function to recover the earth impulse response. The theory and method were first described by Wright et al. (2002, 2005) and Ziolkowski et al. (2007).

The air wave is coupled to the air–earth interface and the energy stays with the wave: it travels at the speed of light without losses or dispersion. Propagation through the earth is slower, diffusive and dispersive; the pulse spreads out in time as it propagates. After deconvolution of the received response for the source time function, the air wave is an impulse at the start time of transmission; the dispersive wave through the earth follows this. Removal of the air wave is straightforward if the complete earth impulse response is recovered: after deconvolution the initial impulse is simply removed by setting it to zero.

Once the air wave has been removed, the remaining earth impulse response resembles the response of a homogeneous half-space and the time to the peak of the response is indicative of the average conductivity down to a depth equal to approximately half the source–receiver separation. This practical approximation is very helpful in the design of data acquisition parameters.

8.2 Acquisition of 2D and 3D CSEM Data

A two-dimensional section of the earth is produced from 2D data. Geophysical data are obtained along a line, say the x-direction, and the processed and interpreted data form a depth section of a geophysical parameter under this line. A point on the section has coordinates (x, z). For CSEM data the parameter to be displayed should be the conductivity σ, or its reciprocal ρ, the resistivity. Since the earth is very often anisotropic, it is desirable to display both horizontal and vertical conductivity.

A plan view of the setup for acquisition of 2D CSEM data is shown in Figure 2.7 and reproduced here in Figure 8.1 for convenience. The voltage between receiver electrodes C and D is

$$V_{CD,AB}(t) = \Delta x_s \Delta x_r I_{AB}(t) * g_{CD,AB}(t) + n_{CD}(t), \tag{8.1}$$

in which $I_{AB}(t)$ is the source current, Δx_s is the distance between the source electrodes A and B, Δx_r is the distance between the receiver electrodes C and D, the asterisk * denotes convolution, $g_{CD,AB}(t)$ is the impulse response of the earth, or

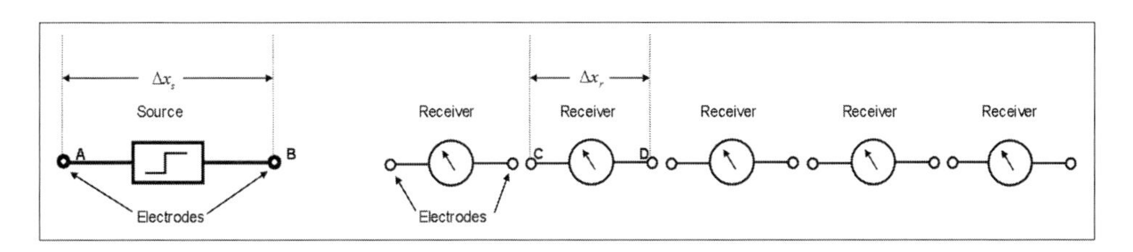

Figure 8.1 Schematic plan view of multi-channel dipole–dipole setup. The separation of the source dipole electrodes is Δx_s; the electric field receivers are in-line, with Δx_r separation of electrodes for each channel.

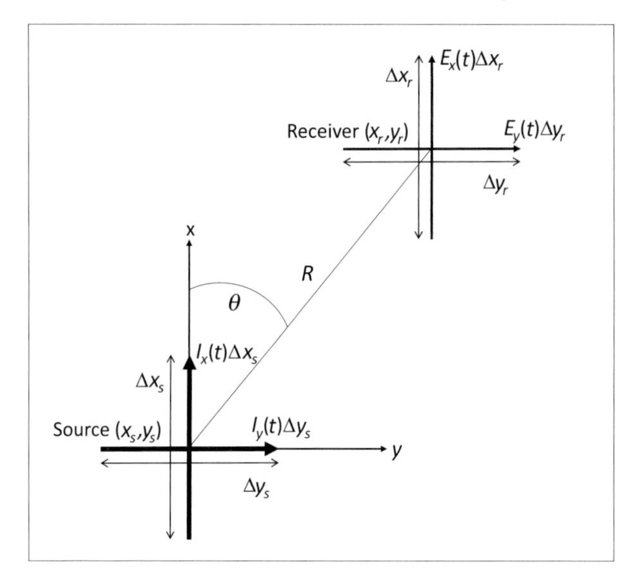

Figure 8.2 Layout of one source and one receiver position in a 3D setup.

Green's function, and $n_{CD}(t)$ is noise. The object is to recover $g_{CD,AB}(t)$ by deconvolution. This is discussed in Section 8.3. The deconvolution can be performed if $V_{CD,AB}(t)$, $I_{AB}(t)$, Δx_s and Δx_r are all known by measurement, if the noise $n_{CD}(t)$ is not too large and if $I_{AB}(t)$ has a broad enough frequency bandwidth.

To simplify the notation, we designate i the source number, corresponding to electrodes A and B in this case, and j the receiver number, corresponding to electrodes C and D. Equation 8.1 can then be written as

$$V_{ji}(t) = \Delta x_s \Delta x_r I_i(t) * g_{ji}(t) + n_j(t). \tag{8.2}$$

For 3D data acquisition receivers are laid out on a grid and sources may also be on a grid. Figure 8.2 shows the layout of one source at (x_s, y_s) and the layout of one receiver at (x_r, y_r). The Cartesian coordinate system is right-handed, with the z-axis pointing downwards. The source consists of two perpendicular horizontal current dipoles which are not operated simultaneously. The dipole moment for the ith x-directed source is $I_{ix}\Delta x_s$ and the dipole moment for the ith y-directed source is $I_{iy}\Delta y_s$. The receiver consists of two perpendicular electric dipoles, each of which measures the voltage between two electrodes. For the x-directed source, the horizontal electric field components at the jth receiver are $E_{jxix}(t)$ and $E_{jyix}(t)$, where the second subscript refers to the receiver orientation and the fourth subscript refers to the source orientation; the voltages measured at the two dipoles are $E_{jxix}(t)\Delta x_r$ and $E_{jyix}(t)\Delta y_r$ Similarly, for the y-directed source, the horizontal electric field components at the jth receiver are $E_{jxiy}(t)$ and $E_{jyiy}(t)$ and the measured voltages are

$E_{jxiy}(t)\Delta x_r$ and $E_{jyiy}(t)\Delta y_r$. As for the 2D case, the subscript i refers to the source number and j refers to the receiver number.

8.3 Deconvolution and Removal of the Air Wave

Deconvolution to recover the complete impulse response, including the air wave, may be illustrated with a 2D example. First, transform equation 8.1 to the frequency domain. The convolution becomes a multiplication. Divide by $\Delta x_s \Delta x_r \hat{I}_i(\omega)$. This yields an estimate $\mathrm{Est}(\hat{g}_{ji}(\omega))$ of the Fourier transform of the impulse response, which is in error by a noise term:

$$\begin{aligned}\mathrm{Est}(\hat{g}_{ji}(\omega)) &= \frac{\hat{V}_{ji}(\omega)}{\Delta x_s \Delta x_r \hat{I}_i(\omega)} = \hat{g}_{ji}(\omega) + \frac{\hat{n}_j(\omega)}{\Delta x_s \Delta x_r \hat{I}_i(\omega)} \\ &= \hat{g}_{ji}(\omega) + \hat{n}_{jm}(\omega),\end{aligned} \tag{8.3}$$

where the noise term $\hat{n}_{jm}(\omega) = \hat{n}_j(\omega)/\Delta x_s \Delta x_r \hat{I}_i(\omega)$. The division is stable for all frequencies if $|\hat{I}_i(\omega)|$ is not zero at any frequency. As discussed in Chapter 6, this condition is satisfied if $I_i(t)$ is, for example, a step function or, even better, a PRBS. The estimated impulse response is obtained by taking the inverse Fourier transform of $\mathrm{Est}(\hat{g}_{ji}(\omega))$:

$$\mathrm{Est}(g_{ji}(t)) = \mathrm{IFT}[\mathrm{Est}(\hat{g}_{ji}(\omega))]. \tag{8.4}$$

The deconvolution has the effect on the response of collapsing the input signal energy into a single instant in time. The energy in the undispersed air wave, therefore, is also collapsed to a single instant in time. The energy in the earth wave is collapsed to the dispersed wave that would have arisen from an impulse at the source with the same energy as the actual source time function. The concentration of source energy in time increases the signal-to-noise ratio by a factor we call the *deconvolution gain*, as discussed in Chapter 6.

An example of recovery of the land impulse response by deconvolution for a PRBS source time function is shown in Figure 8.3. Figure 8.3(a) shows the measured input current for a PRBS of 2047 samples; Figure 8.3(b) shows the measured voltage at the receiver at offset 2050 m; and Figure 8.3(c) shows the recovered earth impulse response on a different time scale, with units of (ohm m^{-2} s^{-1}); in fact, two estimated impulse responses are shown for two consecutive receiver responses, resulting in two slightly different estimates of the earth impulse response, the difference being caused by noise. The time from the start of data to the peak of the earth response we call t_{peak}, which in this case is 0.0009 s. The recording time at source and receiver should be long enough to record the source time function, of duration T_s, plus the duration T_g of the earth impulse response, or Green's function.

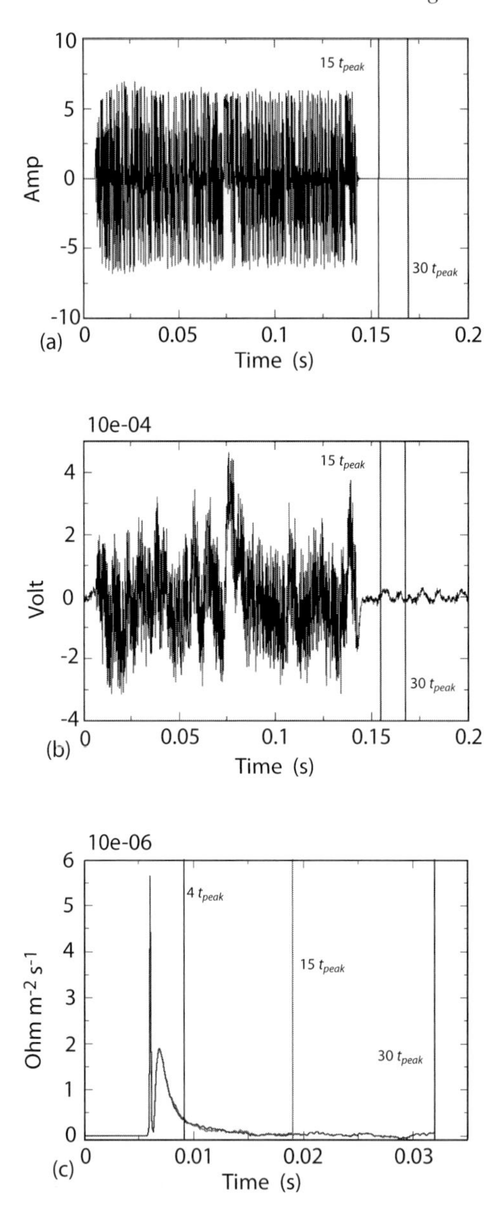

Figure 8.3 Recovery of the earth impulse response by deconvolution: (a) measured source input current (A) for PRBS of 2047 samples; (b) receiver voltage (V) for source–receiver separation of 2050 m; (c) complete earth impulse response (ohm m^{-2} s^{-1}). Redrawn from Ziolkowski, 2007.

The value of T_g is not obvious from the measurement. In this case the data were recorded with an estimate of $T_g = 30t_{peak}$, which may be enough. The cutoff for $T_g = 15t_{peak}$ is shown in Figure 8.3(a)–(c), while the cutoff for $T_g = 4t_{peak}$ is also shown in Figure 8.3(c).

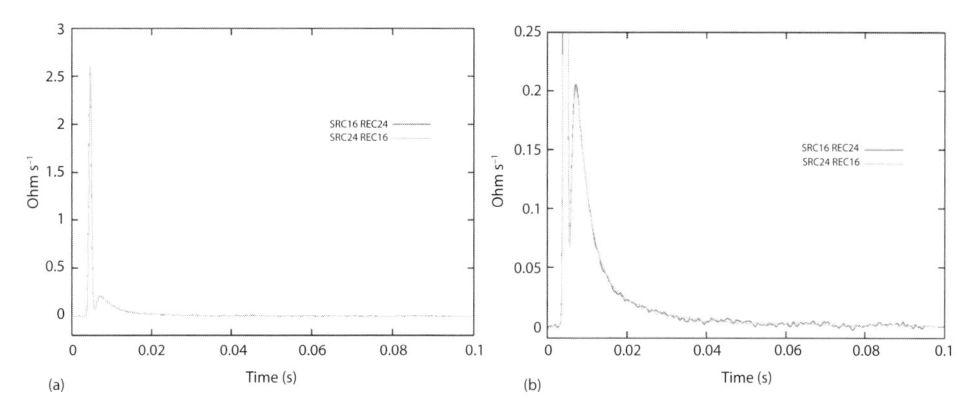

Figure 8.4 Estimates of the earth impulse response, without division by source and receiver dipole lengths; the source and receiver are interchanged for the two responses. (a) and (b) are the same data with different vertical scales (redrawn from Ziolkowski et al., 2007). For the colour version, please refer to the plate section.

In Figure 8.3(c) the air wave is the initial impulse at the start time of data at 0.006 s, corresponding to the beginning of the source time function. This impulse is less than 0.001 s in duration. It is clearly separated from the subsequent response, which is affected only by the subsurface resistivity. It is this part that is of interest. We remove the air wave simply by setting the value of the initial impulse to zero.

It is possible to check that the recovery of the impulse response is correct by using the reciprocity principle, as demonstrated by Ziolkowski et al. (2007): the earth impulse response is the same when the source and receiver are interchanged. In the interchange, the source electrodes occupy the same positions as the receiver electrodes before the interchange and the receiver electrodes occupy the same positions as the source electrodes before the interchange. An example is shown in Figure 8.4.

8.4 Isotropic Half-Space Response

The in-line electric field response at the surface of a vertically transverse isotropic (VTI) half-space to a 1 A-m impulsive horizontal x-directed surface current dipole source at the origin is given by equation 4.140. For an isotropic half-space, equation 4.140 reduces to equation 4.146, which is repeated here:

$$G_{xx}^{ee}(r,0,t) = \frac{\delta(t)}{2\pi\sigma r^3} + \frac{1}{\sigma}\left(\frac{\mu_0\sigma}{4\pi}\right)^{3/2} t^{-5/2}\exp(-\mu_0\sigma r^2/(4t))H(t), \tag{8.5}$$

in which r is the magnitude of the source–receiver distance and the second coordinate is depth z, which is zero. The first term on the right-hand side of equation 8.5 is the air wave, which travels at the speed of light and appears essentially instantaneously at frequencies much less than 10^5 Hz. The second term is the wave that propagates through the half-space, which we regard as the impulse response of the earth:

$$g_{xx}(r,0,t) = \frac{1}{\sigma}\left(\frac{\mu_0\sigma}{4\pi}\right)^{3/2} t^{-5/2}\exp(-\mu_0\sigma r^2/(4t))H(t). \tag{8.6}$$

This function is zero at $t = 0$, rises to a peak, and decays to zero again as $t \to \infty$. The time to the peak is found by differentiating $g_{xx}(r,0,t)$ with respect to time and setting the result to zero. Differentiating with respect to t yields

$$\frac{dg_{xx}(r,0,t)}{dt} = \frac{1}{\sigma}\left(\frac{\mu_0\sigma}{4\pi}\right)^{3/2} \exp(-\mu_0\sigma r^2/(4t))\left[-\frac{5H(t)}{2t^{7/2}} + \frac{\mu_0\sigma r^2 H(t)}{4t^{9/2}} + t^{-5/2}\delta(t)\right]. \tag{8.7}$$

This is zero when $t = t_{peak}$, where

$$t_{peak} = \frac{\mu_0\sigma r^2}{10} = \frac{\mu_0 r^2}{10\rho}. \tag{8.8}$$

Now we define a dimensionless time parameter

$$\tau = t/t_{peak}. \tag{8.9}$$

Substituting for t from equation 8.9 into equation 8.6 gives

$$g_{xx}(r,0,\tau) = \frac{10^{5/2}}{(4\pi)^{3/2}\mu_0}\frac{1}{\sigma^2 r^5}\tau^{-5/2}\exp(-5/(2\tau))H(\tau). \tag{8.10}$$

The factor $10^{5/2}/((4\pi)^{3/2}\mu_0\sigma^2 r^5)$ depends on the medium conductivity σ and the source–receiver separation r. Note that the amplitude decays as r^{-5}. The expression $\tau^{-5/2}\exp(-5/(2\tau))H(\tau)$ is independent of the medium and the source–receiver separation. It is plotted in Figure 8.5. The dimensionless time to the peak is 1, by definition.

The isotropic half-space response shown in this figure is similar in shape to the real earth impulse response shown in Figure 8.3. The time to the peak, after the air wave, gives an indication of the resistivity of the medium under the source–receiver pair. Using equation 8.8 we can define an *equivalent half-space resistivity* ρ_H:

$$\rho_H = \frac{\mu_0 r^2}{10 t_{peak}}, \tag{8.11}$$

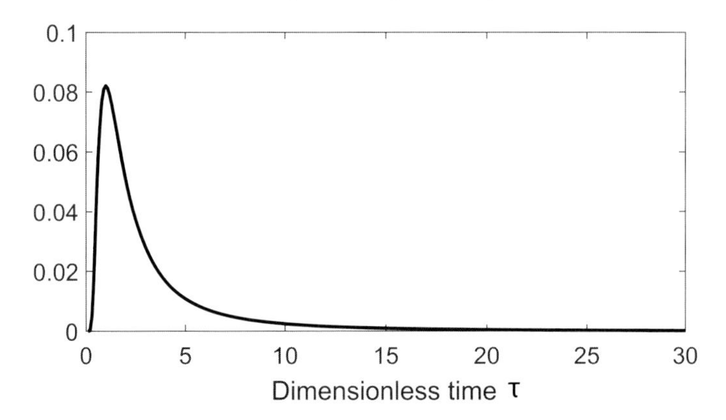

Figure 8.5 Isotropic half-space impulse response in dimensionless time.

which provides an estimate of the subsurface resistivity. Equation 8.11 was also derived in Chapter 4 as equation 4.148. The equation applies to a point dipole source and point dipole receivers. For sources and receivers of finite size it is an approximation, as discussed in Chapter 4. Using the approximation for the case of Figure 8.3, $t_{peak} = 0.0009$ s, $r = 2050$ m, and the equivalent half-space resistivity is 587 ohm-m.

The causal time-domain response of equation 8.6 has the one-sided Laplace transform (cf. equation 4.138 with $\lambda = 1$)

$$\hat{g}_{xx}(r, 0, s) = \frac{1}{2\sigma\pi r^3}(1 + r\sqrt{s\mu_0\sigma})\exp(-r\sqrt{s\mu_0\sigma}). \tag{8.12}$$

Substituting $s = -i\omega$ yields the two-sided Fourier transform

$$\hat{g}_{xx}(r, 0, i\omega) = \frac{1}{2\sigma\pi r^3}(1 + r\sqrt{-i\omega\mu_0\sigma})\exp(-r\sqrt{-i\omega\mu_0\sigma}). \tag{8.13}$$

Following the time domain approach, it is now convenient to define a dimensionless frequency. Let

$$\omega = 2\pi\nu f_0, \tag{8.14}$$

where ν is dimensionless frequency and we choose

$$f_0 = \frac{1}{t_{peak}}. \tag{8.15}$$

Then the dimensionless Fourier transform is

$$\hat{g}_{xx}(r, 0, i\nu) = \frac{1}{2\sigma\pi r^3}(1 + \sqrt{-20i\pi\nu})\exp(-\sqrt{-20i\pi\nu}). \tag{8.16}$$

The amplitude spectrum of the function $(1+\sqrt{-20i\pi\nu})\exp(-\sqrt{-20i\pi\nu})$ is shown in Figure 8.6. From this figure it is clear that the highest dimensionless frequency

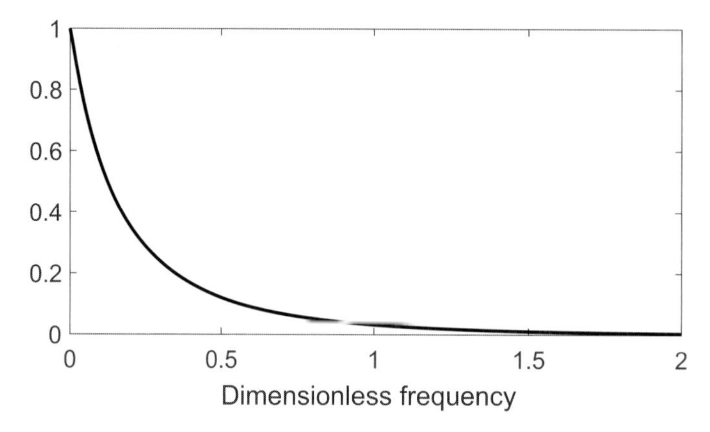

Figure 8.6 Isotropic half-space impulse response in dimensionless frequency.

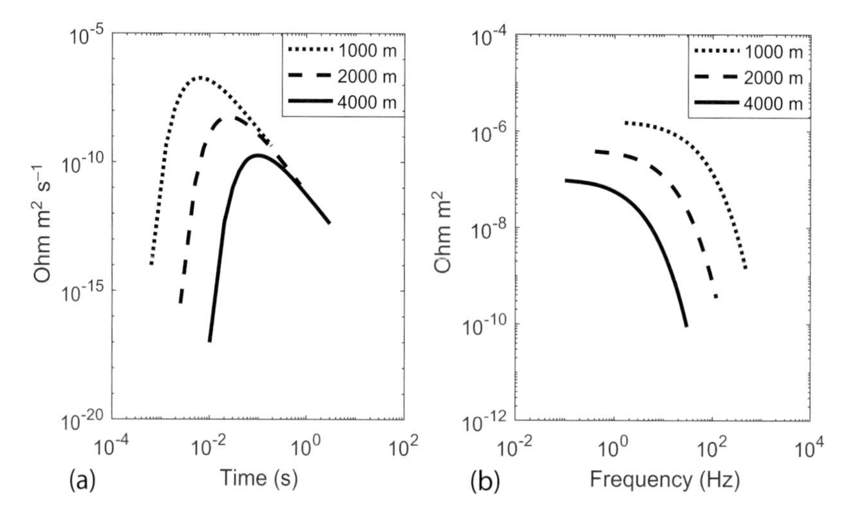

Figure 8.7 20 ohm-m isotropic half-space impulse response for 1000, 2000 and 4000 m offsets: (a) time; (b) amplitude spectrum.

at which there is significant energy is no greater than 2. That is, there is no need to record frequencies higher than $2f_0$ or $2/t_{peak}$.

The effect of offset on the frequency bandwidth and duration of the half-space impulse response function is shown in Figure 8.7 for the example of a 20 ohm-m half-space. Since t_{peak} is proportional to r^2, the highest frequency of interest decreases as r^{-2} with a very steep cutoff, as shown in Figure 8.7(b). The amplitude decreases as r^{-5}, as already noted, and it is obvious that the signal-to-noise ratio must decrease with offset if the noise has a broad bandwidth.

At the low-frequency end of the spectrum there is no cutoff frequency: the response contains DC. This poses a problem: what is the duration T_g of the impulse response and how long, therefore, should we record? This is of great practical interest in setting up the recording parameters. The answer probably requires more analysis than is available with these simple pictures. A reasonable number may be

$$T_g = 30t_{peak}. \tag{8.17}$$

It is also important to observe that both the bandwidth and dynamic range of the responses range over many orders of magnitude. Furthermore, the amplitude of the response is proportional to the square of the resistivity, while t_{peak} is inversely proportional to resistivity. Since the resistivity can vary by at least three orders of magnitude (e.g. from 1 ohm-m to 1000 ohm-m) for rocks on land, the equipment used to acquire transient CSEM data on land must be very flexible. It is a challenge to design equipment that will work with all possible parameters in all geological environments.

From Figure 8.7(b) it is clear that the high-frequency response of the earth decreases dramatically with offset. If a PRBS source time function is used with a smallest time interval Δt_s between polarity reversals, the highest frequency in the source time function is $f_H = 1/(2\Delta t_s)$. For short offsets Δt_s needs to be small and for large offsets Δt_s needs to be large. At large offsets, high frequencies are not observed and it is a waste of time and effort to inject high-frequency energy at the source. To obtain data efficiently at all offsets, a range of source time sample intervals Δt_s should be used.

8.5 Signal-to-Noise Ratio of MTEM Data

The discrete form of equation 8.2 may be written as

$$V_{ji}(k\Delta t) = \Delta x_s \Delta x_r \Delta t \sum_{\tau=0}^{N_{source-1}} I_i(\tau\Delta t) g_{ji}((k-\tau)\Delta t) + n_j(k\Delta t), \tag{8.18}$$

where Δt is the sampling interval of the recording system and N_{source} is the number of time samples of the source time function at a sampling rate of $1/\Delta t$. The raw signal-to-noise ratio of transient CSEM data can then be written as

$$[\mathrm{S{:}N}](k\Delta t) = \frac{\Delta x_s \Delta x_r \Delta t \sum_{\tau=0}^{N_{source}-1} I_i(\tau\Delta t) g_{ji}((k-\tau)\Delta t)}{n_j(k\Delta t)}, \tag{8.19}$$

in which the signal amplitude is directly proportional to each of the following parameters: Δx_s, Δx_r, Δt and $I_i(\tau\Delta t)$. It is also related to N_{source}.

It is clearly important to maximise the source dipole moment $I_i(t)\Delta x_s$, which is independent of the noise. Therefore both the current $I_i(t)$ and the dipole length Δx_s should be maximised. The lateral resolution of the data decreases as Δx_s increases. There is therefore a trade-off between resolution and signal-to-noise ratio to be made in the choice of Δx_s.

For noise which is essentially an electric field, such as magnetotelluric signals and cultural noise, increasing the receiver dipole length Δx_r increases both the signal and the noise in the same proportion, so it has no effect on the signal-to-noise ratio. If electrode noise is a significant fraction of the total noise, this is independent of the length of the receiver dipole and increasing Δx_r can improve the signal-to-noise ratio.

The signal-to-noise ratio is proportional to the sampling interval Δt. Therefore Δt should be as large as possible. We showed in Section 8.4 that the frequency response of a half-space attenuates dramatically with offset and for each offset there is a maximum frequency beyond which it is pointless to transmit energy. In principle this frequency should be the Nyquist frequency of both the recording system and the source time function. They should be the same.

Assuming $I_i(t)$ is a PRBS of amplitude I_s, the sampling interval should be set to Δt and $\Delta t = 1/2f_{Nyquist}$, where $f_{Nyquist}$ is the highest frequency of interest in the data. Then $N_{source} = N = 2^n - 1$, where n is the order of the PRBS. Deconvolution for the source time function replaces $I_i(t)$ by an impulse of amplitude $\sqrt{N}I_s$ and width Δt, as described in Chapter 6. For random noise this results in a gain in signal-to-noise ratio of $\sqrt{N}$.

Another way to improve the signal-to-noise ratio is by vertical stacking. The experiment is repeated a number, say M, times and the impulse response of the earth is recovered for each experiment. Summing the recovered impulse responses and dividing by M gives the average response, which should be identical for every experiment, and reduces the noise if the noise is not coherent from experiment to experiment. If the noise is random, the increase in signal-to-noise ratio by stacking is $\sqrt{M}$. It may be more efficient to increase the length of the PRBS than to increase the number of experiments, because the duration of M experiments is $M(T_s + T_g)$, whereas increasing the length of the PRBS increases only T_s. In practice one uses the longest possible PRBS.

Sometimes it is difficult to get current into the ground in land surveys. Every factor of 2 reduction in source current amplitude results in approximately a factor of 4 increase in the time required to obtain data with adequate signal-to-noise ratio. The survey costs can increase dramatically. We see from this that it is vitally important to maximise the source current and source dipole moment.

8.6 Attenuation of Cultural Noise

Cultural noise is not random: it is organised. We can exploit the organisation to reduce its effect. In general, cultural noise arises because electrical machinery is grounded for a number of reasons, including safety. As a result, electricity goes into the ground wherever there are electrical devices. These operate essentially independently and the noise in the ground comes from different directions, the amplitudes depending on the usage of the devices and the value of near-surface resistivity. A low RMS noise level would be 10 μV between receiver electrodes 100 m apart, or 1 nV/m. Noise levels can be considerably greater than this. We have measured cultural RMS noise levels of 10 mV over 100 m; that is 1000 times, or 60 dB, greater noise level. This extremely high noise level occurred where there was a shallow, highly resistive thick layer, which had a skin depth of many kilometres and prevented the attenuation of cultural noise. The noise conditions and the local resistivity of the subsurface determine how easy it is to obtain data with a good signal-to-noise ratio. Before beginning a survey in a new area, it is a good idea to determine the noise levels and near-surface resistivity.

In this section we discuss two methods of attenuating cultural noise. The first exploits the periodicity of alternating current electricity; the second exploits the symmetry of the field of the dipole current source.

In the first method, we use the knowledge that most machines are driven by alternating current electricity from the grid, which operates at 50 Hz in Europe and 60 Hz in North America. The noise is often at the fundamental frequency (50 Hz and 60 Hz) and at odd harmonics (150, 250, Hz, etc. and 180, 300 Hz, etc.).

Over a period of one second, sometimes longer, both the amplitude and phase of the cultural noise are often fairly constant. Figure 8.8 from Ziolkowski et al. (2007) shows how this can be exploited to attenuate the cultural noise using pairs of measurements. Starting with equation 8.1, the first measurement at a particular receiver is expressed as

$$V_1(t) = \Delta x_s \Delta x_r I(t) * g(t) + nc_1(t) + na_1(t), \tag{8.20}$$

where $nc_1(t)$ is the cultural noise and $na_1(t)$ is all other sources of noise. At a time T_R later, in which T_R is chosen to be an integer number of periods of the fundamental oscillation of the alternating current, the experiment is repeated using a source time function of opposite polarity. The second measurement is

$$V_2(t - T_R) = \Delta x_s \Delta x_r I(t - T_R) * g(t - T_R) + nc_2(t - T_R) + na_2(t - T_R). \tag{8.21}$$

The earth impulse response does not change in this time, so $g(t) = g(t - T_R)$ and if the source time function is repeatable $I(t) = -I(t - T_R)$; that is, the second source time function has opposite polarity. Subtracting equation 8.21 from equation 8.20 yields

$$V_1(t) - V_2(t - T_R) = 2\Delta x_s \Delta x_r I(t) * g(t) + nc_1(t) - nc_2(t - T_R) + na_1(t) - na_2(t - T_R). \tag{8.22}$$

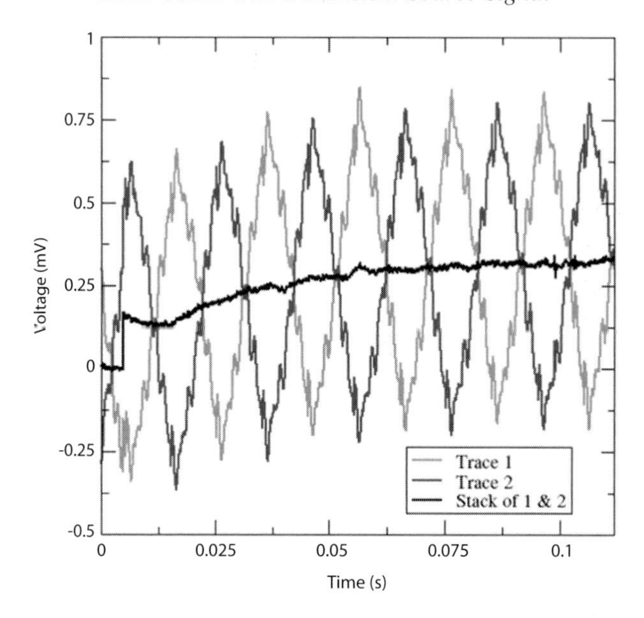

Figure 8.8 The effect of cultural noise cancellation by changing source polarity at integer number of noise cycles (redrawn from Ziolkowski et al., 2007).

If the amplitude and phase of the cultural noise are constant, $nc_1(t) = nc_2(t - T_R)$ and equation 8.22 becomes

$$V_1(t) - V_2(t - T_R) = 2\Delta x_s \Delta x_r I(t) * g(t) + na_1(t) - na_2(t - T_R). \quad (8.23)$$

In Figure 8.8, Trace 1 is $V_1(t)$, the first measurement, Trace 2 is $-V_2(t - T_R)$, minus the second measurement, the source time function was a step, and the repeat time T_R was 1 s. The stack essentially cancels out the noise, and the step response is clear, but still noisy because of other sources of noise. This example shows that the principle works on real data and can be a very powerful way to reduce cultural noise, which in this case had a peak-to-peak amplitude of about 1 mV.

The second method (Ziolkowski and Carson, 2007) exploits the symmetry of the source dipole field (see Figure 2.2). An experiment was conducted using the setup shown in Figure 8.9. The noise was measured between two electrodes on opposite sides of the line through the source dipole axis, equidistant from this line, and on a line perpendicular to it. In a uniform medium the source dipole field is symmetric about its axis and therefore generates the same potential at these two electrodes, so the potential difference between the electrodes due to the source must be zero. This is also the case in horizontally layered media and if there is no significant cross-line dip. If there is no noise, there should be no voltage between these electrodes. Any voltage that is measured must be noise and it is very likely to be correlated with the noise in the adjacent receiver dipole measurement.

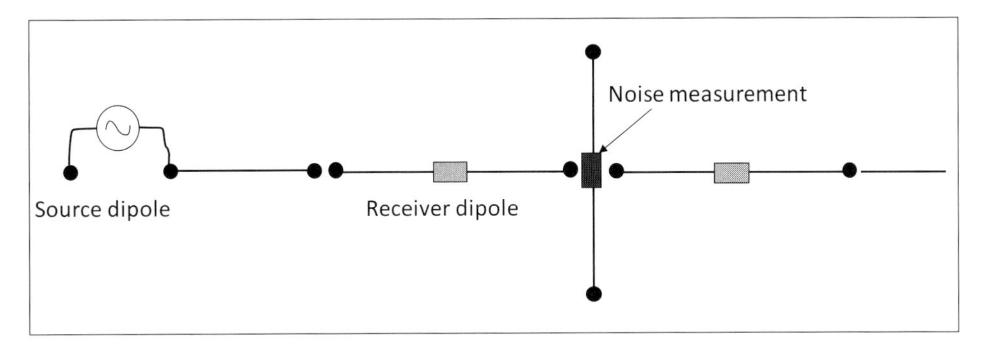

Figure 8.9 Plan view of the setup showing measurement of noise perpendicular to the axis of source dipole (redrawn from Ziolkowski and Carson, 2007). Electrodes are shown as solid black circles.

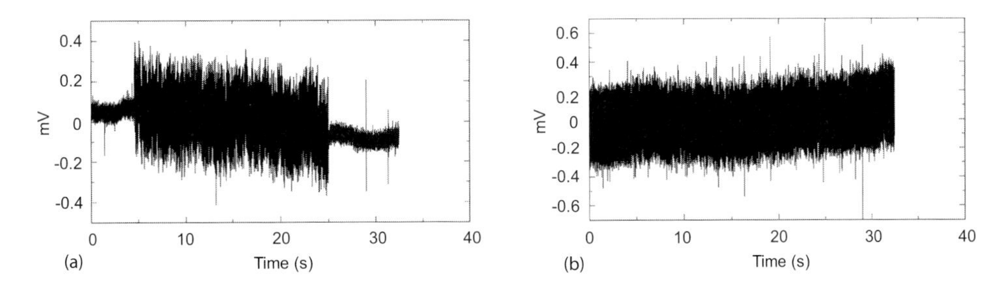

Figure 8.10 Simultaneous in-line and cross-line measurements using setup of Figure 8.9: (a) in-line; (b) cross-line (redrawn from Ziolkowski and Carson, 2007).

Figure 8.10 shows an example of data recorded with the setup shown in Figure 8.9 using 100 m receiver dipoles. Figure 8.10(a) shows the in-line signal response to a PRBS, preceded by noise and followed by noise. Figure 8.10(b) shows the simultaneous cross-line response: it is pure noise with no detectable signal. The cross-line noise happens to have larger amplitude than the in-line noise.

Since the cultural noise is propagating in the ground from diverse sources in different directions, each component can be considered to be a coherent dispersive wave. It is likely that there is a correlation of the noise wavefield in the in-line direction with the noise wavefield in the cross-line direction and there will be a filter that can be applied to the cross-line noise that is a best fit in a least-squares sense to the in-line noise. The filter can be estimated as a Wiener filter (see Section 3.19) from a time window before the arrival of the signal, and applied to the whole of the cross-line noise signal to estimate the in-line noise. The estimated noise can be subtracted from the in-line measurement to reduce the noise and increase the signal-to-noise ratio. The result of this process for the data of Figure 8.10 is shown in Figure 8.11. The noise reduction is about 6 dB, or a factor of 2. This was not a

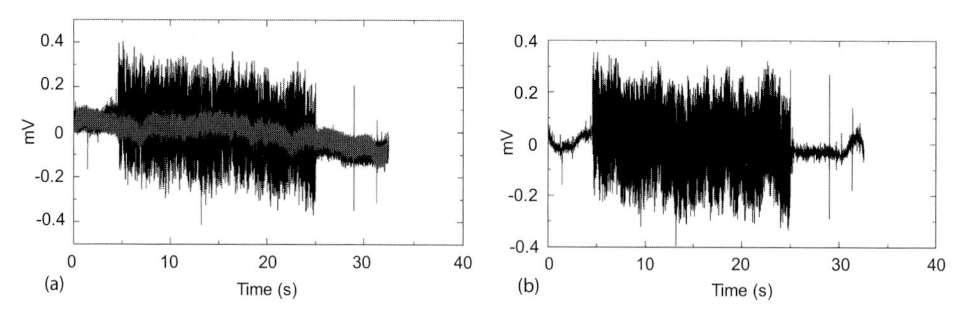

Figure 8.11 (a) In-line measurement with estimated noise in grey; (b) in-line measurement after subtraction of estimated noise (redrawn from Ziolkowski and Carson, 2007).

very noisy environment. Signal-to-noise ratio improvements much greater than this can be obtained in noisier environments.

The noise is not stationary and the filter is locally valid for a short period. A different filter should be calculated for each receiver and for each source transient. In practice, it would be practical to use receivers in the configuration shown in Figure 8.2. For a 3D setup it would be necessary to determine the direction of the source field at the receiver, by correlation with the source time function, and rotate the receiver into in-line and cross-line for that field direction.

8.7 CSEM Survey Over an Underground Gas Storage Site in France

Ziolkowski et al. (2007) describe a transient CSEM survey over a known underground gas storage reservoir in France, using a step function source time function, with a polarity reversal every 1 s cycle. Approximately 1500 cycles were stacked for each source position. The impulse response depends on the source–receiver offset r and on the source–receiver midpoint. The midpoint is defined as the point midway between the centre of the source dipole and the centre of the receiver dipole. A display of impulse responses with the same offset, but with varying midpoint, shows lateral variations if there are lateral variation in subsurface resistivity. Figure 8.12 shows a set of common-offset sections of impulse responses from this survey for offsets 700–2000 m in increments of 100 m. The air wave has been removed from each impulse response.

At the shortest offset, 700 m, the response is confined to times less than about 15 ms, with highest amplitudes in the top-left of the section, corresponding to shallow resistors in the first 1–2 km on the left of the profile. This resistor is most prominent at an offset of 900 m. At greater offsets it becomes weaker and at 1300 m offset it has disappeared. At 1100 m offset a resistor appears at 3 km from the

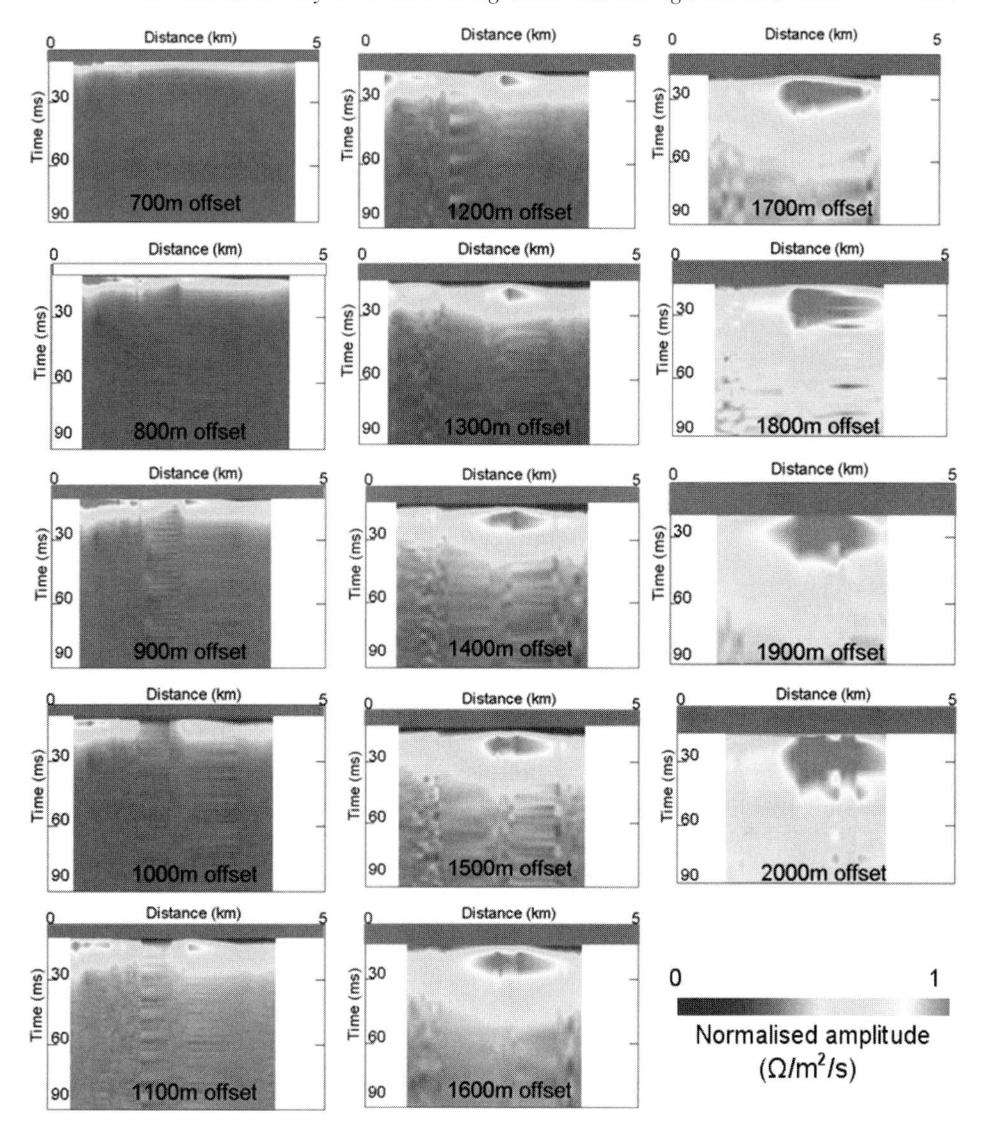

Figure 8.12 Common-offset sections of impulse responses over gas storage site; horizontal coordinate is midpoint; vertical coordinate is time (redrawn from Ziolkowski et al., 2007). For the colour version, please refer to the plate section.

left-hand end of the profile, with a peak at about 15 ms. As the offset increases this peak grows dramatically, both laterally and vertically (in time). The high amplitudes of this anomaly are caused by the known hydrocarbon gas, which is resistive.

This simple display of common-offset sections of recovered impulse responses shows there is a strong resistive anomaly which requires further investigation. This principle can obviously be used as an early check on the data to see if anything interesting has been found.

8.8 Apparent Resistivities from Time to Peak of Impulse Response

One way to investigate the data is by inversion. Ziolkowski et al. (2007) applied collated 1D inversions of common midpoint (CMP) gathers to the data, with a result that agreed with the geology known from seismic data and well information. Inversion takes time and requires care. Often a quick rough result is required before embarking on inversion. Ziolkowski et al. (2007) also developed a simple method for estimating subsurface resistivities using the time to the peak of the earth impulse response.

Equation 8.11 gives the equivalent half-space resistivity for the response for a given source–receiver pair. Using this simple formula on each recovered earth impulse response yields the section shown in Figure 8.13. It shows a shallow resistor, mostly in the upper-left of the section, overlying a conductor, overlying a deeper resistor in the middle of the section at offsets greater than about 1800 m. This deeper resistor is the gas storage reservoir.

The vertical coordinate in the display is offset. We would prefer to have depth. Since the value plotted is a kind of average, it is not possible to obtain the depth without doing much more work. In fact, this is an inversion problem.

8.9 Resistivities from Step Responses

An impulse response may be integrated to give the step response. Figure 8.14 shows a real data example of a step response obtained by integrating a complete impulse response including the air wave. At late times the curve rises slowly and is asymptotic to a value S_∞. An estimate of S_∞ may be made by fitting an appropriate curve to the data at late times. $S_\infty = V_\infty/(\Delta x_s \Delta x_r)$ where V_∞ is the DC value that would be obtained for a 1 A-m source, Δx_s is the distance between the source electrodes and Δx_r is the distance between the receiver electrodes. Thus the apparent resistance is S_∞ ohms. A better way to obtain S_∞ is to use the final value theorem, derived in Appendix D. This says that S_∞ is the DC value of the amplitude spectrum of the time derivative of the step response: that is, the DC value of the amplitude spectrum of the impulse response.

One apparent resistance may be estimated for each source–receiver pair. Using the known positions of the source and receiver electrodes, a standard DC resistivity inversion routine such as the one described by Loke and Barker (1996) may be used to obtain another estimate of subsurface resistivities. This requires more effort than

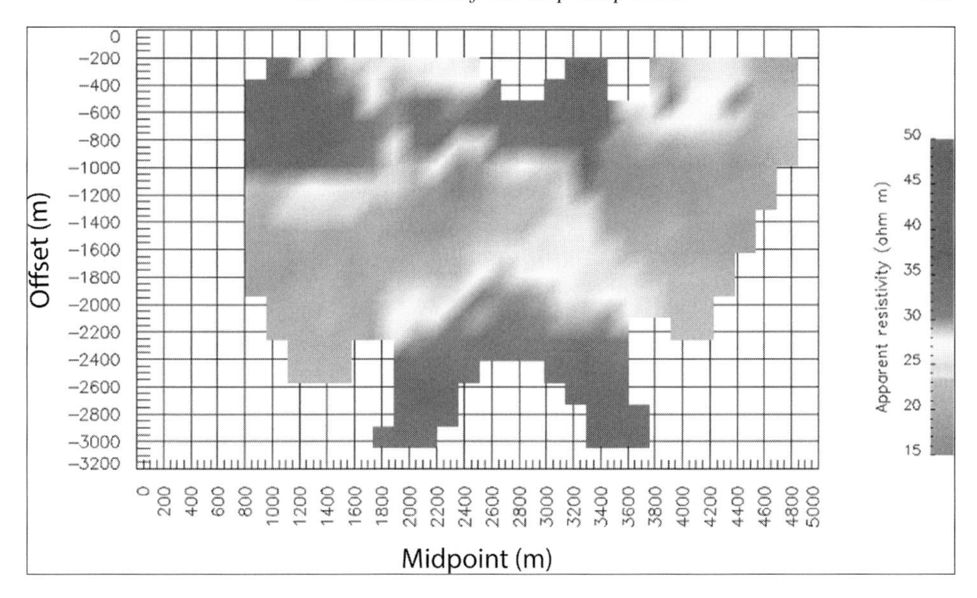

Figure 8.13 Apparent resistivities derived from time to peak of impulse response using equation 8.11 displayed in CMP-offset coordinates (redrawn from Ziolkowski et al., 2007). For the colour version, please refer to the plate section.

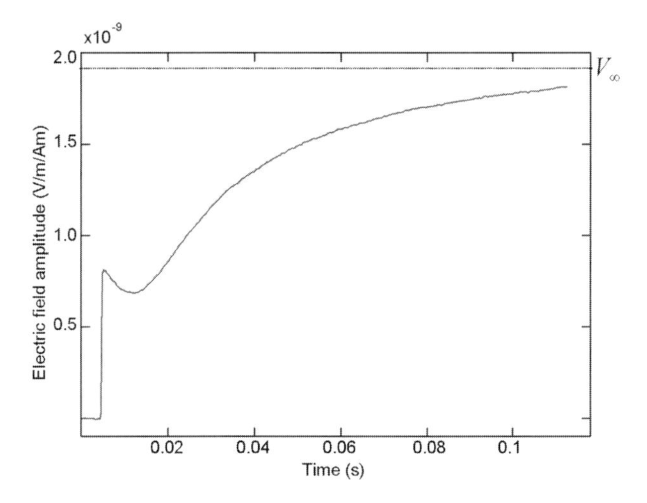

Figure 8.14 Real data example of step response (redrawn from Ziolkowski et al., 2007).

the recovery of apparent resistivities from the peak times of the impulse responses, but is much less effort than full inversion of the complete set of impulse responses. The result for the experiment described by Ziolkowski et al. (2007) is shown in Figure 8.15, in which the vertical coordinate is depth. The top of the deep resistor

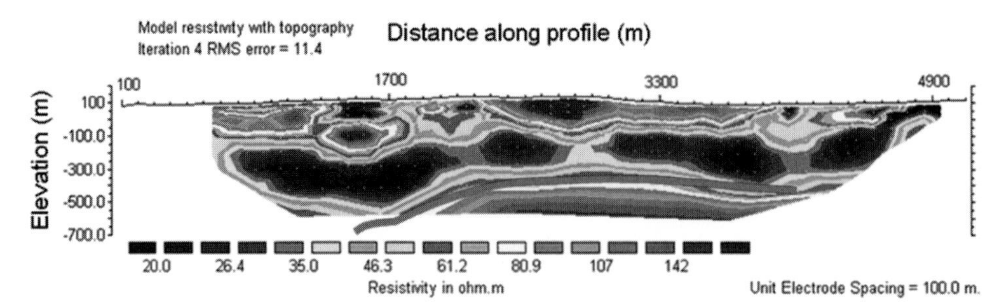

Figure 8.15 Two-dimensional dipole–dipole DC resistivity inversion of the transient CSEM data using the estimated asymptotic values of the step responses. The red curve shows the top of the reservoir (redrawn from Ziolkowski et al., 2007). For the colour version, please refer to the plate section.

matches the known position of the top of the reservoir. This is a very satisfactory result.

The penetration depth of this section is much greater than that of conventional DC resistivity surveys. This is a consequence of the enormous increase in signal-to-noise ratio obtained by the combination of the various techniques described above.

9

Shallow Water CSEM with a Transient Source Signal

The principles of the land controlled-source electromagnetic (CSEM) system with a transient source, described in Chapter 8, can be applied offshore. This chapter describes the marine application, first with an ocean-bottom receiver cable (OBC) and second with a towed streamer receiver. In both cases the data shown are from experimental systems. The advantage of the OBC system is that there is a much smaller time constraint on the data acquisition. The disadvantage is the cost per data point, which is far higher than with the towed streamer system. In both the OBC and towed streamer systems, positioning of the source and receiver is much more precise than with the deep water system described in Chapter 7. Noise levels, however, are much greater in shallow water than in deep water.

9.1 2D Data Acquisition with an OBC

Figure 9.1 shows the setup used for the acquisition of shallow marine CSEM data with an OBC, using a source vessel and a receiver vessel, enabling the source–receiver offset to be varied, as described by Ziolkowski et al. (2010). The distance from the source vessel to the source is much less than for the deep-towed source required in deep water. Losses in the connecting cables are much lower than for the deep-towed source and the shallow water source system is consequently less complicated. It is reasonably straightforward to generate currents of 1000 A because the conductivity of sea water is high, typically 3 S m^{-1}; the voltage required is less than 100 V. Heavy copper cables connect the on-board power supply and signal generator to the sea-bottom current dipole transmitter.

The OBC contains approximately 30 dipole receivers, each 200 m long and consisting of a silver–silver chloride electrode at each end connected to a box in the centre, which performs analogue-to-digital conversion and data transmission back to the receiver vessel. The power for each box is delivered from the vessel.

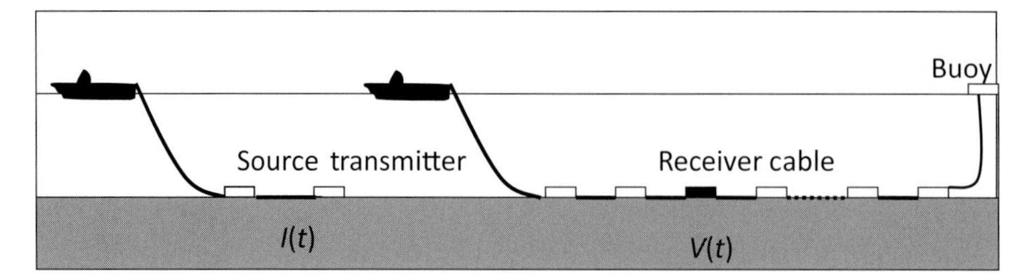

Figure 9.1 Profile of a marine CSEM system showing dipole current source on the sea floor and separate receiver cable (OBC) with dipole receivers, synchronised with GPS, and with real-time quality control; typical dipole source and receiver lengths are 200 m (redrawn from Ziolkowski, 2007).

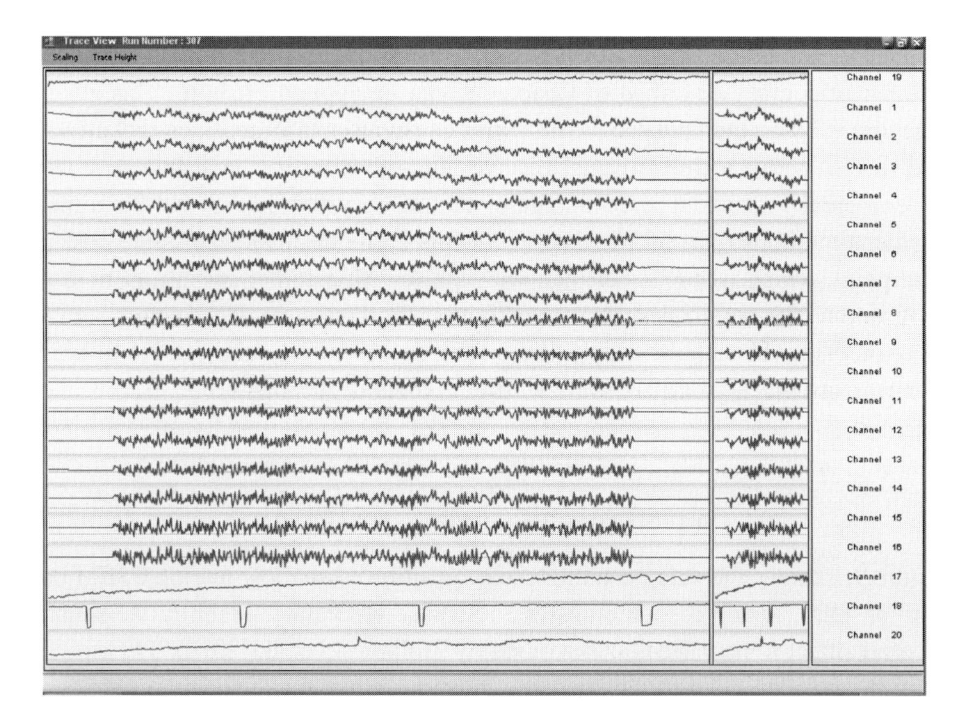

Figure 9.2 Screengrab of a typical marine OBC record: source–receiver offset increases from bottom of the screen to the top. Only 16 of the 30 receivers are in the water; the rest of the cable is on the vessel (redrawn from Ziolkowski, 2007).

A screengrab of typical marine pseudo-random binary sequence (PRBS) data is shown in Figure 9.2.

The transmission of the PRBS signal from the source and recording of the source current time function by the source vessel is synchronised with the recording of the receiver data and the recorded source time function is transmitted by radio to the receiver vessel for deconvolution of the data. Acoustic transponders attached to the source and receiver cables at the electrode positions allow the positions of all the electrodes to be determined with precision of about 2.5 m (Ziolkowski et al., 2010).

9.2 Examples of OBC Data

For the land case it was noted in Section 8.5 that the temporal sampling interval of the data should be as large as possible for optimum signal-to-noise ratio and the source and receiver sample rates should be equal. The situation is the same for the marine case. The bandwidth of the data decreases with increasing offset as higher frequencies are attenuated more rapidly than lower frequencies; consequently, the sample rate should decrease with increasing offset. Examples of OBC data obtained with this principle are shown in Figure 9.3.

Figure 9.3 (a)–(e) shows recovered impulse responses from common shot gathers with offset increasing from 2000 m to 5800 m. At short offsets the air wave is not noticeable. As the offset increases, the amplitude of the air wave increases relative to the subsequent earth response and at offsets greater than 5000 m its peak amplitude is more than twice the amplitude of the earth response. The air wave in water is not a perfect impulse, as it is on land, and it has a tail that overlaps the earth response. The duration of the earth impulse response increases with offset, but the amplitude at longer offsets is always smaller than at shorter offsets. The sampling interval was 25 ms for offsets 2000–3000 m, 100 ms for offsets 3200–4200 m, and 200 ms for offsets 4400–5800 m. If all the data had been acquired with a sampling interval of 25 ms, the data at offsets greater than 3000 m would have been considerably noisier.

Figure 9.3 (b)–(f) shows common-offset impulse response sections with common midpoint (CMP) number as the horizontal coordinate and time as the vertical coordinate. If the earth were laterally invariant, there would be no variation in the response with CMP. The lateral variation that is observed in these displays is caused by lateral variations in subsea resistivity, reflecting variations in subsea geology.

9.3 Removal of Spatially Correlated Noise

The impulse responses in Figure 9.3 were obtained after deconvolution for the PRBS source time function and correlated noise removal. Ziolkowski et al. (2010) developed a method for removal of correlated electromagnetic noise that increases the signal-to-noise ratio of each dataset by as much as 20 dB. The elements of the process are illustrated in Figure 9.4, taken from Ziolkowski et al. (2010).

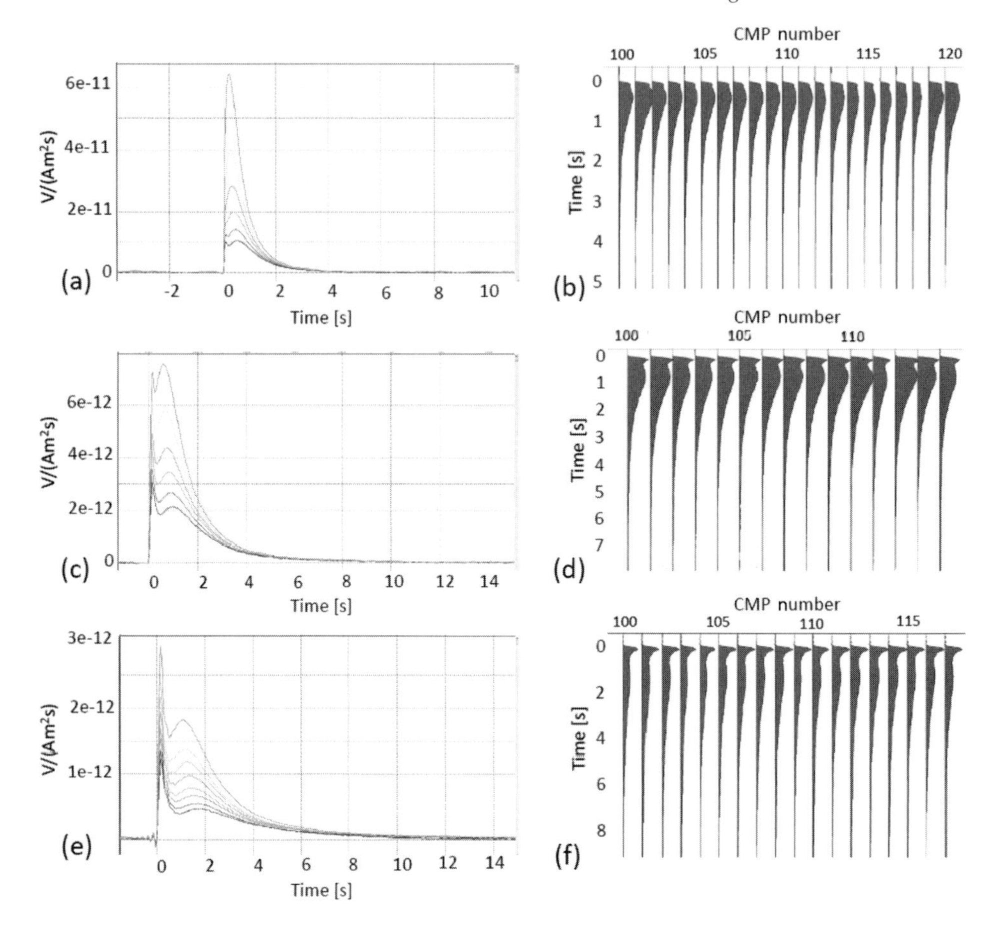

Figure 9.3 OBC data: (a) source gather, offsets 2000–3000 m, Δt = 25 ms; (b) 2500 m common-offset gather, Δt = 25 ms; (c) source gather, offsets 3200–4200 m, Δt = 100 ms; (d) 4000 m common-offset gather, Δt = 100 ms; (e) source gather, offsets 4400–5800 m, Δt = 200 ms; (f) 5000 m common-offset gather, Δt = 200 ms. For the colour version, please refer to the plate section.

Figure 9.4(a) shows a raw common source gather, with source–receiver separation increasing from left to right. The response to the PRBS, of approximately 200 s duration, can be seen clearly as high-frequency oscillations, the amplitude decreasing with offset. On the shortest-offset channels the amplitude is large. It decreases progressively as the offset increases and is so small on the furthest offset as to be imperceptible. The other obvious component has a much lower frequency content, is strongly correlated from channel to channel and its amplitude appears to be about the same on each channel. This correlated low-frequency component is magnetotelluric noise. A typical spectrum of such noise is shown in Figure 2.9.

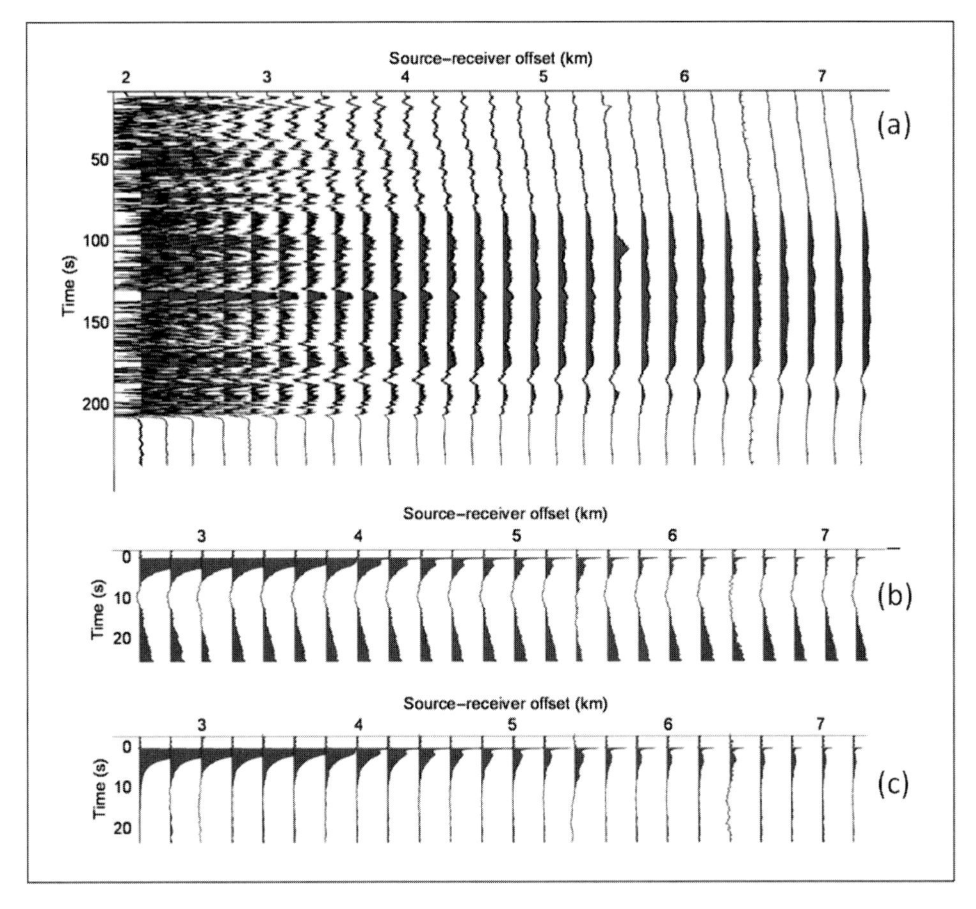

Figure 9.4 (a) Common-source gather, 250 s in duration (vertical axis), with offsets (horizontal axis) increasing from 2200 m on the left to 7000 m on the right in 200 m intervals; (b) result of deconvolving source gather for measured PRBS current, showing only 20 s of data; (c) result of subtracting estimated noise from data in (b) (Redrawn from Ziolkowski et al., 2010) .

Its amplitude increases as the frequency decreases. Deconvolution of the data for the source time function recovers the earth impulse response on every channel, as shown in Figure 9.4(b), but the low-frequency correlated noise is still present. The noise may be estimated, as described in the following and subtracted from the deconvolved data to recover improved responses, as shown in Figure 9.4(c).

Following the notation of the 2D land case, equation 8.2, the voltage in the jth channel of the OBC, in response to the current $I_i(t)$ input at the ith source, may be expressed as

$$V_{ji}(t) = \Delta x_s \Delta x_r I_i(t) * g_{ji}(t) + nc_j(t) + na_j(t), \tag{9.1}$$

in which $nc_j(t)$ is the correlated noise and $na_j(t)$ is all other noise. Deconvolution of the recorded data for the measured system response in the presence of the noise yields

$$Y_{ji}(t) = f_i(t) * V_{ji}(t) = g_{ji}(t) + f_i(t) * nc_j(t) + f_i(t) * na_j(t), \tag{9.2}$$

where $f_i(t)$ is the inverse filter, defined as

$$f_i(t) * [\Delta x_s \Delta x_r I_i(t)] = \delta(t). \tag{9.3}$$

As noted above, the result of this deconvolution is shown in Figure 9.4(b), where the benefit of deconvolution gain is clear: before the deconvolution the response to the PRBS source is imperceptible on the far traces to the right of Figure 9.4(a); after deconvolution the earth impulse response can be seen on all traces. On the short offset traces the signal-to-noise ratio is good and the greater part of the earth impulse response is clearly only a few seconds in duration.

We now consider estimating the noise on a near-offset trace in which the signal-to-noise ratio is good. That is,

$$g_{ji}(t) \gg f_i(t) * nc_j(t) + f_i(t) * na_j(t). \tag{9.4}$$

Normally the nearest trace would be used, but any trace can be used, provided inequality 9.4 is satisfied. Let this be the mth trace. We want to separate the signal from the noise on this trace in which the signal is of short duration. We note that the analytic impulse response of a half-space, equation 8.6, is of the form $A\exp(B/t)t^C$. It decays to zero at late times. We assume the true curve has a similar form and seek to find a curve of this form in which A, B and C are chosen to minimise the misfit between the curve and the measurement over a time window in which inequality 9.4 applies. The curve is calculated for a window beginning at the start of the response and finishing at the end of the trace. The resulting curve is

$$\bar{g}_{mi}(t) = A\exp(B/t)t^C. \tag{9.5}$$

This curve is subtracted from the mth trace to provide an estimate of the noise on the mth trace:

$$\bar{n}_{mi}(t) = Y_{mi}(t) - \bar{g}_{mi}(t). \tag{9.6}$$

That is, $\bar{n}_{mi}(t) \approxeq f_i(t) * nc_m(t) + f_i(t) * na_m(t)$.

Now, a large part of the noise of each trace is correlated with the noise on other traces, as we have observed. We can find a Wiener filter that converts the estimated noise on the mth trace, $\bar{n}_{mi}(t)$, to the estimated noise on the jth trace. To do this we make $\bar{n}_{mi}(t)$ the input to the filter design and $Y_{ji}(t)$ the desired output. The resulting filter, which we can call $f_{mj}(t)$, will give an estimate of the correlated noise on the jth trace, when convolved with $\bar{n}_{mi}(t)$:

$$\bar{n}_{ji}(t) = f_{mj}(t) * \bar{n}_{mi}(t). \tag{9.7}$$

Finally, this estimated noise is subtracted from the jth trace to obtain an estimate of the jth impulse response:

$$\bar{g}_{ji}(t) = Y_{ji}(t) - \bar{n}_{ji}(t). \tag{9.8}$$

This is repeated for all traces and the result is shown in Figure 9.4(c). On the large-offset traces this increases the signal-to-noise ratio by about 20 dB (Ziolkowski et al., 2010).

9.4 Time-Lapse Marine OBC Data Repeatability

Seismic surveys over producing oil fields are often repeated to identify remaining reserves of hydrocarbons that were not swept by the production process. CSEM surveys may also be used. A critical factor in the use of such time-lapse surveys is the repeatability of the data. If the source and receiver are in the same positions for two surveys, the earth responses should be identical if the earth has not changed. It is often difficult to put the source and receiver in exactly the same positions in a repeat survey; furthermore, the noise in the data is different. These two factors ensure that the measured responses in two surveys over the same oil field are not exactly the same. It is important to mitigate the effects of both correlated noise and positioning errors.

Ziolkowski et al. (2010) describe repeat CSEM surveys over the North Sea Harding field in 2007 and 2008 in which these problems were encountered. The problem of correlated noise is illustrated in Figure 9.5. Figure 9.5(a) shows recovered impulse responses for a nominal offset of 6000 m. There is a difference between the responses for 2007 and 2008 that is especially noticeable at times later than about 1 s. The difference is even greater for the step responses, shown in Figure 9.5(b). When the correlated noise is removed, most of these differences are eliminated, as shown in Figure 9.5(c) and (d).

The issue of positioning errors is illustrated in Figure 9.6. Figure 9.6(a) shows recovered impulse responses for a nominal offset of 2000 m. The 2007 offset was 2012 m and the 2008 offset was 1997 m. The amplitude varies with offset r approximately as r^{-5}, as given by equation 8.10. The decrease in offset divided by the offset $(-\Delta r/r)$ is -0.0075. The corresponding increase in amplitude is five times this; that is, 0.0375, or 3.75%. This small increase can be seen in Figure 9.6(a). Reducing the amplitude of the 2008 response by this amount compensates for the slightly

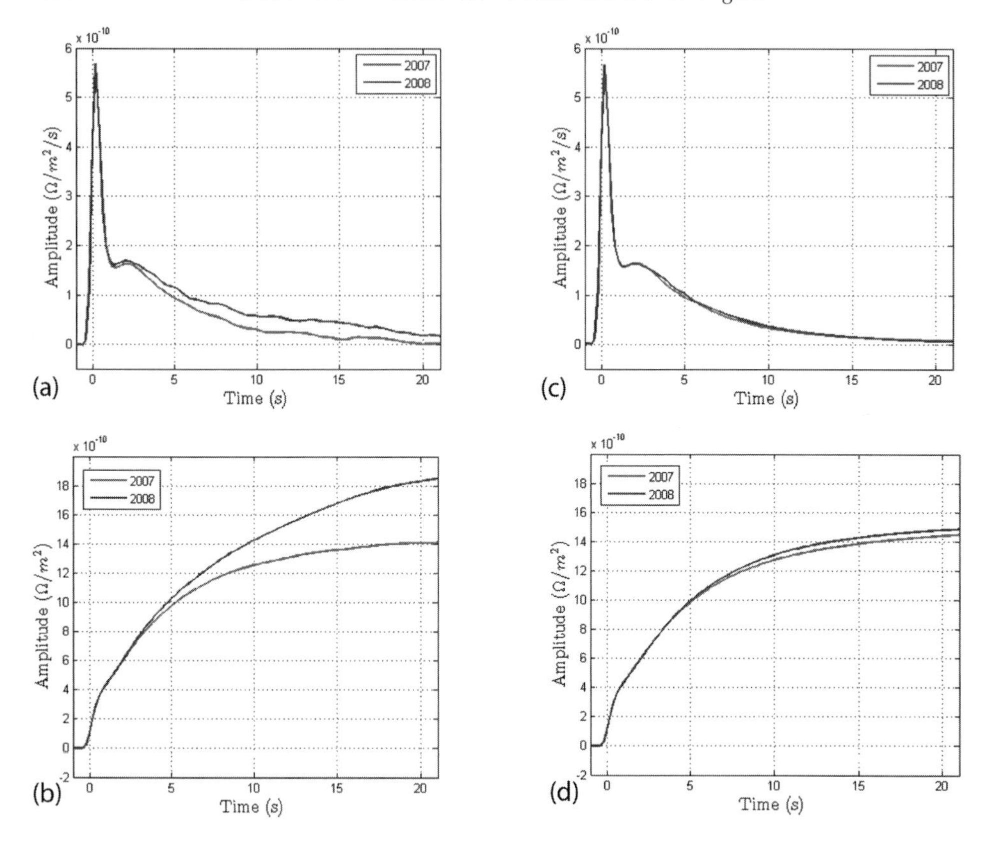

Figure 9.5 Example of correlated noise removal for 6000 m offset for 2007 and 2008 surveys. (a) corresponding impulse responses before noise removal; (b) step response integrals of responses in (a); (c) same as (a) but correlated noise has been removed; (d) step response integrals of responses in (c) (Redrawn from Ziolkowski et al., 2010).

smaller offset and gives the result shown in Figure 9.6(b). The step responses, obtained by integrating the impulse responses, are shown in Figure 9.6(c) and (d)

The effect of these mitigating effects can be measured using the normalised root mean square difference (NRMSD) between the 2007 data a_t and the 2008 data b_t, using the definition of Kragh and Christie (2002):

$$\text{NRMSD} = \frac{200\text{RMS}(a_t - b_t)}{\text{RMS}(a_t) + \text{RMS}(b_t)}, \tag{9.9}$$

where

$$\text{RMS}(x_t) = \left(\frac{1}{N} \sum_{t=1}^{N} x_t^2 \right)^{\frac{1}{2}}. \tag{9.10}$$

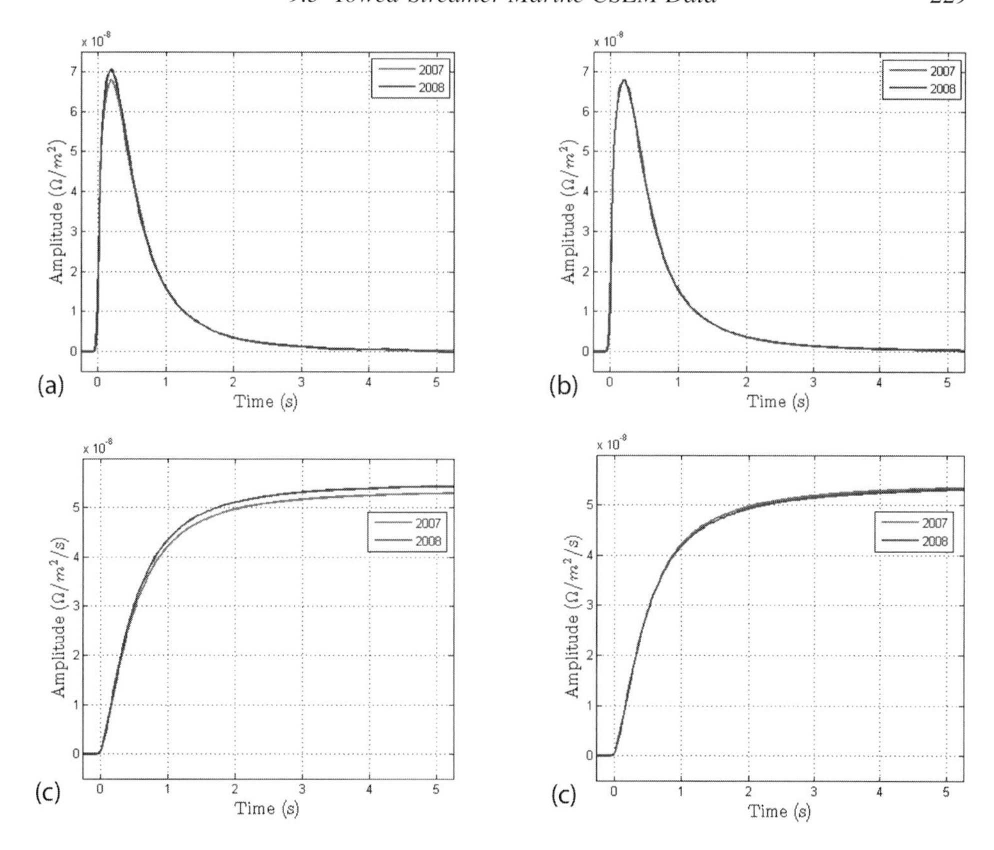

Figure 9.6 Example of positioning error removal for nominal offset of 2000 m. (a) 2007 impulse response at offset of 2012 m and 2008 impulse response at offset of 1997 m; (b) same as (a) but after offset correction for 2008 response; (c) step response integrals of corresponding responses in (a); (d) step response integrals of corresponding responses in (b). (Redrawn from Ziolkowski et al., 2010).

The big improvement is with the correlated noise removal. Before correlated noise removal, NRMSD was 11.9%; after correlated noise removal it was 3.9%. When Ziolkowski et al. (2010) wrote their paper, this was considerably better than the best seismic NRMSD (7%) they were able to find in the literature. CSEM data can be more repeatable than seismic data. A key reason for this is the excellent repeatability of the source time function.

9.5 Towed-Streamer Marine CSEM Data

In 2004 Petroleum Geo-Services (PGS) decided to engineer a towed marine CSEM system that would be able to collect data much faster than the deep water node-based system (Engelmark et al., 2014). The system was built and is being used

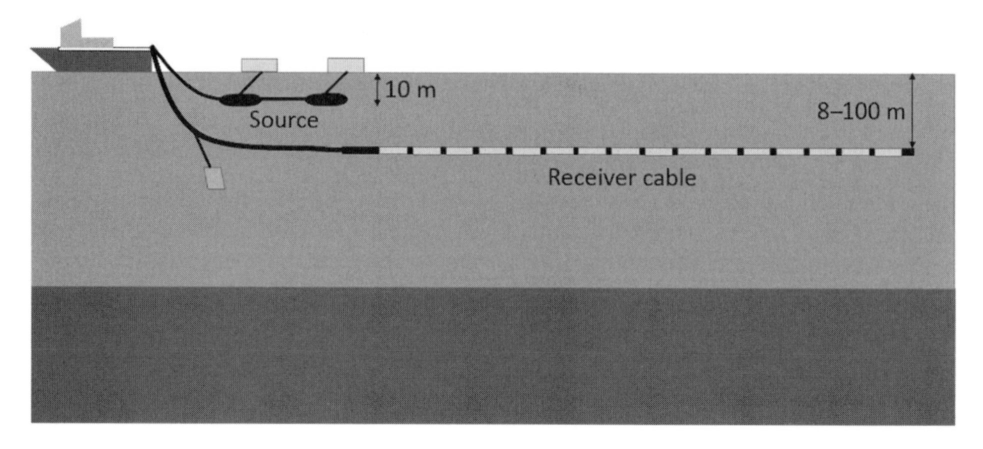

Figure 9.7 The in-line towing configuration of source and receivers along the survey line (redrawn from Mattsson et al., 2012).

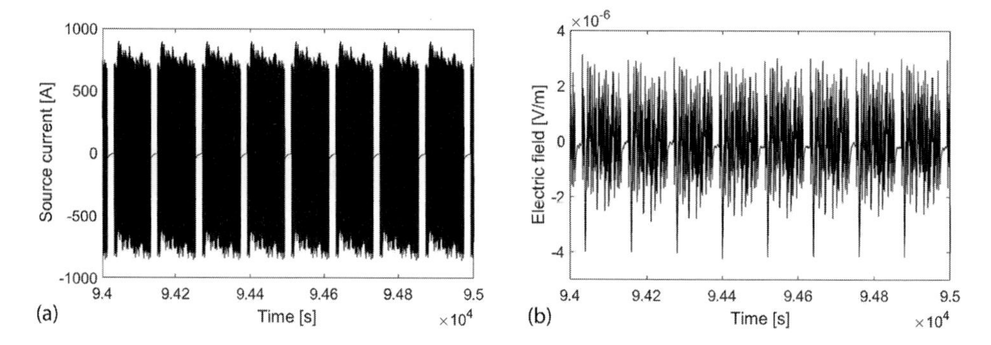

Figure 9.8 1000 s of data from the Peon test: (a) source current; (b) electric field at receiver (redrawn from Ziolkowski et al., 2011).

routinely to collect CSEM data using a streamer with over 70 channels. In an early version of the system, data were obtained in a test over the shallow Peon gas field using square wave and PRBS source time functions (Ziolkowski et al., 2011). The published results obtained with the PRBS may be used to illustrate the benefits of deconvolution gain and the potential for correlated noise removal.

The setup is shown in Figure 9.7. For this test the source–receiver offset was 2025 m. Figure 9.8 shows 1000 s of data along the survey line. Figure 9.8(a) shows the current measured at the source; Figure 9.8(b) shows the electric field measured at the receiver in the same time window. The source time function was a PRBS of order 10 with 1023 samples at a sample interval of 0.1 s. The data are periodic, with a period of 120 s, giving 17.8 s 'listening time' between cycles.

Deconvolution of a single cycle is shown in Figure 9.9. Figure 9.9(a) shows the measured dipole current, which is a PRBS that has been filtered by the recording

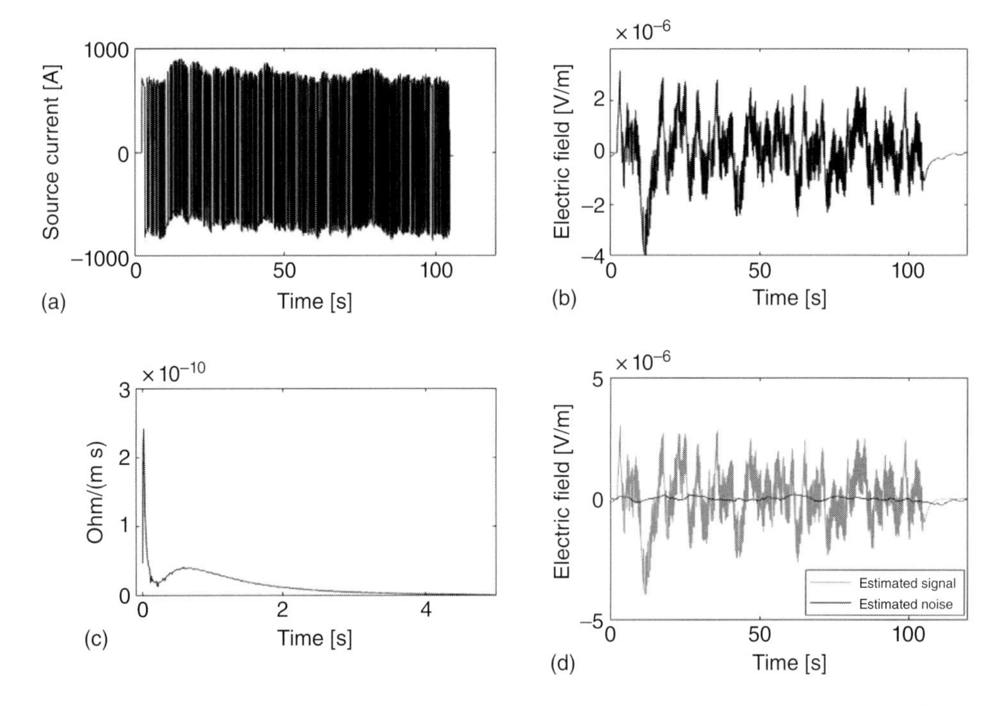

Figure 9.9 Deconvolution of one trace: (a) measured source current; (b) measured electric field; (c) recovered impulse response; (d) estimated received signal and noise (redrawn from Ziolkowski et al., 2011).

system. Figure 9.9(b) shows the measured field response. Figure 9.9(c) shows the result of deconvolving the response (b) for the input signal (a) using least-squares in the time domain. That is, the response (c) is a Wiener filter. The sharp peak at $t = 0$ is the air wave. Note that this impulse response is only a few seconds in duration and considerably less than the 17.8 s listening time. Each 120 s receiver response is therefore a complete convolution. As discussed in Chapter 6, the deconvolution gain is $\sqrt{N}$, in which $N = 1023$ for this order 10 PRBS; that is 31.98, or 30 dB. As discussed by Ziolkowski et al. (2011), this analysis can be taken a little further. The convolution of the recovered impulse response (c) with the input current (a) gives an estimate of the signal in (b) which is correlated with the input (a); this estimated signal is the grey curve in Figure 9.9(d). The measured response (b) minus the estimated signal is the estimated noise, shown as the black signal in (d).

Deconvolution of all the traces is shown in Figure 9.10. The sharp peak at $t = 0$ on each trace is the air wave. The time between traces is 120 s; at a vessel speed of 4.5 knots or about 2.25 m/s the distance between traces is about 280 m and the section is about 13 km long.

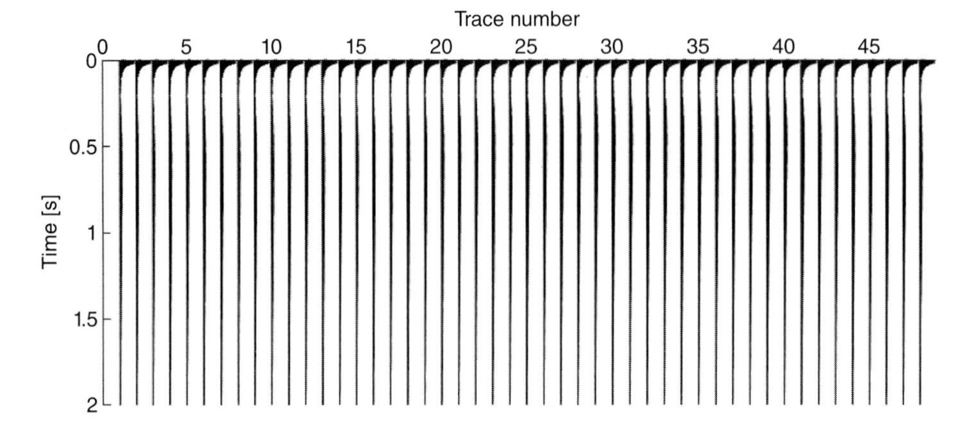

Figure 9.10 Deconvolution of all traces, showing first 2 s of data (redrawn from Ziolkowski et al., 2011).

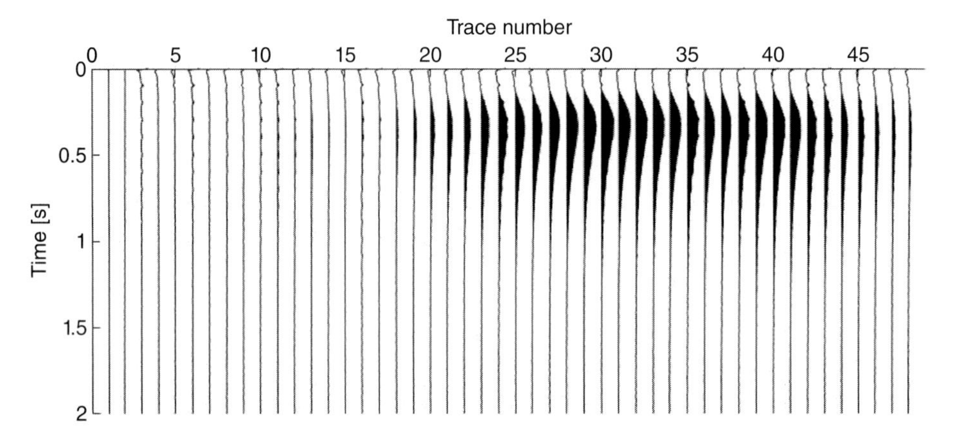

Figure 9.11 Peon anomaly: obtained from Figure 9.10 by subtracting trace 1 from all the traces (redrawn from Ziolkowski et al., 2011).

Figure 1.7 in Chapter 1, shows the effect of a subsurface resistive layer on the dipole–dipole response: the amplitude of the response increases. To see this effect in these data, we subtract the first trace from all the traces and obtain the anomaly due to the shallow resistive gas of the Peon gas field. This is shown in Figure 9.11, where the anomalously high amplitudes are observed on traces after trace 17.

The frequency bandwidth of the data deserves consideration. The source switch rate was 10 Hz, so the highest frequency in the signal is the Nyquist frequency, 5 Hz. Figure 9.12 shows a section of the amplitude spectra of the data of Figure 9.10 and Figure 9.13 shows a section of the amplitude spectra of the data of Figure 9.11. It appears from Figure 9.12 that there is no useful energy above about 1 Hz.

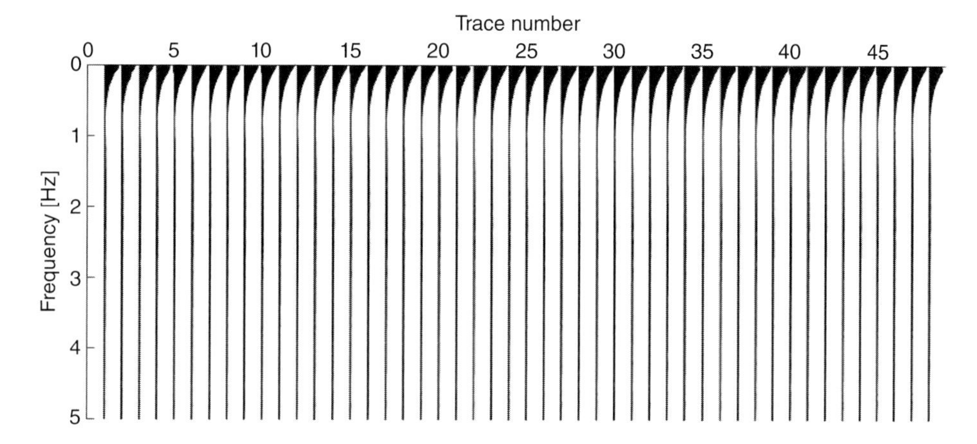

Figure 9.12 Amplitude spectra of traces in Figure 9.12 (redrawn from Ziolkowski et al., 2011).

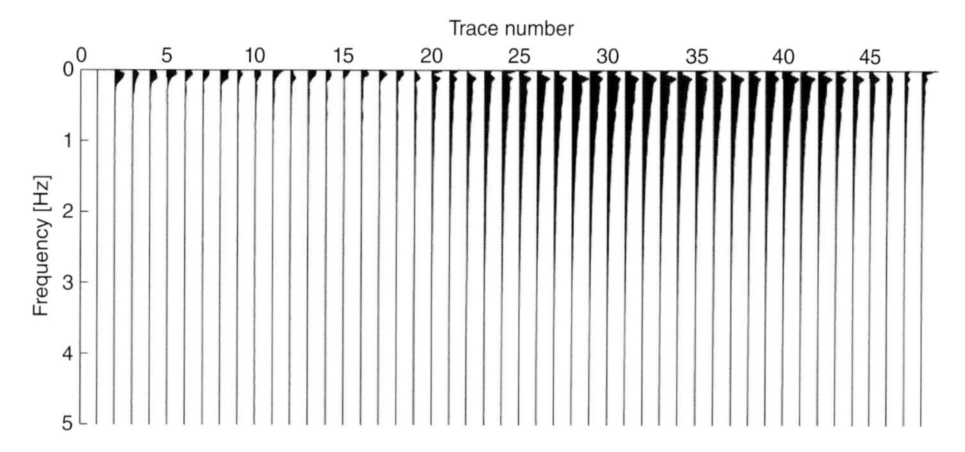

Figure 9.13 Amplitude spectra of traces in Figure 9.13 (redrawn from Ziolkowski et al., 2011).

The spectrum of the anomaly, Figure 9.13, shows that the anomaly has a broader spectrum – possibly up to 3 Hz. Based on Figure 9.12, one might conclude that the 10 Hz switch rate was too high and that the signal-to-noise ratio could be enhanced by reducing the switch rate by a factor of 2, even though that would require the order of the PRBS to be reduced from 10 to 9 to preserve the time window. Looking at the spectrum of the anomaly, it is less obvious that this is the case. Nevertheless, it is such considerations that must be made to get the best-quality data.

Towed streamer CSEM is a fast method to acquire CSEM data in shallow water. For deep targets the main limitation is the short time window available to obtain data with adequate signal-to-noise ratio, especially as towed streamer data are noisier than OBC data. There are two main sources of additional noise. First, the flow of water around the electrodes can increase the noise above the static noise level, as observed in a tank experiment by Djanni et al. (2016). Second, there is noise induced by wave-motion flexure of the cables in the Earth's magnetic field. Fortunately this motionally induced noise is not as large as was originally feared and is reduced when the tension in the streamer is increased, as noted by Engelmark et al. (2014) and Djanni et al. (2016).

Finally, as Engelmark et al. (2014) describe, towed streamer CSEM may be combined with seismic data acquisition on the same vessel. This is a major step in the direction of improved efficiency of marine geophysical data acquisition.

10

Forward and Inverse Modelling of CSEM Data

This chapter briefly discusses 3D modelling methods that are being used today to compute electric and magnetic fields at receiver locations. Forward modelling is the term used for such computations when all model parameters are known and the goal is to compute data given these model parameters. Inverse modelling is the term used for the same computations when the model parameters are not known. The goal is to compute data from an estimated model such that the computed data match the measured data within limits defined by the user. These model estimates are updated by finding adjustments to the conductivity in a numerical 3D model based on the misfit between the modelled and measured data, possibly augmented with additional conditions. Adjustments are usually made in small steps and this leads to iterative forward modelling as the strategy for inverse modelling. Forward modelling can be performed using exact expressions for the electric and magnetic fields in the space–frequency domain and the space–time domain for the whole space and half-space configurations, as described in detail in Chapter 4. For a horizontally layered model with general anisotropy, exact expressions can be found in the horizontal wavenumber–frequency domain and numerical transformations can be used to compute space–frequency domain and space–time domain results. The specific case of a vertically transverse isotropic model is described in Chapter 4, where it is explained how to do the transformation from the horizontal wavenumber domain to the space domain by the so-called fast Hankel transform filter method and to transform from the frequency domain to the time domain by the so-called cosine- or sine-transform filter method. Because these transformations are integral transformations, they can be performed to a high degree of accuracy and we call them semi-exact methods.

For more general models, where the conductivity can be an anisotropic function of position in 3D space, numerical methods have to be used. To adjust model parameters in inversion, a scalar objective function must be formulated and one seeks to minimise this function. When it comes to numerical modelling, the questions that must be answered are whether the solution exists, whether the solution is unique and whether the solution changes in a continuous manner with changing initial conditions. Answering yes to all three questions means the problem is well-posed according to the definition given by Hadamard. If any of these questions cannot be answered positively, the problem is called ill-posed in the Hadamard sense. Forward problems are well-posed, but may still suffer from numerical instabilities. Proper discretisation and proper formulation of the problem together with stable algorithms ensures an accurate solution can be reached numerically. Inverse modelling leads to ill-posed problems because the amount of data is insufficient, the data have finite precision and the parameters to be found are highly sensitive to small changes in the data. This problem is usually dealt with using additional assumptions on smoothness of the parameter distribution in space and/or restrictions on deviation from a starting model (Tikhonov and Arsenin, 1977). This does not remove the non-uniqueness problem and it is important to find a good starting model.

Over recent years the number of codes available to the electromagnetic community has grown and we now have several free modelling software packages available. This will expedite the advances made in computational electromagnetic methods. For 1D isotropic models, forward and inverse modelling code can be found at the Scripps Institution of Oceanography (http://marineemlab.ucsd.edu/resources.html; Key, 2009) and this forward modelling code is used with a Bayesian inversion shell at the Commonwealth Scientific and Industrial Research Organisation (CSIRO) (Gunning et al., 2010). For 1D vertical transverse isotropy (VTI) models, forward and inverse modelling code is available at the *Geophysics* source-code archive of SEG, (http://software.seg.org; Hunziker et al., 2015, 2016). At the same website, new Python code is available for 1D VTI models and several options for computing space–frequency and space–time domain results (Werthmüller, 2017).

For 2D modelling and inversion, the MARE2DEM code is available from the website of Kerry Key. Because his working address may change, we recommend searching for his name on the web (Key and Ovall, 2011; Key, 2012b).

For 3D modelling based on a fast and accurate integral equation solver, the free code can be found at GIEM2G (https://github.com/DarthLaran/GIEM2G; Kruglyakov and Bloshanskaya, 2017).

One of the important aspects is mesh generation, and a free software package is available as TetGen (Si, 2015). This program generates tetrahedral meshes of any set of 3D polyhedral domains suitable for numerical modelling independent of the method that is used.

10.1 Forward Modelling

In the diffusive approximation it is possible to express the electric and magnetic fields as the impulse response of simple models in terms of ordinary and special functions. For a whole space and a half-space this can be done in the space–frequency domain and in the space–time domain, as we show in Chapter 4 and Appendix A. When the model contains more than one conductive layer, it is not possible to describe the electric and magnetic fields in the space–frequency domain and in the space–time domain using ordinary and special functions, but it is possible to express the fields as integrals of ordinary and special functions provided the boundaries are flat and parallel surfaces of infinite extent, as we show in Chapter 4 and Appendix B. For more complicated models we must find numerical solutions.

For general three-dimensional models we need to find approximate solutions to Maxwell's equations. This requires discretisation of the space–time domain, or of the space domain in frequency domain modelling. We need to do this in such a way that the discrete equations produce solutions as close as possible to solutions of the true Maxwell equations. In forward modelling the first choice to be made is whether to use the coupled first-order Maxwell equations or the second-order equation for either the electric or magnetic field. The second choice is whether the full solution is going to be computed numerically or whether the primary field is obtained analytically and the secondary field numerically. The third choice is how to discretise the model and set up a linear matrix-vector equation that can be solved efficiently with a computer. This requires choices to be made on the boundary conditions at the outside of the computational domain. For large-scale problems the choices of computer language, algorithm and data structures, and possibly the distribution of the computational load, are important. The last question is how to verify or validate the numerical results that are obtained.

Layered-earth models have a finite number of horizontal interfaces and are bounded by half-spaces at the top and at the bottom. Each layer is unbounded in the horizontal directions. The fields decay towards infinity in all spatial directions and this can be seen as the radiation conditions. These conditions are physical and are incorporated in the physical solutions we want to find. For a general three-dimensional model we want to model a finite 3D region and assume that the world outside this region is non-reflecting or that it does not influence our modelling results. This can be done by enforcing radiation conditions at the outside of the computational domain. This method is used in integral equation modelling, where the system matrix is already full. It can also be done by putting the outside boundary of the computational domain sufficiently far away so that we can enforce non-physical boundary conditions. This is used with finite difference, finite element and finite integration modelling techniques, where the system matrix is large but sparse. The second problem is the discretisation problem. In forward modelling we know the distribution of conductivity values in 3D space and can accurately follow the boundaries between regions where these values differ. In finite element, finite volume and integral equation techniques these can be used in the generation of the discrete system of equations. If we now assume that the boundaries coincide with the changes in the conductivity values, the choice must be made where the unknown discrete field values are positioned in the grid. Because the electric and magnetic fields have continuous tangential components across a boundary, it makes sense to locate them on edges connecting the grid points in the mesh and to orient them along those edges. In finite difference methods a rectangular mesh is usually used and the three vector components are located on nodal points in the grid. This reduces the ability to follow boundaries in the model and thereby to satisfy boundary conditions. Effective medium theories have been developed to circumvent this problem.

Finite integration technique. Let us begin by transforming equations 4.3 and 4.4 without a magnetic source to the frequency domain to give

$$\nabla \times \hat{\mathbf{E}} - \mathrm{i}\omega\hat{\mathbf{B}} = 0, \tag{10.1}$$

$$\nabla \times \hat{\mathbf{H}} - \hat{\mathbf{J}} = \hat{\mathbf{J}}^e, \tag{10.2}$$

and assume we have defined a grid with connecting cells. A 3D rectangular cell has six rectangular faces, each with a unique constant unit normal vector $\mathbf{n}$. We perform scalar multiplication by $\mathbf{n}$ on equations 10.1 and 10.2, and integrate over one face to obtain

$$\oint_{\mathcal{C}} \tau \cdot \hat{\mathbf{E}} dl - \mathrm{i}\omega \int_{\mathcal{S}} \mathbf{n} \cdot \hat{\mathbf{B}} dA = 0, \tag{10.3}$$

$$\oint_{\mathcal{C}} \tau \cdot \hat{\mathbf{H}} dl - \int_{\mathcal{S}} \mathbf{n} \cdot \hat{\mathbf{J}} dA = \int_{\mathcal{S}} \mathbf{n} \cdot \hat{\mathbf{J}}^e dA, \tag{10.4}$$

where τ denotes the unit tangent vector along the edges, $\mathcal{C}$, of the face $\mathcal{S}$. We can then connect all faces in a cell and connect all cells to build 3D space. Equations 10.3 and 10.4 are exact representations of Maxwell's equations. The next steps involve approximations. For second-order accuracy it is customary to define the electric and magnetic fields in the line integrals as edge averages on each line segment of each face. Second, we need to connect the fields in the surface integrals to the electric and magnetic fields using equations 4.9 and 4.10. We then need face averages of the electric and magnetic fields. To be consistent within second-order accuracy, dual cells are introduced whose edges coincide with the centre of the faces of the original cells. Third, we want to solve the equations in 3D space and must scale to the volumes within the defined cells. Because we have assumed the x-component of the electric field to be an edge average along an edge in the x-direction, we assume this value can be used along the whole edge and in y- and z-directions up to half the cells adjacent to the edge where it is defined. Let the edge under consideration connect the two nodes given by (x_k, y_l, z_m) and (x_{k+1}, y_l, z_m). The corresponding edge averaged value for $E_{x;klm}$ is then assumed constant in the rectangle defined by $x_{k+1} < x < x_k, y_{l+1/2} < y < y_{l-1/2}, z_{m+1/2} < z < z_{m-1/2})$. This can be viewed as a finite-volume generalisation of Yee's staggered grid scheme (Yee, 1966). The computational domain is truncated by assuming the electric or magnetic field is zero at the edges of the outer faces of the cells on the boundary. This is usually implemented with stretched grids to move this boundary far away from the region of interest. Based on this discretisation procedure, a matrix vector problem is obtained that can be written as

$$Lu = f, \tag{10.5}$$

where L is the system matrix of known coefficients, u is the vector containing all unknown discrete edge averages of the electric and/or magnetic fields and f is the vector containing all discrete source elements. Details of this technique can be found in Clemens and Weiland (2001) and in combination with a multigrid solver in Mulder (2006). The finite integration technique is not restricted to the use of rectangular grid cells and can be used on unstructured grids with tetrahedral cells.

Finite element methods are very similar to the finite integration technique but are based on variational principles and not further discussed with regard to reducing Maxwell's equations to a discrete matrix-vector equation. Direct and indirect solvers have been developed recently for single computer and parallel implementations (Schwarzbach et al., 2011; da Silva et al., 2012; Cai et al., 2017; Li and Han, 2017). Transient data can be obtained by computing models at a sufficient number of frequencies using the fast sine transform, as we discuss in Chapter 4, by optimised irregular logarithmic frequency axis sampling with interpolation and fast Fourier transform (FFT) (Mulder et al., 2008), or using logarithmic FFT (Hamilton, 2000). Note that the method of optimised irregular sampling of the logarithmic frequency axis with interpolation can also be used to minimise the number of frequencies for which full models have to be computed for the fast sine transform and logarithmic FFT methods.

Finite difference methods. Finite difference was, and possibly is, the first method of choice for computing 3D transient CSEM data for land models. Druskin and Knizhnerman (1994) published the spectral Lancsoz decomposition method, which has the feature that the computation of the transient solution is hardly more expensive than computation for a single frequency. The authors began the development of this method in the second half of the 1980s. To reduce the effect of the regular stretched grid, Davydycheva et al. (2003) improved the method by conductivity averaging and spectral optimal grid refinement. It is implemented on GPUs (graphics processing units) by Sommer et al. (2013). An alternative method was given by Maaø (2007), who modified the basic equations and introduced an artificial wave propagation velocity, thereby reducing the number of time steps necessary for a stable and accurate solution. As a consequence a correction has to be computed afterwards. Yet another approach was given by Mittet (2010), who used the diffusive-to-wave-field transform to convert the diffusive Maxwell equations to a lossless wave equation that can be solved with standard finite-difference time-domain (FDTD) algorithms used for lossless wave propagation. The solution must then be integrated using the numerically stable wave to diffusive field transformation. Parallel implementations of frequency domain finite difference can be found in Alumbaugh et al. (1996) and Streich (2009).

Integral equation methods. Integral equation methods are constructed by rewriting Maxwell's equations in integral form with matrix Green's functions after creating so-called secondary or scattering sources. Because Maxwell's equations are linear in the source, we can write the electric and magnetic fields as sums of background and scattered fields, e.g. the electric field can be expressed as $\hat{\mathbf{E}} = \hat{\mathbf{E}}_b + \hat{\mathbf{E}}_s$, $\hat{\mathbf{E}}_b$ denoting the background field and $\hat{\mathbf{E}}_s$ the scattered field. The basic equations take the form

$$\nabla \times \hat{\mathbf{E}}_b - \mathrm{i}\omega\mu_b\hat{\mathbf{H}}_b = 0, \tag{10.6}$$

$$\nabla \times \hat{\mathbf{H}}_b - \sigma_b\hat{\mathbf{E}}_b = \hat{\mathbf{J}}^e, \tag{10.7}$$

$$\nabla \times \hat{\mathbf{E}}_s - \mathrm{i}\omega\mu_b\hat{\mathbf{H}}_s = -\mathrm{i}\omega(\mu_b - \mu)\hat{\mathbf{H}}, \tag{10.8}$$

$$\nabla \times \hat{\mathbf{H}}_s - \sigma_b\hat{\mathbf{E}}_s = (\sigma - \sigma_b)\hat{\mathbf{E}}. \tag{10.9}$$

By adding equations 10.6 and 10.8, we find equation 10.1 and likewise adding equations 10.7 and 10.9 yields equation 10.2. The background model distribution of the conductivity must be taken such that the electric and magnetic fields can be computed quickly and accurately everywhere in space. A horizontally layered model such as we discussed in Chapter 4 is a suitable background model. The scattered field equations contain right-hand sides that are known as scattering sources. These depend on the difference in permeability and conductivity between the background and real models and on the total fields in the domain where the background and real models have non-zero differences in their parameters. Equations (10.8) and

10.9 can be rewritten as a pair of coupled integral equations. For simplicity we assume that $\mu = \mu_b$ everywhere and we find

$$\hat{\mathbf{E}}(\mathbf{r},\omega) - \int_{\mathbb{D}_s} \hat{\mathbf{G}}_b(\mathbf{r},\mathbf{r}',\omega) \cdot [\sigma(\mathbf{r}') - \sigma(\mathbf{r}')]\hat{\mathbf{E}}(\mathbf{r}',\omega)d\mathbf{r}' = \hat{\mathbf{E}}_b(\mathbf{r},\omega). \qquad (10.10)$$

In this equation the scattering domain $\mathbb{D}_s$ is the volume in which the conductivity in the background and real models differ. The objective is to make this volume as small as possible. The reason is that to find the electric field at the receivers from equation 10.10, first the total electric field inside the scattering domain must be determined. For all points $\mathbf{r} \in \mathbb{D}_s$ equation 10.10 constitutes a Fredholm integral equation of the second kind. The two biggest problems with this formulation are the fact that the resulting system matrix is small but completely filled and that finding accurate values for the matrix elements is not a trivial task because the matrix Green's function is singular when $\mathbf{r} = \mathbf{r}'$. In a horizontally layered background model the matrix Green's function is a spatial convolutional operator in horizontal space which means the system matrix exhibits a block Toepplitz structure that can be exploited in computing solutions with either iterative schemes or direct solution methods. This structure also helps to reduce the number of elements that must be stored in memory or on disc. The second helpful property is reciprocity, stated as $\hat{\mathbf{G}}_b(\mathbf{r}',\mathbf{r},\omega) = \left(\hat{\mathbf{G}}_b(\mathbf{r},\mathbf{r}',\omega)\right)^t$, which can be retained for the corresponding elements in the system matrix. This method was implemented in the last two decades of the previous century and an important development has been the formulation of the contracting kernel technique that ensures the solution of the resulting integral equation has a convergent Neumann series expansion (Pankratov et al., 1995; Singer, 1995). In this century the method is still of interest (Zhdanov et al., 2006), although recently a shift to hybrid methods combining finite difference and integral equations has been reported (Yoon et al., 2016). Interest may be revived with the recent implementation exploiting symmetry properties and the use of parallel computing (Kruglyakov and Bloshanskaya, 2017).

A final note is that equations 10.8 and 10.9 are also used with finite integration, finite element and finite difference techniques in the so-called primary–secondary formulation, e.g. Streich (2009).

10.2 Inverse Modelling

Finding medium parameters from measured data can be done in various ways. Usually the amount of data from which this has to be done is insufficient to use data-driven methods. For this reason most work is carried out on model-driven methods. This leads to the formulation of finding the medium parameters based on simulating the data from a model and trying to fit the measured data as closely as possible. The misfit between the simulated and measured data has to be minimised in some sense. This is inverse modelling by iterative forward modelling. The objective is to find the model that results in simulated data that fit the measured data the best in the least number of iterations. The first task is then to define an objective function to be minimised. Once the objective function is defined, the method of solution to the minimisation problem has to be chosen. Global methods exist that search the entire solution space and avoid getting trapped in a local minimum. This can be a solution to the non-uniqueness problem, but assumes that the global minimum defines the desired solution. For 2D and 3D problems the solution space is far too big to be searched exhaustively. Geophysicists tend to regard feasible methods as gradient-based methods that are usually used with a regularised objective function (Tikhonov and Arsenin, 1977; Tarantola, 1987; Parker, 1994) given by

$$2F(\mathbf{m}) = \left([\mathbf{d} - \mathbf{g}(\mathbf{m})]^t \mathbf{C}_d^{-1}[\mathbf{d}^* - \mathbf{g}^*(\mathbf{m})] + \lambda[\mathbf{m} - \mathbf{m}_0]^t \mathbf{C}_m^{-1}[\mathbf{m} - \mathbf{m}_0]\right). \quad (10.11)$$

assuming the parameters are real numbers and * denotes complex conjugate. The simulated data are given by $\mathbf{d}_{\text{sym}} = \mathbf{g}(\mathbf{m})$ in terms of a non-linear operator function $\mathbf{g}$ that converts the model vector $\mathbf{m}$ to simulated data, which can be achieved using finite integration, finite element, finite difference and integral equation methods. The measured data are denoted $\mathbf{d}$ and the initial or prior model is represented by $\mathbf{m}_0$. The matrices $\mathbf{C}$ are covariance matrices: $\mathbf{C}_d$ represents the uncertainties in the data and $\mathbf{C}_m$ represents uncertainties in the model. λ is a regularisation parameter that can be used to adjust the contribution of the deviation of the obtained model from the initial model. The objective function $F(\mathbf{m})$ has a minimum where its gradient to the model is zero, which can be expressed as

$$\boldsymbol{\gamma} = \partial_m F(\mathbf{m}) = \left(\partial_{m_1} F, \partial_{m_2} F, \ldots, \partial_{m_M} F\right)^t = \mathbf{0} \quad (10.12)$$

for a parameter vector of length M. For the given objective function the gradient can be written as

$$\boldsymbol{\gamma} = \Re\{\mathbf{J}^t \mathbf{C}_d^{-1}(\mathbf{g}^*(\mathbf{m}) - \mathbf{d}^*)\} + \lambda \mathbf{C}_m^{-1}(\mathbf{m} - \mathbf{m}_0), \quad (10.13)$$

where $*$ $\Re$ denotes the real part. The sensitivity matrix or Jacobian is given by

$$\mathbf{J} = \begin{pmatrix} \partial_{m_1} g_1 & \cdots & \partial_{m_M} g_1 \\ \cdots & \cdots & \cdots \\ \partial_{m_1} g_N & \cdots & \partial_{m_M} g_N \end{pmatrix}, \quad (10.14)$$

for a data vector of length N. The technique to minimise equation 10.11 is Newton's method, which can be written as

$$\mathbf{m}_{n+1} = \mathbf{m}_n - \mu_n \mathbf{H}_n^{-1} \boldsymbol{\gamma}_n, \quad (10.15)$$

given the model vector $\mathbf{m}_n$ after the nth iteration and μ_n is a parameter determining the step length and can be found by a line search in the gradient direction. In equation 10.15 $\mathbf{H}$ is the Hessian matrix of second-order derivatives that can be written as

$$\mathbf{H} = \begin{pmatrix} \partial^2_{m_1} F & \cdots & \partial_{m_1}\partial_{m_M} F \\ \cdots & \cdots & \cdots \\ \partial_{m_M}\partial_{m_1} F & \cdots & \partial^2_{m_M} F \end{pmatrix}. \tag{10.16}$$

It can be computed from

$$\mathbf{H}_n = \Re\{\mathbf{J}_n^\dagger \mathbf{C}_d^{-1} \mathbf{J}_n + (\partial_{\mathbf{m}} \boldsymbol{\gamma}\ ) \mathbf{C}_d^{-1} [\mathbf{d} - \mathbf{g}(\mathbf{m}_n)]\} + \mathbf{C}_m^{-1}. \tag{10.17}$$

The second term inside the bracketed term in the right-hand side of equation 10.17 is usually not taken into account as it involves second derivatives. This choice leads to the Gauss–Newton method in which we can write the model update equation as

$$\mathbf{m}_{n+1} = \mathbf{m}_n - \mu_n \left(\Re\{\mathbf{J}_n^\dagger \mathbf{C}_d^{-1} \mathbf{J}_n\} + \mathbf{C}_m^{-1}\right)^{-1} \left(\Re\{\mathbf{J}_n^t \mathbf{C}_d^{-1} [\mathbf{g}^*(\mathbf{m}_n) - \mathbf{d}^*]\} \right.$$
$$\left. + \lambda \mathbf{C}_m^{-1} [\mathbf{m}_n - \mathbf{m}_0]\right). \tag{10.18}$$

Nguyen et al. (2016) show that this method yields much better results than the Broyden–Fletcher–Goldfarb–Shanno (BFGS) method, which initially behaves as a steepest descent method with the intention to arrive approximately at the Newton method near convergence (Plessix and Mulder, 2008). Key (2012b) shows very good results in a 2.5D inversion with the same method. He applied it to a slightly different objective function to be minimised, which involves controlling the model roughness and a penalty term inside the data misfit (Constable et al., 1987), and it is implemented with bounds on the model parameters.

11

Recovery of Resistivities from CSEM Data

Geophysicists have not yet found a way to determine resistivities directly from controlled-source electromagnetic (CSEM) data. This is in stark contrast to the processing of seismic data in which the idea of lining up arrivals to find velocities that fit the recovered travel-time curves is central. There is no corresponding key idea in the analysis of CSEM data. It appears that the diffusion process is inherently more difficult to understand and analyse than propagating waves. We need to use inversion to estimate subsurface resistivities. As described in Chapter 10, inversion aims to minimise the error – defined by a scalar objective function – between the real data and synthetic data generated from a model.

This chapter looks at three elements in the inversion process: the need to weight the data as a function of source–receiver offset; the reduction of resolution with depth and its effect on model parameterisation; and the use of rock physics to estimate earth conductivity models from seismic data and well logs.

11.1 Effect of Offset

The common-offset sections of Chapter 8, particularly Figures 8.12 and 8.13, show that deeper structure is seen with greater offsets. The effect is somewhat similar to seismic refraction. We also know that the amplitude of the response attenuates rapidly with source–receiver offset r. For a half-space the amplitude of the impulse response decreases as r^{-5}, as shown in equation 8.10. Both the real data and the synthetic data have smaller amplitudes at larger offsets. This emphasises the shallower structure at the expense of the deeper structure. If there are large errors in the model for deeper bodies, these are hard to see with the r^{-5} attenuation. If we want the deeper structure to be as important in the model as the shallower structure, both the real and synthetic impulse responses need to have an offset-dependent gain applied before the error is calculated. For a half-space the gain is r^5. This is an approximate guide. The response of a dipole in a non-conducting medium decays as r^{-3}. The attenuation in conducting media increases the rate of decay and stretches out the time function by dispersion. The best gain function will certainly be greater than r^3, but may not be as great as r^5; it will depend on the subsurface resistivity distribution.

If the inversion is performed in the frequency domain, the gain required is less dramatic: it is r^3, as seen in equation 8.16.

The effect of the proposed offset-scaling is illustrated in Figure 11.1 using the example of a 20 ohm-m half-space and omitting the air wave, which propagates in air as a wave with r^{-3} decay. Figure 11.1(a) shows the true amplitude time response concentrated at short times and short offsets; Figure 11.1(b) shows the amplitude spectrum of the corresponding frequency response. The effect of the r^5 gain for the time domain responses is shown in Figure 11.1(c), in which it is seen that the maximum amplitude is the same at each offset; the effect of the r^3 gain for the corresponding frequency domain responses is shown in Figure 11.1(d). The effect of attenuation is seen very clearly in 11.1(d); attenuation increases with frequency and reduces the frequency bandwidth towards the lower frequencies with offset. Figure 11.1(e) has the same r^5 offset scaling as Figure 11.1(c), but the vertical axis is now the logarithm of time and it is seen that the response on each offset is the same and is shifted vertically. To give equal weight to each offset requires not only offset scaling, but equi-spaced samples in log(time). Figure 11.1(f) shows the corresponding plot in log(frequency) with r^3 offset scaling and the frequency attenuation with offset is clear and linear in this domain.

11.2 Attenuation and Model Parameterisation

Deeper structure is seen with greater offsets. Attenuation increases with increasing offset, with the highest frequency decreasing as r^{-3}, as shown in Figure 11.1. The smallest body that can be resolved in a uniform medium is inversely proportional to the highest frequency. Deeper bodies can be resolved less well than shallower bodies. The effect depends on the resistivity distribution as well as depth, but the r^{-3} dependence is correct for uniform media. If a body of volume $\Delta V = \Delta x \Delta y \Delta z$ can just be resolved at a depth z, then a body $2^3 \Delta V = (2\Delta x)(2\Delta y)(2\Delta z)$ can just be resolved at a depth $2z$.

This simple result is the key to model parameterisation. The number of independent model parameters should decrease as the cube of the depth. If $m(z)$ is the number of model parameters at depth z, then

$$m(z) \propto \frac{1}{z^3}. \tag{11.1}$$

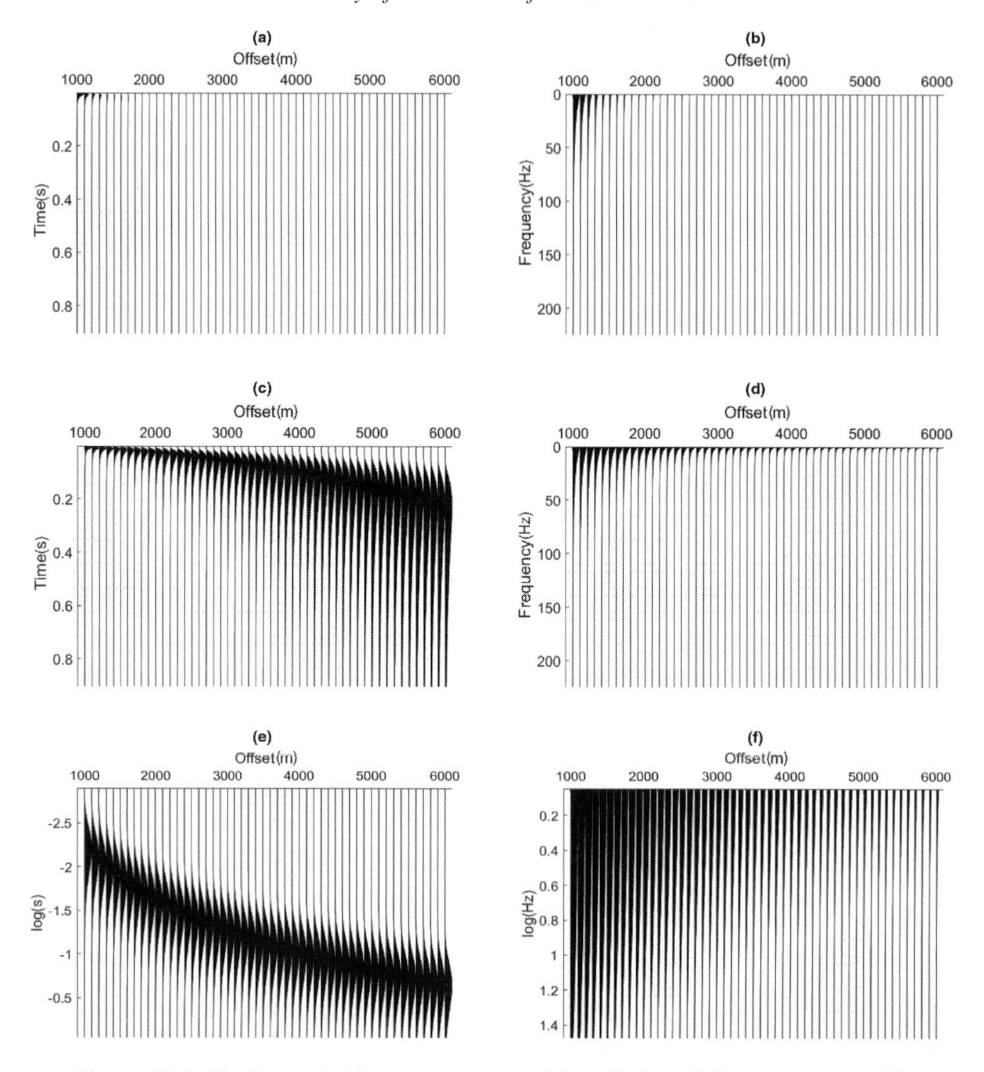

Figure 11.1 20 ohm-m half-space response: (a) variation of time response with offset; (b) variation of amplitude of frequency response with offset; (c) same as (a), but scaled as $(\text{offset})^5$; (d) same as (b), but scaled as $(\text{offset})^3$; (e) same as (c), but vertical scale is log(s); (f) same as (d), but vertical scale is log(Hz).

11.3 Resistivities from Seismic Velocities

Section 1.10 and Chapter 10 emphasise the importance of a good starting model of subsurface resistivity. It is necessary to have a method for finding the model from pre-existing geological, geophysical and well data. CSEM data are complementary to seismic data. Seismic data yield a velocity model of subsurface structure; CSEM data provide information about the fluid content of the rocks. Very often seismic

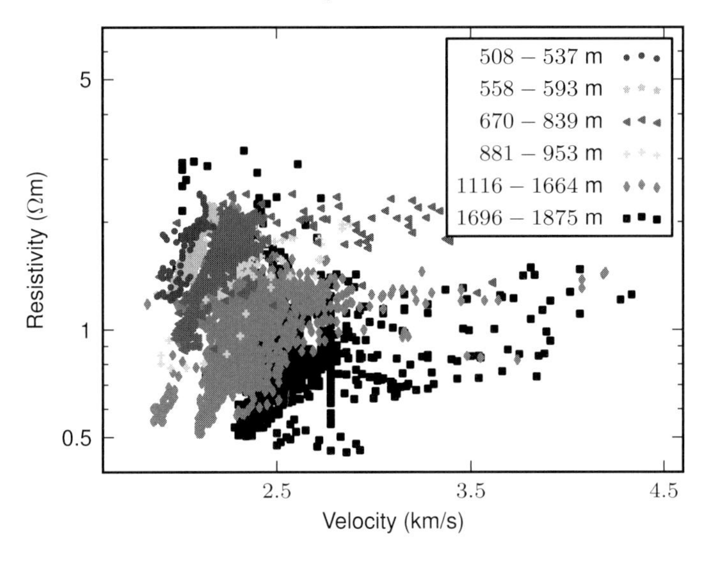

Figure 11.2 Resistivity versus velocity plot for shales in North Sea well 9/23b-8 (reproduced from Werthmüller et al., 2013). For the colour version, please refer to the plate section.

data have been obtained before CSEM surveys are considered, and they can provide constraints on the range of compatible resistivity models.

The theory of seismic wave propagation is derived from Newton's laws of mechanics and Hooke's law of linear elasticity; the theory of electromagnetic wave propagation is derived from Maxwell's equations and Ohm's law. Seismic waves share no physical medium properties with electromagnetic waves, so it is not straightforward to relate them.

The problem can be reduced to the determination of the electrical resistivities of the rocks from the corresponding seismic velocities. Rocks are mixtures of mineral grains and fluids and are often very heterogeneous. Figure 11.2 illustrates the complexity for shales in a single well; there may be a weak relationship between velocity and resistivity for shales in this well. There is a clear depth trend in these data: velocity tends to increase with depth, while resistivity tends to decrease.

Relating velocity to resistivity is usually done using theoretical or empirical relations between velocity and porosity and between porosity and resistivity. Carcione et al. (2007) present an overview of common relations. Engelmark (2010) emphasises the importance of background shale modelling for electromagnetic inversions and especially the inclusion of depth dependence of rock physics models. A velocity-resistivity transform should take the complexity into account and should work for a range of rocks in a field, and a range of depths.

Werthmüller et al. (2013) propose a method that includes these requirements and apply it to the Harding field in the North Sea. Their method may use a variety of models for the velocity–resistivity transform. It is illustrated with the Gassmann equation (Gassmann, 1951) relating velocity to porosity and the self–similar relation (Sen et al., 1981) relating porosity to resistivity. They begin with the following expression for the P-wave velocity in a rock:

$$v_P = \sqrt{\frac{K + 4G/3}{(1-\phi)\varrho_s + \phi\varrho_f}}, \tag{11.2}$$

in which K is the bulk modulus of the saturated rock, G is the shear modulus of the saturated rock, ϕ is the porosity, ϱ_s is the density of the solid matrix and ϱ_f is the density of the fluid. The Gassmann equation expresses the saturated rock moduli in terms of the unsaturated rock moduli, porosity and fluid density and compressibility. Equation 11.2 is solved for porosity ϕ in an iterative way, given v_P, K and G.

Werthmüller et al. (2013) stress that the assessment of the uncertainty of any rock physics model is more important than the choice of the model. All the models oversimplify the aggregate properties of the rocks. There are three sources of uncertainty in the velocity–resistivity transform: the uncertainty of the value of the velocity; the uncertainty of each of the parameters in the model; and the uncertainty of the model itself. These are expressed as probability density functions.

Werthmüller et al. (2013) show that velocity uncertainty can be estimated from a well log, as shown in Figure 11.3. The grey curve v in Figure 11.3(a) is the original log; the black curve v^s, is obtained from the grey curve by smoothing with a Hanning window over 320 samples. The P-wave probability density function, shown in Figure 11.3(b) is obtained from the well data by taking the difference between the original and smoothed data: $v(z) - v^s(z)$.

The parameter uncertainty is obtained by applying the velocity–resistivity transform to the given velocity and a value for each of the parameters in the transform, repeating this many times, choosing the values at random from a uniform distribution for each parameter with a range of $\pm 5\%$ around the best estimate. Werthmüller et al. (2013) used a Markov chain Monte Carlo (MCMC) sampler to do this.

Following Chen and Dickens (2009), the model uncertainty is described as a gamma distribution with error E:

$$f(\rho|v,\theta) = \frac{\beta^\alpha \rho^{\alpha-1}}{\Gamma(\alpha)}\exp(-\beta\rho), \tag{11.3}$$

where ρ is the estimated resistivity, θ is a vector containing all model parameters, $\Gamma(\alpha)$ is the gamma function, the shape parameter $\alpha = 1/E^2$ and the scale parameter

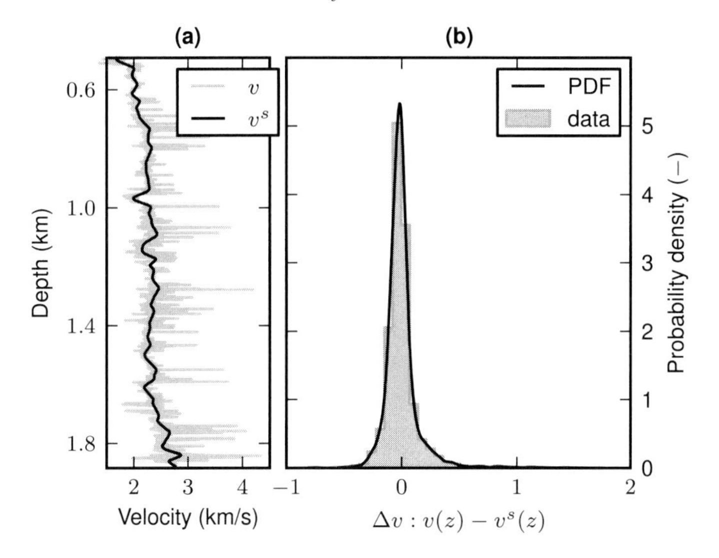

Figure 11.3 Velocity log $v(z)$ of North Sea well 9/23b-8. (a) Grey curve: the log; black curve $v^s(z)$ is a smoothed version of the log; (b) P-wave probability density function is computed from $v(z)-v^s(z)$ (reproduced from Werthmüller et al., 2013).

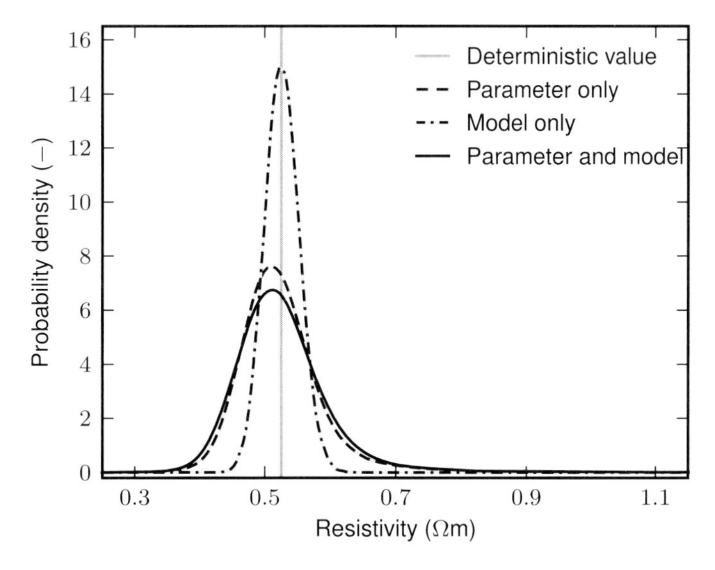

Figure 11.4 Example of uncertainty analysis (reproduced from Werthmüller et al., 2013).

$\beta = (\alpha - 1)/\rho$. An example of this uncertainty analysis for a particular set of parameters is shown in Figure 11.4, with model error E estimated to be 5%.

The rock-physics model is calibrated with a well in the vicinity of the target prospect, perhaps in a nearby field. The transform is checked using the velocity and resistivity logs. Then the transform is applied to a seismic velocity section or volume to obtain a model of background resistivities.

In vertical wells with horizontal laminations, normal induction tools measure the horizontal component of resistivity (Ellis and Singer., 2007). Therefore the velocity–resistivity transform calculated in this way normally provides an estimate of the horizontal component of resistivity.

11.4 Example from North Sea Harding Field

Werthmüller et al. (2013) applied this methodology to the North Sea Harding field. The calibration was performed using well 9/23b-8 in the Harding South field, a few kilometres to the south of Harding Central. To fit the data for the whole log, they found it necessary to include linear depth trends for the model parameters and structural constraints in the form of steps in the model parameters at the boundaries of a sandstone layer at about 1 km depth, identifiable on the seismic data.

A comparison of the rock physics model analysis and well data is shown in Figure 11.5. The plotted data are from well 9/23b-8 in the depth interval 1424-1524 m. The depth was set to $z = 1.474$ km for the depth-dependent parameters. The deterministic result of the transform is shown in grey and the outcome of the uncertainty analysis is in black. The well data fall within two standard deviations of the mode, apart from some high-velocity outliers originating from thin limestone layers.

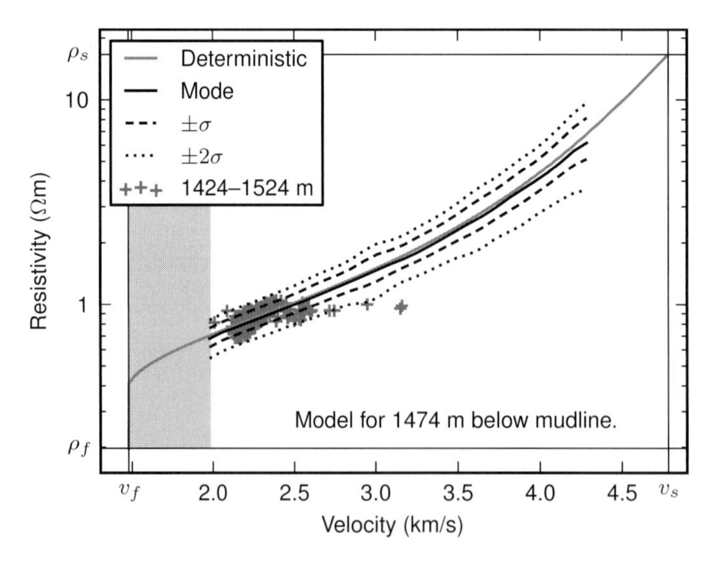

Figure 11.5 Comparison of the rock physics model and uncertainty analysis with well data (reproduced from Werthmüller et al., 2013).

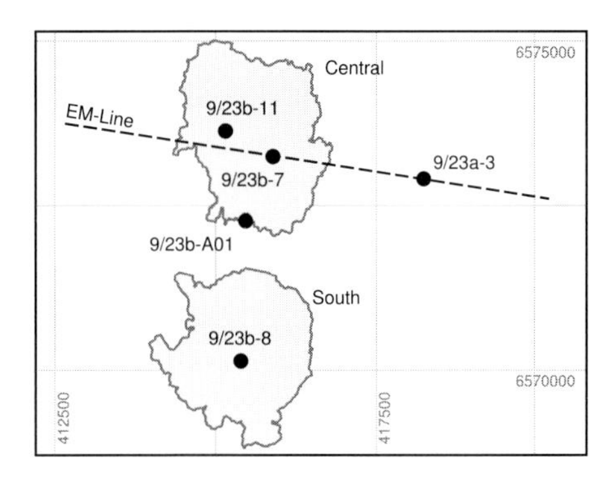

Figure 11.6 Map showing location of Harding Central, Harding South, wells 9/23a-3, 9/23b-7, 9/23b-8, 9/23b-11 and a CSEM line through 9/23b-7 and 9/23a-3. The field outlines are the oil–water contacts. (reproduced from Werthmüller et al., 2013).

The velocity–resistivity transform calibrated at well 9/23b-8 in Harding South was used to estimate the background resistivities from known velocities at Harding Central, a few kilometres north of Harding South. Figure 11.6 shows the location of the fields.

BP had obtained a velocity cube at Harding from the wells and seismic data. A section through this cube along the electromagnetic line shown in Figure 11.6 is shown in Figure 11.7(b). Figure 11.7(a) shows the mode resistivity section calculated from (b) using the velocity–resistivity transform. Figures 11.7(c) and (d) show the background resistivity sections minus one standard deviation and plus one standard deviation, respectively.

11.5 Test of Methodology Using Real CSEM Data

Once there is a subsurface resistivity model, it can be used to model synthetic CSEM data. Werthmüller (2017) compared synthetic and real data for the Harding field surveys discussed in Chapter 9. There were three major issues. First, there was no measurement of the sea-water resistivity profile. A resistivity of 0.3 ohm-m was assumed. Second, the shallowest section of the well was not logged. Third, they suspected that resistive anisotropy was important.

The shallow seismic data appeared to be horizontal, so common midpoint (CMP) gathers were inverted for the shallow resistivities, assuming horizontally layered vertical transverse isotropy (VTI) media. This recovered a layered structure down to 700 m with an average anisotropy factor $\lambda = \sqrt{\rho_v/\rho} = 1.5$. There were no data against which to check the inversion. An average anisotropy factor $\lambda = 1.5$ was used for inversion of the complete dataset using the 3D frequency domain

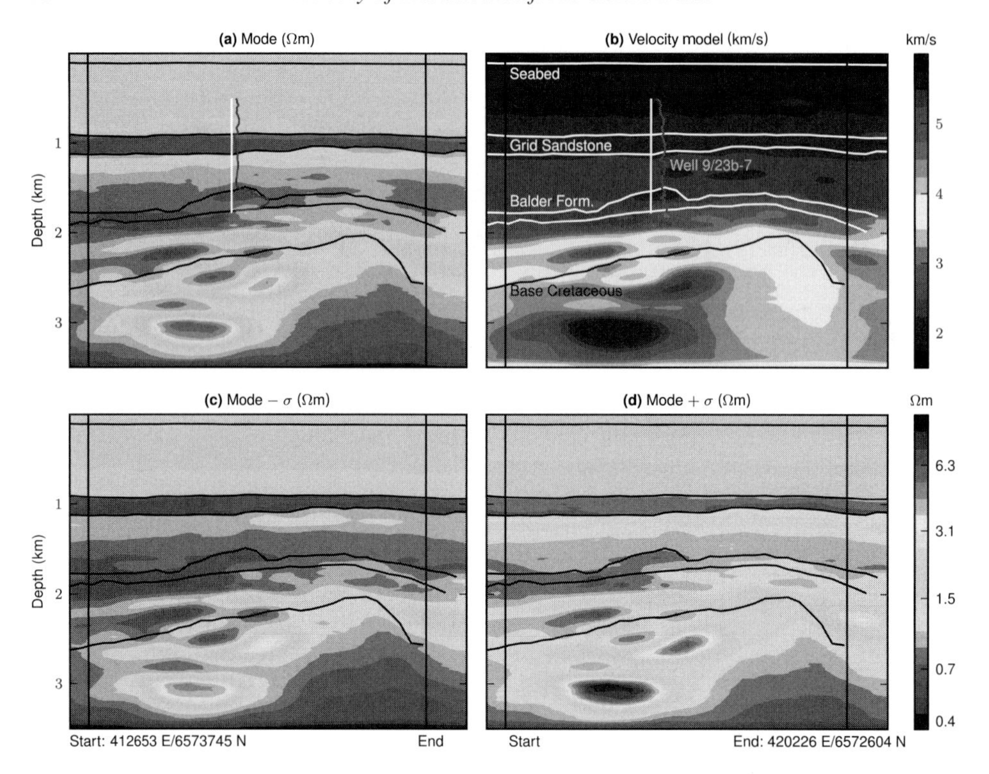

Figure 11.7 Results of velocity–resistivity transformation: (a) resistivity log at 9/23b-7 and mode background resistivity model obtained from velocity model shown in (b); (b) velocity section along electromagnetic line, showing major formations and velocity log; (c) mode resistivity minus one standard deviation; (d) mode resistivity plus one standard deviation (reproduced from Werthmüller et al., 2013). For the colour version, please refer to the plate section.

integral equation forward modeller of Hursán and Zhdanov (2002). Figure 11.8 illustrates how the synthetic data compared with the measured data. The measured data are in black; the synthetic data are colour-coded by offset, with the filled areas representing plus and minus one standard deviation. The prediction of the responses is good. Werthmüller (2017) points out that amplitudes are predicted with the right magnitude, and the peaks of the acquired data are within plus and minus one standard deviation: 'This is achieved without any scaling, shifting, or fudge factors.'

Sometimes there is no well control and no seismic data. In that case one has to guess. For instance, one can assume the resistivity of the water to be 0.3 ohm-m and one can start with a guess for the sediments below the sea floor to be a half-space with resistivity 1 ohm-m. One can always check such starting models against synthetic data computed from more complicated models.

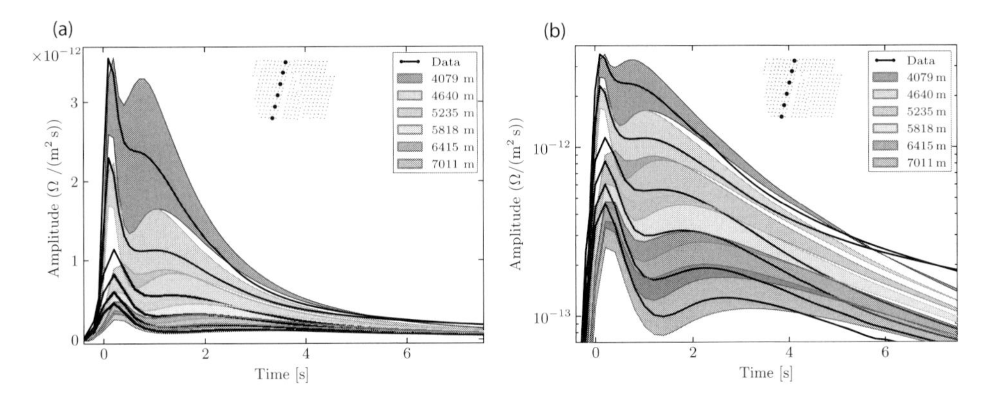

Figure 11.8 Common-source responses for real and synthetic data: (a) linear amplitude scale; (b) logarithmic amplitude scale. (reproduced from Werthmüller et al., 2014). For the colour version, please refer to the plate section.

11.6 Implications for Electromagnetic Survey Planning

In situations in which seismic data exist and there are no CSEM data, it is of course possible to compute synthetic 3D data. One can compare synthetic data with and without hydrocarbon reservoirs and determine whether a reservoir is detectable using CSEM. Such a feasibility study must be performed using 3D modelling, not 1D or 2D, because these over-emphasise small differences in amplitude.

CSEM also has the potential to monitor production and locate bypassed hydrocarbons. Whether this is likely to be beneficial can also be estimated by 3D modelling.

11.7 Example Inversion of Deep Water 3D CSEM Data

The ultimate goal is to recover resistivities from CSEM data and, because we do not have a method of obtaining the resistivities directly from the data, we are forced to use inversion. As we say in Section 1.10, the key issues are the origin of the model and the method of updating it. Section 11.3 describes one approach to the determination of a plausible model using seismic data and well logs, and Section 10.2 describes non-stochastic methods of updating a model. We have selected a case history presented by Nguyen et al. (2016) to illustrate the methodology. Nguyen et al. (2016) present inversion of deep water 3D CSEM data over the Snøhvit field in the central part of the Hammerfest Basin in the Barents Sea. They compare the BFGS algorithm with the Gauss–Newton algorithm, first on synthetic 3D CSEM data and then on the real data.

The synthetic model, named Circus, is shown in Figure 11.9(a) and (b) and has two targets:

> A shallow target about 850 m below the seabed (2400 m below sea level), which is 100 m thick and has a resistivity 12 Ωm, and a deeper target at about 1700 m below the seabed (3300 m below sea level), having a resistivity 50 Ωm and a thickness of 50 m. The deeper target is placed mainly below the shallower target, and this relative position of the targets is chosen since experience shows that stacked targets are challenging to image in a BFGS inversion. The water depth varies from 1000 m to 1800 m, and there are several layers of resistivity increasing with depth. Artificial noise was added to the simulated data with a multiplicative noise standard deviation of 0.02, and this noise level was also used in calculating the data weights in the inversion. (Nguyen et al., 2016)

The four lines of receiver nodes each had 15 receivers. Source–receiver offsets up to 12 km were included, giving 901 source points along the tow lines. The horizontal E_x and E_y electric fields were inverted for four frequencies between 0.1 and 1.0 Hz; 142,000 data points were used.

The initial model for the inversion of the synthetic data is shown in Figure 11.9(c). The sea-water resistivity is presumably identical to that in the model and the sea-bottom topography appears to be identical to that in the model. Below the seabed is a simple resistivity profile increasing with depth, but with no structural information. The boundaries between the layers appear smooth and the layer resistivities are given by the colour bar on the side.

The BFGS inversion shown in Figure 11.9(d) images one anomaly at a depth of approximately 3000 m below sea level, while the Gauss–Newton inversion images both targets very close to the correct locations. Both inversions had an RMS misfit of 0.92. The BFGS inversion took 77 iterations, while the Gauss–Newton inversion took 12. This result demonstrates the superiority of the Gauss–Newton inversion, which, as Nguyen et al. (2016) explain, leads the inversion to a better local minimum.

The inversion of the 3D Snøhvit dataset is shown in Figure 11.10. Figure 11.10(a) shows the receiver positions over the three discoveries, Albatross, Snøhvit and Snøhvit North, and the yellow line of the section shown in (c), (d) and (e) below. Figure 11.10(b) shows the geographic location of these discoveries.

The initial model, shown in Figure 11.10(c), consists of three layers: the water layer, the upper formation above the prominent reflection of the Base Cretaceous Unconformity (BCU) and the lower formation below the BCU. The vertical and horizontal resistivities of the upper formation are 6 and 3 Ωm, an anisotropy factor λ of $\sqrt{2}$, while the corresponding resistivities for the lower formation are 20 and 5 Ωm, an anisotropy factor λ of 2. The model was then smoothed before being used in the inversion. The authors do not disclose how these parameters were derived. We know from Chapter 10, however, that the choice of starting model is critical,

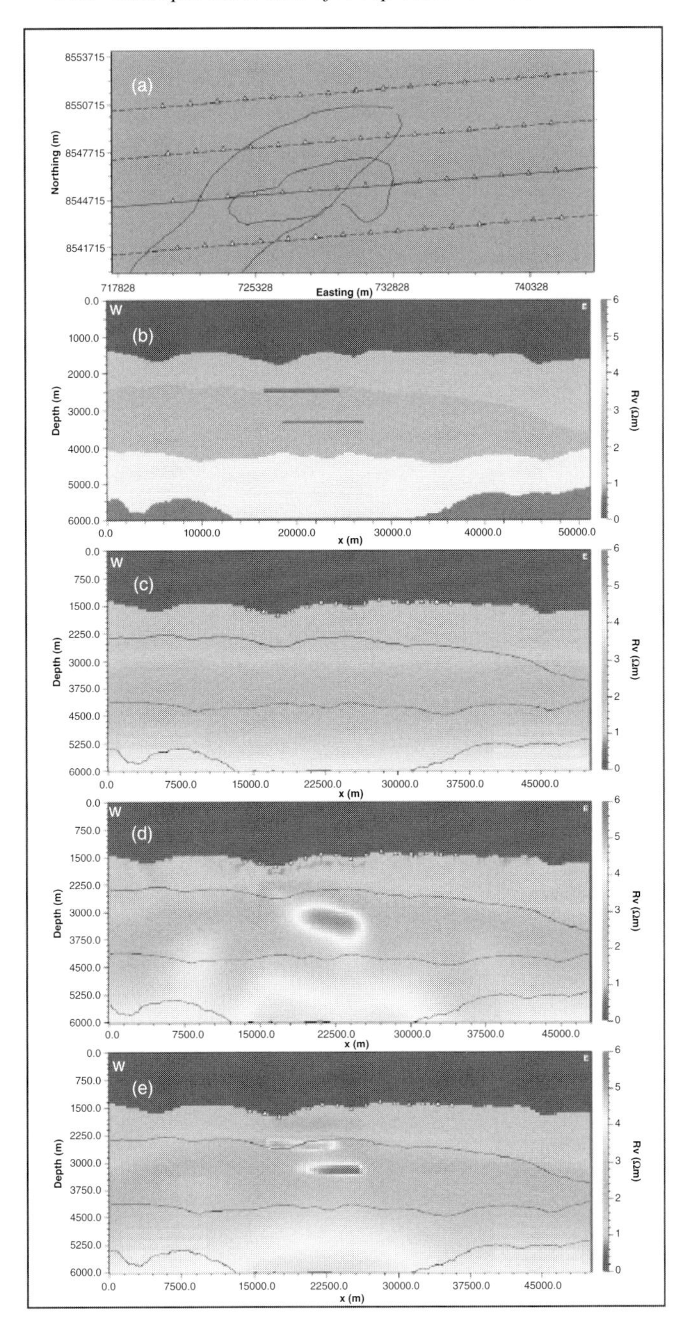

Figure 11.9 Synthetic data example. (a) Survey layout and target outlines; the towlines are along the lines of seabed nodes, the solid line is the section shown in (b)–(e). (b) The Circus model used to generate the synthetic data. (c) The initial model used for the inversions. (d) BFGS inversion. (e) 3D Gauss–Newton inversion (reproduced from Nguyen et al., 2016). For the colour version, please refer to the plate section.

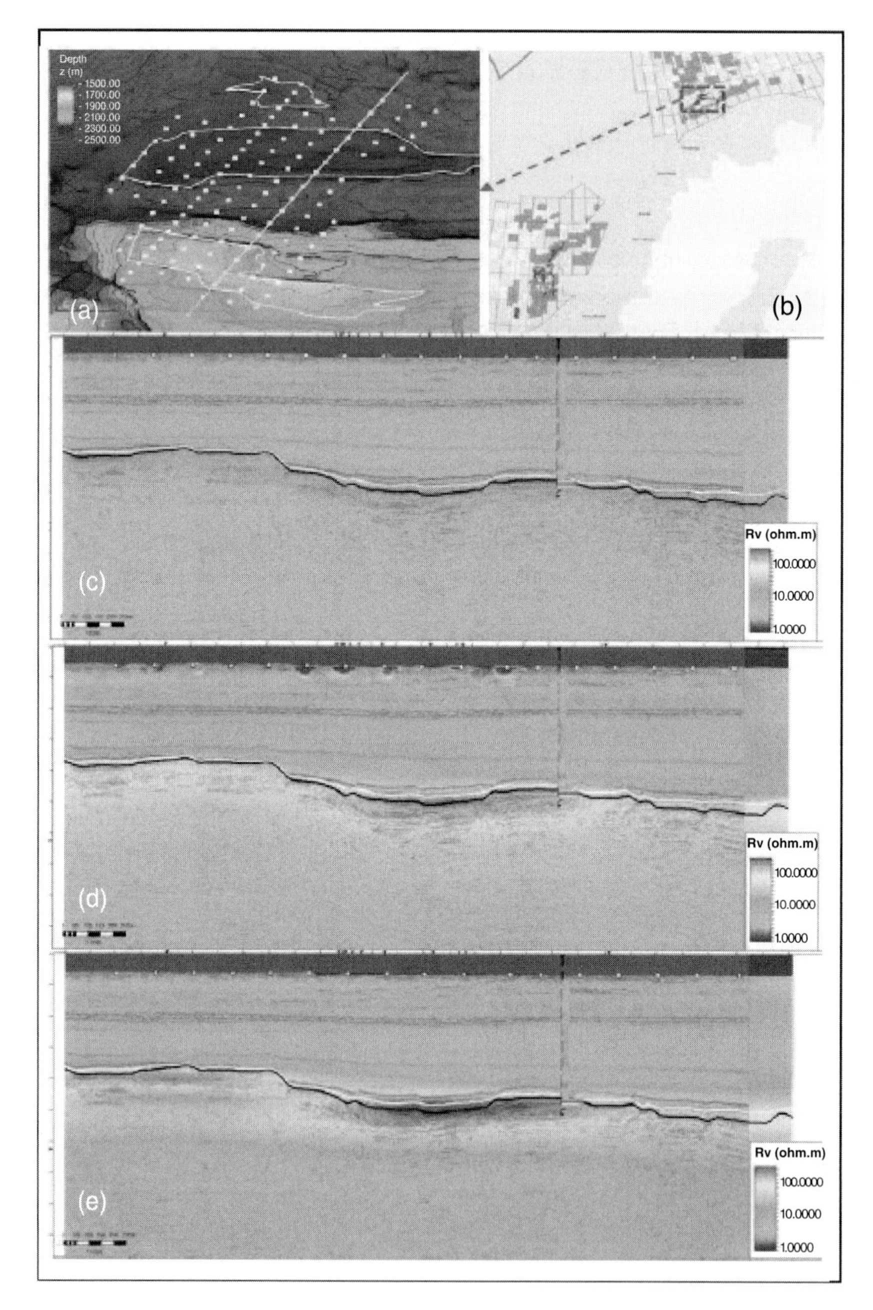

Figure 11.10 Real data example. (a) Seabed receiver locations and outlines of discoveries from bottom to top: Albatross, Snøhvit and Snøhvit North. The yellow line shows the positions of sections (c), (d) and (e). (b) Geographic location of Snøhvit field. (c) Initial resistivity model, (d) BFGS inversion and (e) 3D Gauss–Newton inversion all on a log scale, overlain on the seismic section (reproduced from Nguyen et al., 2016). For the colour version, please refer to the plate section.

since the inversion scheme drives the inversion to a local minimum close to the starting model. For the inversions both horizontal receiver components and three frequencies, 0.5, 1.0 and 2.0 Hz, were used.

Comparing the BFGS inversion of the data in Figure 11.10(d) with the Gauss–Newton inversion in Figure 11.10(e), Nguyen et al. (2016) point out that the BFGS inversion creates many artificial resistive anomalies right under the receivers, which the Gauss–Newton inversion does not. Furthermore, the 3D Gauss–Newton inversion maps the Snøhvit anomaly as a stronger event, which is more consistent with the resistivity logs. That is, the Gauss–Newton inversion is better, as confirmed by independent data.

12

Efficient CSEM

This final chapter draws together some of the themes in the book and considers how to make controlled-source electromagnetic (CSEM) data acquisition more efficient to enable it to be used as an exploration tool that is truly complementary to seismic exploration.

There are three simple questions an exploration geophysicist will always ask about CSEM:

1. How much does it cost?
2. What is the depth of penetration?
3. What is the resolution?

These questions are not easy to answer. For marine surveys the costs are now reasonably well known. The depth of penetration and the resolution are not so well known. It is tempting to reply that the answers have to be determined by modelling. This is not satisfactory. There have to be straight answers and we should probably consider the way the seismic exploration industry has developed to find these answers. Seismic exploration has been developed as an exploration tool over six or seven decades and is mature. The seismic exploration industry knows what to expect in the field and knows what data acquisition parameters are required to obtain data with adequate signal-to-noise ratio and adequate resolution for targets at a range of depths down to about 10 km. And both the client and the contractor understand what is meant by 'adequate' in this context. The corresponding knowledge in CSEM is much less extensive and there is much less shared knowledge between the client and the contractor. And, because Maxwell's equations appear to be much more difficult to understand than the acoustic or elastic wave equations, there is much less common understanding of the problems.

Thus there are difficulties, but progress has been made on certain problems. We attempt to summarise here the present status of CSEM and see where more progress is required. On land we can separate the incident air wave from the ground wave using coded source time functions and signature deconvolution. For 3D data acquisition, seismic methods should be adopted and research is required on the development of inexpensive reliable electric receivers that can be rapidly deployed and retrieved in their thousands. For shallow water (<500 m depth) marine exploration, a towed source and receiver cable have already been demonstrated to be effective and the extension to many receiver cables is obvious. For deep water, development of an efficient method for deployment and retrieval of the sea-floor receiver nodes will have the greatest impact on reducing costs. Precise positioning of the deep-towed source and measurement of the source time function will improve data quality. In all applications, the use of coded source signals and signature deconvolution will increase the signal-to-noise ratio by deconvolution gain, independent of the domain in which the subsequent inversion is carried out.

12.1 General Considerations

In Chapters 1 and 5 it was established that resistive or conductive targets can be detected at a given depth provided the source–receiver offsets extend to at least twice that depth. To give confidence in the detection of the target, source–receiver offsets should extend even further, to perhaps three times the depth. The short answer to question 2 above, therefore, is that there is no depth limitation to the method in principle, provided the source–receiver offsets extend out to about three times the target depth. This answer ignores the impact of noise.

It has also been shown that the strength of the signal decreases rapidly with source–receiver separation, even for 1D layered media. Since noise is always present and is independent of source–receiver separation, the problem of detection at large offsets is therefore a problem of increasing the signal-to-noise ratio. Several techniques have been developed for increasing signal and reducing noise, as discussed in earlier chapters. One very important factor is time. The more time there is available to inject signal and repeat measurements, the more the signal-to-noise ratio may be increased by summing the results, or *stacking*, to use the seismic exploration term. For random noise, the signal-to-noise ratio increases as the square-root of the number of measurements stacked. In principle, therefore, a doubling of the time can increase the signal-to-noise ratio by $\sqrt{2}$. When everything else has been optimised, this means that costs can double to obtain $\sqrt{2}$ increase in signal-to-noise ratio. This is unacceptable and is therefore a severe limitation of the method. We note that the marine methods using towed sources already have tight constraints on the time available to inject signal without introducing unacceptable lateral smearing.

Turning to seismic exploration for answers to this problem, we note that tremendous increases in signal-to-noise ratio are routinely obtained by common midpoint

(CMP) stacking and by migration, especially multi-azimuth migration. These processing methods are not easy to adapt to diffusive electromagnetic propagation. Nevertheless, we suggest that this could be not only a fruitful area of research, but also of tremendous benefit to the CSEM method and the detection of subsurface resistivity variations.

From Figures 1.8 and 1.11, it is clear that differences in the response from a resistive or conductive target at 2 km depth are easily detected at horizontal intervals of 100 m. For these cases we can claim that the *lateral resolution* for targets at 2 km depth is, in principle, less than 100 m, or less than 5% of the depth. This number may increase in the presence of noise. We do not know the *vertical resolution*: this depends on the quality of the inversion.

12.2 Land

In land CSEM, the source is a horizontal current dipole at the earth surface and the receivers are horizontal dipoles, also on the surface, each of which measures the voltage between the dipole electrodes. The electric field at a receiver is the gradient of the potential and is a vector. The vertical component of the electric field is zero at the earth surface, so two orthogonal horizontal components define the vector. The magnitude of one component is the voltage between the electrodes divided by the distance between them. For an impulse source time function $\delta(t)$, the total electric field response $g(t)$ consists of an impulse at $t = 0$, the incident air wave, followed by the scattered field, which is a response that travels through the ground, consisting of propagation from the source and scattering of the air wave, acting as an impulsive plane wave at $t = 0$. The ground response, or scattered field, contains information about the earth; the incident air wave does not. We are, therefore, interested in the ground wave and not in the incident air wave. If we can obtain the complete impulse response, we can cut out the incident air wave at $t = 0$ and isolate the ground wave, or scattered field, for further analysis.

In practice, it is impossible to generate an impulse. To obtain the impulse response we must use a broad bandwidth source time function $I(t)$ whose energy is spread out over a period of time T_s. The received data are then deconvolved for the source time function $I(t)$. The deconvolution step compresses the energy of the source into a single instant in time and recovers the earth impulse response. For data sampled at discrete intervals Δt, the incident air wave is compressed into a single time sample at $t = 0$. Removal of the incident air wave consists of setting the value of this sample to zero. Unless this process is performed, the scattered field cannot be isolated: the data are contaminated by the incident air wave.

Since the deconvolution step compresses all the source signal energy into a single instant in time, it results in an increase in signal-to-noise ratio, which we call *deconvolution gain*.

Suppose a land CSEM survey is required for geothermal purposes over a given area, say 100 km^2, to a maximum depth of, say, 2.5 km. Source–receiver separations out to three times the depth, or 7.5 km, would be required. If adequate horizontal resolution is obtainable with a spatial sampling interval of 100 m, a line of 75 receivers each side of the source dipole axis, or 150 receivers, would be required. Let this be the x-direction. As noted in Chapter 8, a uniform (x, y) grid of receivers would be used for an areal 3D survey and 22,500 receiver units would therefore be required for a given source position. Such a CSEM survey has never been performed. Unless such a survey can be performed on a routine basis, however, and the data processed for a reasonable cost, there is little prospect of land CSEM being used for such a survey.

Exploration geophysicists are used to seismic surveys and seismic exploration costs. Unless the costs of CSEM surveys are comparable, they will not be considered. A 'reasonable' cost in this context is the cost of a comparable seismic survey.

The cost of installing a current dipole source should be similar to the cost of a dynamite shot hole, or a vibroseis sweep with a group of seismic vibrators. Vibroseis surveys are generally less costly than dynamite, so the cost of the dynamite shot hole can be considered an upper limit. The problem, however, is not the source, but the receivers.

Emplacing electric receivers has its origins with magnetotelluric receivers in which non-polarising electrodes are carefully emplaced by hand and used for periods of days, weeks or even months. A single magnetotelluric station represents a considerable investment in time and labour. A survey using thousands of such receivers would be exorbitantly expensive and the concept is automatically rejected. The land CSEM system must use simpler receivers.

For CSEM on land, Ziolkowski et al. (2007) found that simple copper electrodes worked well, provided the electric field from the source had alternating polarity. Ziolkowski et al. (2007) obtained good data, but the receiver dipoles were 100 m long. It would be good if the receivers could be made small, say 50 cm between electrodes. The issue is then the noise level of individual electrodes.

Each electrode generates its own noise, which adds to the signal to contribute to the measured voltage. The electrode noise is essentially independent of the distance between the electrodes. The signal-to-electrode-noise ratio of the electric field measurement is therefore proportional to the distance between electrodes. We do not know how small this distance can be. Experimental research is required to determine this and, perhaps, for the development of low-noise electrodes.

The seismic exploration industry has developed systems for the rapid deployment of thousands of multi-component seismic receiver units that are able to determine their own positions and record data for long periods. The CSEM industry must catch up and borrow the seismic exploration technology. Development of small electric field receivers is an essential component of the design of a low-cost land CSEM data acquisition system.

12.3 Shallow Water Marine

Towed streamer CSEM in shallow (<500 m deep) water has been established since about 2010. The CSEM vessel moves at the same speed as seismic vessels; in principle, therefore, the CSEM data acquisition costs are comparable with towed streamer marine seismic data acquisition costs. Since the CSEM acquisition and seismic acquisition can be performed simultaneously on the same vessel, there is a potential to reduce the costs of both methods.

The source is towed close to the sea surface, typically at a depth of about 10 m. The receiver cable can be towed, in principle, at any depth and is typically at a depth of 100 m. In principle many receiver cables can be used, as in seismic reflection surveying. Orientation and positioning of source and receivers can be performed using compasses and underwater positioning acoustic systems, as in marine seismic surveying.

In this configuration, where the source and receivers are in the water, the incident field is the direct wave from the source to the receiver. The scattered field includes scattering from the sea floor, the sea surface and bodies below the sea floor. The air wave originates as any upcoming wave at the sea surface, extending essentially instantaneously across the sea surface, and propagating downwards into the water again. Inversion of the data to determine the resistivity of subsea bodies requires precise forward modelling, which, in general, must be 3D.

The principal limitation of towed streamer CSEM in the current configuration is the water depth. The sea water below the source attenuates the downgoing signal exponentially with depth. The obvious solution is to tow the source closer to the sea floor. This is not so easy, as discussed in Chapter 7. To minimise power losses down the cable, the current must be minimised and the voltage maximised. Therefore, high-voltage AC power must be transmitted from the vessel to the head fish of the source. To deliver a large current, a transformer is used to convert from high-voltage low-current AC power to low-voltage high-current AC power. This is then rectified and fed into a waveform generator. Finally the outgoing current signal must be measured and transmitted back to the vessel. The technology to do all this exists.

Chapter 7 describes the problems that occur with the control of a deep-towed source when the towing speed is low. The position of the source and its orientation can be very variable and must be measured and transmitted to the vessel. Processing of the data needs to take account of these variables. If the receiver cable is also deep-towed, these same problems occur. The problems are reduced if the towing speed can be increased.

12.4 Deep Water Marine

The principal attraction of the deep water marine system is the very low noise level of the ocean-bottom nodes and, consequently, the potential to obtain data with very good signal-to-noise ratio. The principal component of the high cost of data acquisition is the deployment and retrieval of the nodes.

To reduce this cost we should look at the seismic industry, where several competing systems have been developed for rapid deployment and retrieval of ocean-bottom nodes. The seismic nodes are much smaller than the CSEM nodes, because the seismic measurement is of motion or pressure at a point. CSEM nodes measure the magnetic field in two orthogonal horizontal directions and the electric field in the same two orthogonal horizontal directions. The size of the node is determined by the distance between the electric field electrodes, as in the land case. If this distance can be made small, say 50 cm, the seismic deployment methods can be used for deep water CSEM. As for the land case, this requires low-noise electrodes. It may be possible to do without the magnetic field measurements to reduce the cost and conserve battery life. It may further be possible to combine seismic sensors with the electric field sensors to enable seismic and CSEM surveys to be performed with the same vessel and the same nodes.

Once the nodes are deployed, the principal consideration is the deep-towed source. When the tow speed is low, the position and orientation of the source are very variable and determination of these parameters is critical. The faster the source can be towed, the lower the cost of the survey and the more stable are the position and orientation of the source. These are great motivations for increasing tow speed. The downside is the reduced time for injecting source signal into the water.

Improvements to this system are being made. In 2017 Statoil, Shell and EMGS announced a 100-fold increase in signal-to-noise ratio of the deep-water marine acquisition system (Hanssen et al., 2017). This consists of increasing the source current from 1000 A to 10,000 A and reducing the noise floor of the receiver system by a factor of 10. The authors estimate that this increases the depth of detectable targets to 4.5 km below the sea floor.

12.5 Source Time Function

Chapter 6 discusses source time functions. If it is desired to obtain the complete earth impulse response in the time domain, Chapter 6 shows that a broad bandwidth source time function, such as a pseudo-random binary sequence (PRBS), is required, for then the complete earth impulse response can be recovered by deconvolution. The deconvolution gain improves the signal-to-noise ratio. After the deconvolution has been performed and, in the land case, the scattered field has been isolated by eliminating the incident air wave, the subsequent analysis, including inversion, can be performed in the time domain or the frequency domain.

Suppose it is known prior to data acquisition that the inversion will be performed in the frequency domain: is there a case for not using a PRBS and using another signal with energy concentrated in fewer frequencies? We argue, first, that it is not known prior to surveying which are the few best frequencies to use, and second, that by not using a PRBS, you lose the benefit of deconvolution gain. Section 6.7 compares the performance of a PRBS with a special periodic function of about the same length. With the special periodic function, the signal-to-noise ratio at a given frequency is unchanged by the frequency domain division of the spectrum of the measured response by the spectrum of the source time function. With the PRBS this is also true, but then the transformation back to time gives the deconvolution gain. In the case considered in Section 6.7 the gain was a factor of 12.4, or 22 dB. After this gain has been obtained, the impulse responses may be inverted in the time domain or the frequency domain.

12.6 Conclusions

Many of the developments required to improve efficiency are extensions of existing technology and we can expect them to occur if sufficient resources are made available. The biggest limitation on the method is the depth of penetration. This is a problem of obtaining good signal-to-noise ratio at large offsets. Incremental improvements in raw signal strength and noise reduction will doubtless come, but a real breakthrough is required in the processing of CSEM data to combine data from different offsets and azimuths for imaging of subsurface resistivity contrasts. We suggest that this could be a very rewarding area for research.

A barrier to rapid deployment of land and deep marine receivers is the large size of the receiver dipoles used to measure the electric field vector. The magnitude of a component of the electric field vector is presently obtained by measuring the voltage between two electrodes and dividing by the distance between them. The voltage includes the noise of each electrode. The signal-to-electrode-noise ratio is proportional to the distance between the electrodes. If the electrode noise level can be reduced such that electric dipoles of 50 cm have acceptably low noise levels, the costs of land and deep marine CSEM can be reduced to the level of equivalent seismic surveys.

Finally, for marine data, we suggest that surveys can be less expensive and more precise if the deep-towed source can be towed more rapidly.

Appendix A

The Electric Field in a VTI Whole Space

All elements in the Green's functions of equations 4.37–4.42 can be transformed back to the space domain, which is done here. The expressions obtained are equivalent to ones given by Weiglhofer (1990) and Abubakar and Habashy (2006), but we use the reduced expressions in which the apparent singularity is removed. We limit ourselves to the electric field generated by an electric current dipole source, because then the others can be obtained without difficulty. We start with the general procedure for the inverse spatial Fourier transformation of a scalar field $\tilde{G}$, which is given by

$$\hat{G}(\mathbf{r}, \mathbf{r}_s, s) = \frac{1}{4\pi^2} \int_{(k_x,k_y)\in\mathbb{R}^2} \tilde{G}(k_x, k_y, z, z_s, s) \exp(\mathrm{i}k_x x + \mathrm{i}k_y y) dk_y dk_x. \tag{A.1}$$

This integral can be rewritten using radial parameters, with $x = r\cos(\phi)$, $y = r\sin(\phi)$, $0 < r < \infty$ and $0 \leq \phi < 2\pi$, together with $k_x = \kappa\cos(\psi + \phi)$ and $k_y = \kappa\sin(\psi + \phi)$ with $0 < \kappa < \infty$ and $0 \leq \psi < 2\pi$. Changing the integration variables from (k_x, k_y) to (κ, ψ) gives the Jacobian as $dk_x dk_y = \kappa d\kappa d\psi$. The integral therefore becomes

$$\hat{G}(\mathbf{r}, \mathbf{r}_s, s) = \frac{1}{2\pi} \int_{\kappa=0}^{\infty} \tilde{G}(k_x, k_y, z, z_s, s) \mathrm{J}_0(\kappa r) \kappa d\kappa, \tag{A.2}$$

in which $\mathrm{J}_0(\kappa r)$ denotes the Bessel function of the first kind and order zero,

$$\mathrm{J}_0(\kappa r) = \frac{1}{2\pi} \int_{\psi=0}^{2\pi} \exp[-\mathrm{i}\kappa r \cos(\psi)] d\psi. \tag{A.3}$$

When $\tilde{G} = \tilde{G}_H$, as given in equation 4.55, the integral in equation A.2 is well known and given by

$$\hat{G}_H(\mathbf{r}, \mathbf{r}_s, s) = \frac{\exp(-\gamma R)}{4\pi R}, \tag{A.4}$$

where $R = \sqrt{x^2 + y^2 + (h^-)^2}$ is the distance from source to receiver. The TM mode Green's function in equation 4.54 can be written as

$$\tilde{G}_V(k_x, k_y, z, z_s, s) = \frac{\exp(-\sqrt{\kappa^2 + \gamma_v^2}\lambda|h^-|)}{2\lambda\sqrt{\kappa^2 + \gamma_v^2}}, \tag{A.5}$$

and the space domain Green's function is then given by

$$\hat{G}_V(\mathbf{r}, \mathbf{r}_s, s) = \frac{\exp(-\gamma_v R_v)}{4\pi\lambda R_v}, \tag{A.6}$$

where $R_v = \sqrt{x^2 + y^2 + (h_v^-)^2}$, is the TM mode distance from source to receiver, with $h_v^- = \lambda h^-$ being the scaled height difference between source and receiver and $\lambda = \sqrt{\sigma/\sigma_v}$ is introduced as the coefficient of anisotropy.

Using this procedure for the TE mode electric field Green's function for an electric current source as given in equation 4.61 results in

$$\hat{\mathbf{G}}_H^{ee}(\mathbf{r} - \mathbf{r}_s, s) = -\begin{pmatrix} -\frac{\partial^2}{\partial y^2} & \frac{\partial^2}{\partial x \partial y} & 0 \\ \frac{\partial^2}{\partial x \partial y} & -\frac{\partial^2}{\partial x^2} & 0 \\ 0 & 0 & 0 \end{pmatrix} \frac{\zeta}{2\pi} \int_{\kappa=0}^{\infty} \tilde{G}_H \mathrm{J}_0(\kappa r)\kappa^{-1} d\kappa. \tag{A.7}$$

The integral in the right-hand side of equation A.7 is denoted $\mathcal{I}_H$ and can be evaluated in closed form by taking one derivative, e.g.

$$-\frac{\partial \mathcal{I}_H}{\partial x} = \frac{x}{2\pi r} \int_{\kappa=0}^{\infty} \tilde{G}_H(k_x, k_y, z, z_s, s)\mathrm{J}_1(\kappa r) d\kappa. \tag{A.8}$$

The integral in equation A.8 exists, but we take one more derivative to facilitate the evaluation of the matrix elements in equation A.7 and we arrive at

$$-\frac{\partial^2 \mathcal{I}_H}{\partial x^2} = \frac{x^2}{r^2}\hat{G}_H - \frac{x^2 - y^2}{r^2}\frac{1}{2\pi r} \int_{\kappa=0}^{\infty} \tilde{G}_H(k_x, k_y, z, z_s, s)\mathrm{J}_1(\kappa r) d\kappa, \tag{A.9}$$

where the derivatives on the Bessel functions can be found in Appendix C. This example is the solution for the yy-component of the electric field, except for the scaling factor ζ, as can be seen from the $(2, 2)$-element of the matrix in equation A.7. The first term on the right-hand side of equation A.9 is a propagating diffusive wave as we would expect physically. The directional term $(x/r)^2$ for the yy-component of the TE mode field is what we would expect from the physics of the problem: the TE mode field propagating in the (x, z)-plane should be a y-component field. The part of the expression containing the integral in equation A.9 represents an artefact that arises due to the mode decomposition. It can be found in Gradshteyn and Ryzhik (1996:1098, formula 6.637 1) and can be written as (Abramowitz and Stegun, 1972)

$$\frac{1}{2\pi r}\int_{\kappa=0}^{\infty}\tilde{G}_H(k_x,k_y,z,z_s,s)\mathrm{J}_1(\kappa r)d\kappa=\frac{\exp(-\gamma|h^-|)-\exp(-\gamma R)}{4\pi\gamma r^2}. \tag{A.10}$$

This result is regular at $r = 0$ when $h^- \neq 0$, which can be seen by taking the first three terms of the Taylor series expansion around $r = 0$ of the exponentials. Unfortunately it comes at the cost of an exponential function that depends only on depth and not on offset. This creates a strong field at large offsets that is unphysical. This contribution also occurs in the TM mode, but with opposite sign, for which reason we subtract this term from both modes. We do this for all terms in the TE mode Green's function and the final result is given in equations 4.69 and 4.70.

The TM mode electric field Green's function for an electric current source of equation 4.62 can be transformed to the space domain, and we write the expression as

$$\hat{\mathbf{G}}_V^{ee}(\mathbf{r},\mathbf{r}_s,s)=\begin{pmatrix}-\gamma_v^2\frac{x^2}{r^2}+\frac{\partial^2}{\partial x^2} & -\gamma_v^2\frac{xy}{r^2}+\frac{\partial^2}{\partial x\partial y} & \frac{\partial^2}{\partial x\partial z}\\ -\gamma_v^2\frac{xy}{r^2}+\frac{\partial^2}{\partial x\partial y} & -\gamma_v^2\frac{y^2}{r^2}+\frac{\partial^2}{\partial y^2} & \frac{\partial^2}{\partial y\partial z}\\ \frac{\partial^2}{\partial x\partial z} & \frac{\partial^2}{\partial y\partial z} & -\gamma^2+\frac{\partial^2}{\partial z^2}\end{pmatrix}\frac{\hat{G}_V(\mathbf{r},\mathbf{r}_s,s)}{\sigma_v}$$
$$+\begin{pmatrix}\frac{x^2-y^2}{r^2} & 2\frac{xy}{r^2} & 0\\ 2\frac{xy}{r^2} & \frac{y^2-x^2}{r^2} & 0\\ 0 & 0 & 0\end{pmatrix}\frac{\zeta}{2\pi r}\int_{\kappa=0}^{\infty}\tilde{G}_V\mathrm{J}_1(\kappa r)d\kappa. \tag{A.11}$$

The integral is given by

$$\frac{1}{2\pi r}\int_{\kappa=0}^{\infty}\tilde{G}_V(k_x,k_y,z,z_s,s)\mathrm{J}_1(\kappa r)d\kappa=\frac{\exp(-\gamma|h^-|)-\exp(-\gamma_v R_v)}{4\pi\gamma r^2}, \tag{A.12}$$

and is also regular at $r = 0$ when $h^- \neq 0$. We subtract again the depth dependent part of the result when $r \neq 0$ to eliminate the unphysical event. With this function, the angular singularity problems have been dealt with without introducing unphysical fields. Carrying out all derivatives, as detailed in equations C.1–C.4, results in the expressions given in equations 4.71 and 4.72.

Appendix B

The Electromagnetic Field in a VTI Layered Medium

We derive closed form expressions in the horizontal wavenumber–frequency domain for the electromagnetic field generated by an arbitrary point source located inside a vertical transverse isotropic (VTI) layered medium and integral expressions for the electric and magnetic fields in the space–frequency and space–time domains. In a multi-layered medium, space–frequency and space–time domain results must be obtained numerically. For the special cases of a homogeneous whole space and one conductive homogeneous half-space adjacent to a non-conductive half-space, exact expressions in the space–frequency and space–time domain are found. Equations 4.44 and 4.43 form the starting point for the analysis and procedure to the solution. Section 4.1.3 explains that we need only solve for the TM mode and the substitutions given in Table 4.1 may then be used to obtain the desired expressions for the TE mode from the TM mode solution.

Let us assume a layered model with $N+1$ interfaces at depth levels z_n with $n = 0, 1, 2, \ldots, N-1, N$, as depicted in Figure B.1. Because zero depth can be chosen anywhere, we take $z_0 = 0$. Each layer $\mathbb{D}_n$ is characterised by the isotropic magnetic permeability μ_n, the vertical conductivity $\sigma_{v;n}$, horizontal conductivity σ_n and layer thickness $d_n = z_n - z_{n-1}$ with the understanding that the upper and lower half-spaces, $\mathbb{D}_0$ and $\mathbb{D}_{N+1}$, respectively, have an infinite thickness. Let the source be located at z_s with $z_{\varsigma-1} < z_s < z_\varsigma$. Then the TM mode electric field can be written in terms of fields that diffuse in upgoing and downgoing directions. In the source layer the direct field generated by the source is known, while the up- and downgoing reflected fields generated by the interfaces above and below the source are unknown. This gives the following expressions for the field inside the layer that contains the source,

$$\tilde{E}_{V;\varsigma} = X_V(z)\exp(-\Gamma_{v;\varsigma}|h^-|) + A_\varsigma^+ \exp(-\Gamma_{v;\varsigma}(z - z_{\varsigma-1})) + A_\varsigma^- \exp(-\Gamma_{v;\varsigma}(z_\varsigma - z)), \tag{B.1}$$

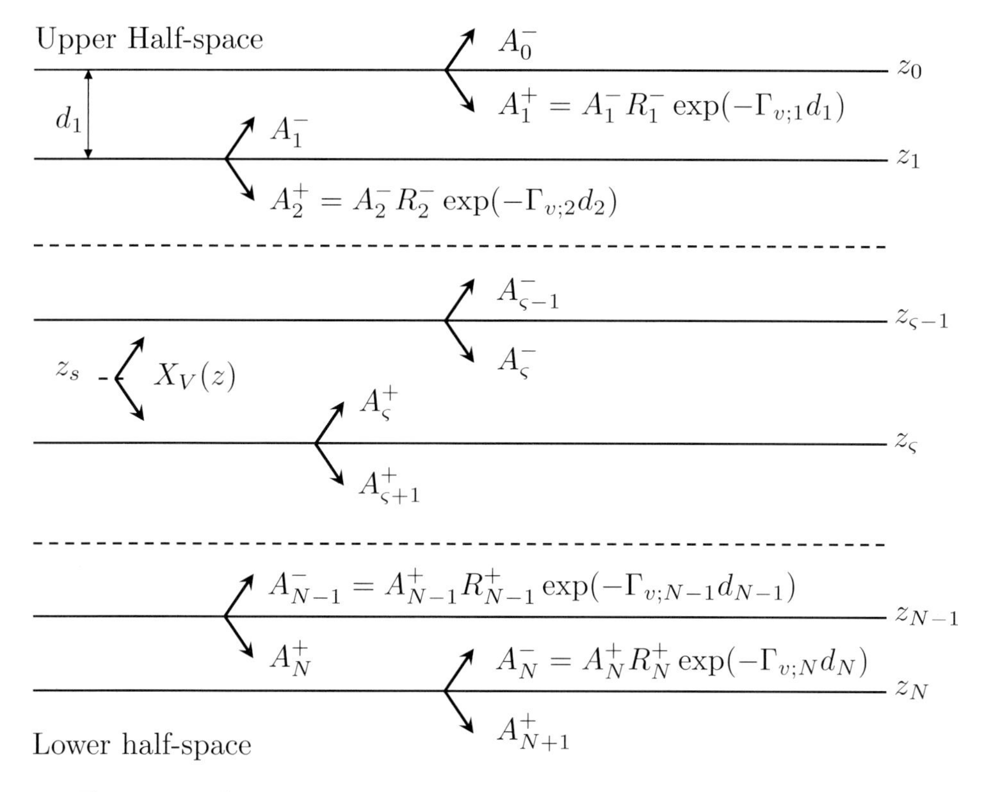

Figure B.1 The TM mode representation for the source signal and up- and downward diffusing fields in a layered medium.

$$
\begin{aligned}
\tilde{H}_{V;\varsigma} = \frac{\sigma_\varsigma}{\Gamma_{v;\varsigma}} \big[& X_V(z)\text{sign}(h^-) \exp(-\Gamma_{v;\varsigma}|h^-|) + A_\varsigma^+ \exp(-\Gamma_{v;\varsigma}(z - z_{\varsigma-1})) \\
& - A_\varsigma^- \exp(-\Gamma_{v;\varsigma}(z_\varsigma - z)) \big],
\end{aligned} \tag{B.2}
$$

in which A_ς^+ is the unknown amplitude of the downward-reflected electric field at the boundary $z_{\varsigma-1}$ and A_ς^- is the unknown amplitude of the upward reflected electric field at the boundary z_ς. The source term X_V is given by

$$
X_V(z) = \frac{\Gamma_{v;1}}{2\sigma_1} \hat{I}_V^e + \frac{1}{2} \left(\frac{\kappa^2}{\sigma_{v;1}} \hat{I}_z^e + \hat{I}_V^m \right) \text{sign}(z - z_s). \tag{B.3}
$$

The variable depth z is where we evaluate the fields. In the end it will be receiver depth z_r. In any layer with finite thickness that is source free we can write the fields as

$$
\tilde{E}_{V;n} = A_n^+ \exp(-\Gamma_{v;n}(z - z_{n-1})) + A_n^- \exp(-\Gamma_{v;n}(z_n - z)), \tag{B.4}
$$

$$
\tilde{H}_{V;n} = \frac{\sigma_n}{\Gamma_{v;n}} \left[A_n^+ \exp(-\Gamma_{v;n}(z - z_{n-1})) - A_n^- \exp(-\Gamma_{v;n}(z_n - z)) \right]. \tag{B.5}
$$

In the upper half-space we have only an upgoing field and the electric and magnetic fields can be written as

$$\tilde{E}_{V;0} = A_0^- \exp(\Gamma_{v;0} z), \tag{B.6}$$

$$\tilde{H}_{V;0} = -A_0^- \frac{\sigma_0}{\Gamma_{v;0}} \exp(\Gamma_0 z). \tag{B.7}$$

Similarly, in the lower half-space we have only a downgoing field and the electric and magnetic fields can be written as

$$\tilde{E}_{V;N+1} = A_{N+1}^+ \exp(-\Gamma_{v;N+1}(z - z_N)), \tag{B.8}$$

$$\tilde{H}_{V;N+1} = A_{N+1}^+ \frac{\sigma_{N+1}}{\Gamma_{v;N+1}} \exp(-\Gamma_{v;N+1}(z - z_N)). \tag{B.9}$$

Note that all coefficients $A_n^{\pm}$ are defined at the depth level at which they are generated, which accommodates causality and ensures only damped exponentials occur in the expressions. The concept of the reflection response has been used in Figure B.1 and we give the physical reasoning behind it now. The plane (diffusive) wave reflection response at, but just above, an interface inside a layered medium for a downward incoming field above that interface is equal to the ratio of the upward travelling field to the downward travelling field at that interface. The reflection response defined in this way only contains information about all primary and multiple reflections at and below the interface where it is defined. This means that all the information about the layered medium above that interface and the interactions between the two zones is contained in the downgoing field. When the incoming wavefield is in the upward direction the same principle applies, but in reversed order. For all layers above the layer containing the source, the upgoing wavefield is the incoming wavefield. Therefore, the downgoing fields can be expressed in terms of the upgoing fields and the reflection response of the layered medium above the layer in question. These downgoing wavefields are given by

$$A_n^+ = A_n^- R_{V;n}^- \exp(-\Gamma_{v;n} d_n), \tag{B.10}$$

for $1 \leq n \leq \varsigma - 1$. Likewise, for all layers below the source layer we write

$$A_n^- = A_n^+ R_{V;n}^+ \exp(-\Gamma_{v;n} d_n), \tag{B.11}$$

for $\varsigma + 1 \leq n \leq N$. These expressions are indicated in Figure B.1. The TM mode electric and magnetic fields are continuous across any interface, which for $z = z_{n-1}$ leads to

$$\begin{aligned} &A_n^- \left(1 + R_{V;n}^-\right) \exp(-\Gamma_{v;n} d_n) \\ &\quad = A_{n-1}^- \left(1 + R_{V;n-1}^- \exp(-2\Gamma_{v;n-1} d_{n-1})\right), \end{aligned} \tag{B.12}$$

$$\frac{\sigma_n}{\Gamma_{v;n}} A_n^- \left(1 - R_{V;n}^-\right) \exp(-\Gamma_{v;n} d_n)$$
$$= \frac{\sigma_{n-1}}{\Gamma_{v;n-1}} A_{n-1}^- \left(1 - R_{V;n-1}^- \exp(-2\Gamma_{v;n-1} d_{n-1})\right), \tag{B.13}$$

for $1 \leq n \leq \varsigma - 1$, and for $z = z_n$ to

$$A_n^+ \left(1 + R_{V;n}^+\right) \exp(-\Gamma_{v;n} d_n)$$
$$= A_{n+1}^+ \left(1 + R_{V;n+1}^+ \exp(-2\Gamma_{v;n+1} d_{n+1})\right), \tag{B.14}$$

$$\frac{\sigma_n}{\Gamma_{v;n}} A_n^+ \left(1 - R_{V;n}^+\right) \exp(-\Gamma_{v;n} d_n)$$
$$= \frac{\sigma_{n+1}}{\Gamma_{v;n+1}} A_{n+1}^+ \left(1 - R_{V;n+1}^+ \exp(-2\Gamma_{v;n+1} d_{n+1})\right), \tag{B.15}$$

for $\varsigma + 1 \leq n \leq N$.

First, we recognise that there is no downward-diffusing wavefield in $\mathbb{D}_0$, hence $R_{V;0}^- = 0$, and there is no upward-diffusing wavefield in $\mathbb{D}_{N+1}$, hence $R_{V;N+1}^+ = 0$. Evaluating equations B.12 and B.13 for $n = 1$ results in

$$R_{V;1}^- = r_{V;1}^- = \frac{\sigma_1 \Gamma_{v;0} - \sigma_0 \Gamma_{v;1}}{\sigma_1 \Gamma_{v;0} + \sigma_0 \Gamma_{v;1}}, \tag{B.16}$$

which implies that the global reflection, $R_{V;1}^-$, at the top interface is the local reflection, $r_{V;1}^-$, at that interface. Evaluating equations B.14 and B.15 for $n = N$ results in

$$R_{V;N}^+ = r_{V;N}^+ = \frac{\sigma_N \Gamma_{v;N+1} - \sigma_{N+1} \Gamma_{v;N}}{\sigma_N \Gamma_{v;N+1} + \sigma_{N+1} \Gamma_{v;N}}. \tag{B.17}$$

Solving for all other values of $n \neq \varsigma$ results in the general expressions

$$R_{V;n}^\pm = \frac{r_{V;n}^\pm + R_{V;n\pm1}^\pm \exp(-2\Gamma_{v;n\pm1} d_{n\pm1})}{1 + r_{V;n}^\pm R_{V;n\pm1}^\pm \exp(-2\Gamma_{v;n\pm1} d_{n\pm1})}, \tag{B.18}$$

$$r_{V;n}^\pm = \frac{\sigma_n \Gamma_{v;n\pm1} - \sigma_{n\pm1} \Gamma_{v;n}}{\sigma_n \Gamma_{v;n\pm1} + \sigma_{n\pm1} \Gamma_{v;n}}, \tag{B.19}$$

where the minus-sign in super- and subscript is used for $1 \leq n \leq \varsigma$ and the plus-sign for $\varsigma \leq n \leq N$.

Now we must evaluate the field inside the source layer, which is where the amplitudes are defined. Evaluating at $z = z_{\varsigma-1}$ gives, after some algebraic manipulation,

$$A_\varsigma^+ = R_{V;\varsigma}^- \left[X_V(z_{\varsigma-1}) \exp(-\Gamma_{v;\varsigma} d^-) + A_\varsigma^- \exp(-\Gamma_{v;\varsigma} d_\varsigma)\right], \tag{B.20}$$

with $d^- = z_s - z_{\varsigma-1}$. Evaluating at $z = z_\varsigma$ yields

$$A_\varsigma^- = R_{V;\varsigma}^+ \left[X_V(z_\varsigma) \exp(-\Gamma_{v;\varsigma} d^+) + A_\varsigma^+ \exp(-\Gamma_{v;\varsigma} d_\varsigma)\right], \tag{B.21}$$

with $d^+ = z_\varsigma - z_s$. We can see that the principle of the reflection response also applies in the source layer. The term in brackets on the right-hand side of equation B.20 is the total upgoing field and that is reflected by the reflection response $R^-_{V;\varsigma}$ defined at $z_{\varsigma-1}$ to give the total downward-reflected response A^+_ς. Similarly, the term in brackets on the right-hand side of equation B.21 is the total downgoing field and that is reflected by the reflection response $R^+_{V;\varsigma}$ defined at z_ς to give the total upward reflected response A^-_ς. These two equations can be solved for $A^\pm_\varsigma$ to find

$$A^-_\varsigma = R^+_{V;\varsigma}\frac{X_V(z_\varsigma)\exp(-\Gamma_{v;\varsigma}d^+) + R^-_{V;\varsigma}X_V(z_{\varsigma-1})\exp(-\Gamma_{v;\varsigma}(d^- + d_\varsigma))}{1 - R^+_{V;\varsigma}R^-_{V;\varsigma}\exp(-2\Gamma_{V;\varsigma}d_\varsigma)}, \tag{B.22}$$

$$A^+_\varsigma = R^-_{V;\varsigma}\frac{X_V(z_{\varsigma-1})\exp(-\Gamma_{v;\varsigma}d^-) + R^+_{V;\varsigma}X_V(z_\varsigma)\exp(-\Gamma_{v;\varsigma}(d^+ + d_\varsigma))}{1 - R^+_{V;\varsigma}R^-_{V;\varsigma}\exp(-2\Gamma_{V;\varsigma}d_\varsigma)}. \tag{B.23}$$

For the layer directly above the source layer, we connect the fields at $z = z_{\varsigma-1}$ and find

$$A^-_{\varsigma-1} = \frac{X_V(z_{\varsigma-1})\exp(-\Gamma_{v;s}d^-) + X_V(z_\varsigma)R^+_{V;\varsigma}\exp[-\Gamma_{v;s}(d^+ + d_s)]}{\left[1 + R^-_{V;\varsigma-1}\exp(-2\Gamma_{v;\varsigma-1}d_{\varsigma-1})\right]\left[1 - R^+_{V;\varsigma}R^-_{V;\varsigma}\exp(-2\Gamma_{v;\varsigma}d_\varsigma)\right]} \times \left(1 + R^-_{V;\varsigma}\right). \tag{B.24}$$

For the layer directly below the source layer, we connect the fields at $z = z_\varsigma$ and find

$$A^+_{\varsigma+1} = \frac{X_V(z_\varsigma)\exp(-\Gamma_{v;s}d^+) + X_V(z_{\varsigma-1})R^-_{V;\varsigma}\exp[-\Gamma_{v;s}(d^- + d_s)]}{\left[1 + R^+_{V;\varsigma+1}\exp(-2\Gamma_{v;\varsigma+1}d_{\varsigma+1})\right]\left[1 - R^+_{V;\varsigma}R^-_{V;\varsigma}\exp(-2\Gamma_{V;\varsigma}d_\varsigma)\right]} \times \left(1 + R^+_{V;\varsigma}\right). \tag{B.25}$$

With these results and equations B.12 and B.14, the amplitudes in every other layer can be determined recursively and this solves the problem as

$$A^-_{n-1} = A^-_n \frac{\left(1 + R^-_{V;n}\right)\exp(-\Gamma_{v;n}d_n)}{1 + R^-_{V;n-1}\exp(-2\Gamma_{v;n-1}d_{n-1})}, \tag{B.26}$$

for $1 \le n \le \varsigma - 1$ and

$$A^+_{n+1} = A^+_n \frac{\left(1 + R^+_{V;n}\right)\exp(-\Gamma_{v;n}d_n)}{1 + R^+_{V;n+1}\exp(-2\Gamma_{v;n+1}d_{n+1})}, \tag{B.27}$$

for $\varsigma + 1 \le n \le N$. The TE mode field is found by direct substitution and with this the electromagnetic field is determined in all space.

For a physical interpretation of the equations, let us look at the field inside layer $\mathbb{D}_{N-1}$, where the downward-diffusing part is given by A^+_{N-1}, the upward-diffusing part is given by $A^-_{N-1} = A^+_{N-1} R^+_{V;N-1} \exp(-\Gamma_{v;N-1} d_{N-1})$ and the global reflection coefficient is given by

$$R^+_{V;N-1} = \frac{r^+_{V;N-1} + r^+_{V;N} \exp(-2\Gamma_{v;N} d_N)}{1 + r^+_{V;N-1} r^+_{V;N} \exp(-2\Gamma_{v;N} d_N)}. \tag{B.28}$$

This is a convenient mathematical form that can be interpreted physically by writing the denominator as a Taylor series expansion given by

$$\begin{aligned} R^+_{V;N-1} =& \left[r^+_{V;N-1} + r^+_{V;N} \exp(-2\Gamma_{v;N} d_N) \right] \left[1 - r^+_{V;N-1} r^+_{V;N} \exp(-2\Gamma_{v;N} d_N) \right. \\ & \left. + (r^+_{V;N-1} r^+_{V;N})^2 \exp(-4\Gamma_{v;N} d_N) - \cdots \right], \\ =& \, r^+_{V;N-1} + \left[1 - (r^+_{V;N-1})^2 \right] r^+_{V;N} \exp(-2\Gamma_{v;N} d_N) \\ & - \left[1 - (r^+_{V;N-1})^2 \right] r^+_{V;N-1} (r^+_{V;N})^2 \exp(-4\Gamma_{v;N} d_N) + \cdots, \\ =& \, r^+_{V;N-1} + \left[1 - (r^+_{V;N-1})^2 \right] r^+_{V;N} \exp(-2\Gamma_{v;N} d_N) \sum_{m=0}^{\infty} \\ & \left(-r^+_{V;N-1} r^+_{V;N} \right)^m \exp(-2m\Gamma_{v;N} d_N). \end{aligned} \tag{B.29}$$

Equation B.29 states that the reflection response to a unit amplitude diffusive plane wavefield incident at $z = z_{N-1}$ can be written as the primary local reflection at $z = z_{N-1}$, a primary local reflection at $z = z_N$ that has passed twice through the boundary at $z = z_{N-1}$ with a two-way local transmission coefficient of $1 - (r^+_{V;N-1})^2$, plus all reverberations inside layer $\mathbb{D}_N$ that arrive as upgoing diffusive wavefields transmitted through and at depth level $z = z_{N-1}$. Each term $\exp(-2m\Gamma_{v;N} d_N)$ accounts for the delay and propagation attenuation of traversing layer $\mathbb{D}_N$ down and up m times, with $m = 0$ describing the primary reflection, $m = 1$ the first-order multiple and each larger value describes the mth order multiple. We can now see that equation B.28 has a denominator that generates all reverberations inside layer $\mathbb{D}_N$. If the downgoing field is not a single plane wavefield but a general downgoing plane wavefield A^+_{N-1}, the upgoing wavefield is the product of the global reflection response and the downgoing wavefield as stated in equation B.11. The general equation B.18 has such a denominator that reverberates all primary and multiple reflections from below depth level z_n in layer $\mathbb{D}_n$. From equation B.19 we can see that $r^+_{V;n-1} = -r^-_{V;n}$ so that we can write equation B.28 as

$$R^+_{V;N-1} = \frac{r^+_{V;N-1} + r^+_{V;N} \exp(-2\Gamma_{v;N} d_N)}{1 - r^-_{V;N} r^+_{V;N} \exp(-2\Gamma_{v;N} d_N)}. \tag{B.30}$$

From this result we can see that the denominator in equations B.22 and B.23 generates all reverberations inside the source layer of all primary and multiple reflections from above the source layer and from below the source layer. In this layer all reverberations are connected throughout the whole layered model. These are then used in all layers above the source layer to compute the upgoing fields and in the layers below the source to compute the downgoing fields, and the whole field is then determined everywhere in the model. The global reflection responses are recursively built from the outer boundaries inward to the source layer, where they are connected. This requires one trip from the outer boundaries to the source layer. In terms of numerical complexity this means that the layered medium is traversed once. To compute the field inside any layer where there is a receiver, the total field is recursively built by updating from the receiver level to the source level also the downgoing field if the receiver is below the source, or the upgoing field if the receiver is above the source. This means that when the numerical scheme connects the upgoing and downgoing fields in the source layer, the total field at the receiver level is immediately known as well. This is the most efficient scheme to compute the total electromagnetic field inside a layered model.

For a configuration in which the receiver is in the same layer as, but below, the source, we can take the direct field from source to receiver together with the reflection response of the field that leaves the source in the downgoing direction and arrives at the receiver in the downgoing direction. The fields are separated in terms of direction in which they leave the source and at which they arrive at the receiver. These are denoted $\tilde{E}^{\pm\pm}_{V,\varsigma}(\kappa, z_r, z_s, s)$ where the first superscript denotes the direction in which the field diffuses towards the receiver and the second superscript denotes the direction in which the field diffuses away from the source. For both the plus-sign indicates the field diffuses in the downward direction and the minus-sign indicates the field diffuses in the upward direction. The total TM mode parts of the electric and magnetic fields are given by

$$\tilde{E}_{V,\varsigma} = \tilde{E}^{++}_{V,\varsigma} + \tilde{E}^{-+}_{V,\varsigma} + \tilde{E}^{+-}_{V,\varsigma} + \tilde{E}^{--}_{V,\varsigma}, \tag{B.31}$$

$$\tilde{H}_{V,\varsigma} = \tilde{H}^{++}_{V,\varsigma} + \tilde{H}^{-+}_{V,\varsigma} + \tilde{H}^{+-}_{V,\varsigma} + \tilde{H}^{--}_{V,\varsigma}. \tag{B.32}$$

The four parts in each field can be obtained using the above results of equations B.22 and B.23 in equation B.1 for the electric field and in equation B.2 for the magnetic field, and these parts are given by

$$(\tilde{E}^{\pm+}_{V,\varsigma}, \tilde{H}^{\pm+}_{V,\varsigma}) = \left(1, \pm\frac{\sigma_\varsigma}{\Gamma_{v;\varsigma}}\right)\left[\frac{\Gamma_{v;\varsigma}}{\sigma_\varsigma}\hat{I}^e_V + \frac{\kappa^2}{\sigma_{v;\varsigma}}\hat{I}^e_z + \hat{I}^m_V\right]\tilde{G}^{\pm+}_{V;\varsigma}, \tag{B.33}$$

$$(\tilde{E}^{\pm-}_{V,\varsigma}, \tilde{H}^{\pm-}_{V,\varsigma}) = \left(1, \pm\frac{\sigma_\varsigma}{\Gamma_{v;\varsigma}}\right)\left[\frac{\Gamma_{v;\varsigma}}{\sigma_\varsigma}\hat{I}^e_V - \frac{\kappa^2}{\sigma_{v;\varsigma}}\hat{I}^e_z - \hat{I}^m_V\right]\tilde{G}^{\pm-}_{V;\varsigma}, \tag{B.34}$$

in which the directional scalar Green's functions are given by

$$\tilde{G}^{++}_{V;\varsigma} = \frac{\exp(-\Gamma_{v;\varsigma}(z_r - z_s))}{2M_{V;\varsigma}}, \tag{B.35}$$

$$\tilde{G}^{-+}_{V;\varsigma} = \frac{R^{+}_{V;\varsigma} \exp(-\Gamma_{v;\varsigma}(2z_{\varsigma} - z_r - z_s))}{2M_{V;\varsigma}}, \tag{B.36}$$

$$\tilde{G}^{+-}_{V;\varsigma} = \frac{R^{-}_{V;\varsigma} \exp(-\Gamma_{v;\varsigma}(z_r + z_s - 2z_{\varsigma-1}))}{2M_{V;\varsigma}}, \tag{B.37}$$

$$\tilde{G}^{--}_{V;\varsigma} = \frac{R^{-}_{V;\varsigma} R^{+}_{V;\varsigma} \exp(-\Gamma_{v;\varsigma}(2d_{\varsigma} - z_r + z_s))}{2M_{V;\varsigma}}, \tag{B.38}$$

and the reverberation operator is given by

$$M_{V;\varsigma} = 1 - R^{+}_{V;\varsigma} R^{-}_{V;\varsigma} \exp(-2\Gamma_{v;\varsigma} d_{\varsigma}). \tag{B.39}$$

We use Table 4.1 to write down the solutions for the TE mode fields as

$$\left(\tilde{E}^{\pm+}_{H,\varsigma}, \tilde{H}^{\pm+}_{H,\varsigma}\right) = \left(\pm 1, \frac{\Gamma_{\varsigma}}{\zeta_{\varsigma}}\right) \left[\hat{I}^{m}_{H} + \frac{\kappa^2}{\Gamma_{\varsigma}} \hat{I}^{m}_{z} + \frac{\zeta_{\varsigma}}{\Gamma_{\varsigma}} \hat{I}^{e}_{H}\right] \tilde{G}^{\pm+}_{H;\varsigma}, \tag{B.40}$$

$$\left(\tilde{E}^{\pm-}_{H,\varsigma}, \tilde{H}^{\pm-}_{H,\varsigma}\right) = \left(\pm 1, \frac{\Gamma_{\varsigma}}{\zeta_{\varsigma}}\right) \left[\hat{I}^{m}_{H} - \frac{\kappa^2}{\Gamma_{\varsigma}} \hat{I}^{m}_{z} - \frac{\zeta_{\varsigma}}{\Gamma_{\varsigma}} \hat{I}^{e}_{H}\right] \tilde{G}^{\pm-}_{H;\varsigma}, \tag{B.41}$$

with

$$\tilde{G}^{++}_{H;\varsigma} = \frac{\exp(-\Gamma_{\varsigma}(z_r - z_s))}{2M_{H;\varsigma}}, \tag{B.42}$$

$$\tilde{G}^{-+}_{H;\varsigma} = \frac{R^{+}_{H;\varsigma} \exp(-\Gamma_{\varsigma}(2z_{\varsigma} - z_r - z_s))}{2M_{H;\varsigma}}, \tag{B.43}$$

$$\tilde{G}^{+-}_{H;\varsigma} = \frac{R^{-}_{H;\varsigma} \exp(-\Gamma_{\varsigma}(z_r + z_s - 2z_{\varsigma-1}))}{2M_{H;\varsigma}}, \tag{B.44}$$

$$\tilde{G}^{--}_{H;\varsigma} = \frac{R^{-}_{H;\varsigma} R^{+}_{H;\varsigma} \exp(-\Gamma_{\varsigma}(2d_{\varsigma} - z_r + z_s))}{2M_{H;\varsigma}}, \tag{B.45}$$

$$M_{H;\varsigma} = 1 - R^{+}_{H;\varsigma} R^{-}_{H;\varsigma} \exp(-2\Gamma_{\varsigma} d_{\varsigma}), \tag{B.46}$$

$$R^{\pm}_{H;n} = \frac{r^{\pm}_{H;n} + R^{\pm}_{H;n\pm1} \exp(-2\Gamma_{n\pm1} d_{n\pm1})}{1 + r^{\pm}_{H;n} R^{\pm}_{H;n\pm1} \exp(-2\Gamma_{n\pm1} d_{n\pm1})}, \tag{B.47}$$

$$r^{\pm}_{H;n} = \frac{\zeta_n \Gamma_{n\pm1} - \zeta_{n\pm1} \Gamma_n}{\zeta_n \Gamma_{n\pm1} + \zeta_{n\pm1} \Gamma_n}. \tag{B.48}$$

The electric and magnetic field vectors of the earth response to an electric current dipole source are now composed from equations B.33–B.48 using equations 4.51–4.53, and we express them explicitly for each component of the electric current dipole source with the aid of equations 4.49 and 4.46 as

$$\tilde{\mathbf{E}}^{\pm+}_{V;\varsigma} = \begin{pmatrix} \frac{(ik_x)^2\Gamma_{v;\varsigma}}{\kappa^2\sigma_\varsigma} & \frac{ik_x ik_y\Gamma_{v;\varsigma}}{\kappa^2\sigma_\varsigma} & -\frac{ik_x}{\sigma_{v;\varsigma}} \\ \frac{ik_x ik_y\Gamma_{v;\varsigma}}{\kappa^2\sigma_\varsigma} & \frac{(ik_y)^2\Gamma_{v;\varsigma}}{\kappa^2\sigma_\varsigma} & -\frac{ik_y}{\sigma_{v;\varsigma}} \\ \mp\frac{ik_x}{\sigma_{v;\varsigma}} & \mp\frac{ik_y}{\sigma_{v;\varsigma}} & \pm\frac{\sigma_\varsigma\kappa^2}{(\sigma_{v;\varsigma})^2\Gamma_{v;\varsigma}} \end{pmatrix} \begin{pmatrix} \hat{I}^e_x \\ \hat{I}^e_y \\ \hat{I}^e_z \end{pmatrix} \tilde{G}^{\pm+}_{V;\varsigma}, \tag{B.49}$$

$$\tilde{\mathbf{E}}^{\pm-}_{V;\varsigma} = \begin{pmatrix} \frac{(ik_x)^2\Gamma_{v;\varsigma}}{\kappa^2\sigma_\varsigma} & \frac{ik_x ik_y\Gamma_{v;\varsigma}}{\kappa^2\sigma_\varsigma} & \frac{ik_x}{\sigma_{v;\varsigma}} \\ \frac{ik_x ik_y\Gamma_{v;\varsigma}}{\kappa^2\sigma_\varsigma} & \frac{(ik_y)^2\Gamma_{v;\varsigma}}{\kappa^2\sigma_\varsigma} & \frac{ik_y}{\sigma_{v;\varsigma}} \\ \mp\frac{ik_x}{\sigma_{v;\varsigma}} & \mp\frac{ik_y}{\sigma_{v;\varsigma}} & \mp\frac{\sigma_\varsigma\kappa^2}{(\sigma_{v;\varsigma})^2\Gamma_{v;\varsigma}} \end{pmatrix} \begin{pmatrix} \hat{I}^e_x \\ \hat{I}^e_y \\ \hat{I}^e_z \end{pmatrix} \tilde{G}^{\pm-}_{V;\varsigma}, \tag{B.50}$$

$$\tilde{\mathbf{E}}^{\pm+}_{H;\varsigma} = \pm\begin{pmatrix} \frac{(ik_y)^2\zeta_\varsigma}{\kappa^2\Gamma_\varsigma} & -\frac{ik_x ik_y\zeta_\varsigma}{\kappa^2\Gamma_\varsigma} & 0 \\ -\frac{ik_x ik_y\zeta_\varsigma}{\kappa^2\Gamma_\varsigma} & \frac{(ik_x)^2\zeta_\varsigma}{\kappa^2\Gamma_\varsigma} & 0 \\ 0 & 0 & 0 \end{pmatrix} \begin{pmatrix} \hat{I}^e_x \\ \hat{I}^e_y \\ \hat{I}^e_z \end{pmatrix} \tilde{G}^{\pm+}_{H;\varsigma}, \tag{B.51}$$

$$\tilde{\mathbf{E}}^{\pm-}_{H;\varsigma} = \pm\begin{pmatrix} -\frac{(ik_y)^2\zeta_\varsigma}{\kappa^2\Gamma_\varsigma} & \frac{ik_x ik_y\zeta_\varsigma}{\kappa^2\Gamma_\varsigma} & 0 \\ \frac{ik_x ik_y\zeta_\varsigma}{\kappa^2\Gamma_\varsigma} & -\frac{(ik_x)^2\zeta_\varsigma}{\kappa^2\Gamma_\varsigma} & 0 \\ 0 & 0 & 0 \end{pmatrix} \begin{pmatrix} \hat{I}^e_x \\ \hat{I}^e_y \\ \hat{I}^e_z \end{pmatrix} \tilde{G}^{\pm-}_{H;\varsigma}. \tag{B.52}$$

Here we restrict ourselves to an electric dipole source, because the electric field generated by a magnetic dipole source can be obtained by reciprocity from the magnetic field generated by an electric dipole source. The magnetic fields generated by electric or magnetic sources are given by

$$\begin{aligned}\tilde{\mathbf{H}}^{\pm+}_{V;\varsigma} = \pm&\begin{pmatrix} -\frac{ik_x ik_y}{\kappa^2} & -\frac{(ik_y)^2}{\kappa^2} & \frac{ik_y\sigma_\varsigma}{\sigma_{v;\varsigma}\Gamma_{v;\varsigma}} \\ \frac{(ik_x)^2}{\kappa^2} & \frac{ik_x ik_y}{\kappa^2} & -\frac{ik_x\sigma_\varsigma}{\sigma_{v;\varsigma}\Gamma_{v;\varsigma}} \\ 0 & 0 & 0 \end{pmatrix} \begin{pmatrix} \hat{I}^e_x \\ \hat{I}^e_y \\ \hat{I}^e_z \end{pmatrix} \tilde{G}^{\pm+}_{V;\varsigma} \\ \pm&\begin{pmatrix} \frac{(ik_y)^2\sigma_\varsigma}{\kappa^2\Gamma_{v;\varsigma}} & -\frac{ik_x ik_y\sigma_\varsigma}{\kappa^2\Gamma_{v;\varsigma}} & 0 \\ -\frac{ik_x ik_y\sigma_\varsigma}{\kappa^2\Gamma_{v;\varsigma}} & \frac{(ik_x)^2\sigma_\varsigma}{\kappa^2\Gamma_{v;\varsigma}} & 0 \\ 0 & 0 & 0 \end{pmatrix} \begin{pmatrix} \hat{I}^m_x \\ \hat{I}^m_y \\ \hat{I}^m_z \end{pmatrix} \tilde{G}^{\pm+}_{V;\varsigma},\end{aligned} \tag{B.53}$$

$$\begin{aligned}\tilde{\mathbf{H}}^{\pm-}_{V;\varsigma} = \pm&\begin{pmatrix} -\frac{ik_x ik_y}{\kappa^2} & -\frac{(ik_y)^2}{\kappa^2} & -\frac{ik_y\sigma_\varsigma}{\sigma_{v;\varsigma}\Gamma_{v;\varsigma}} \\ \frac{(ik_x)^2}{\kappa^2} & \frac{ik_x ik_y}{\kappa^2} & \frac{ik_x\sigma_\varsigma}{\sigma_{v;\varsigma}\Gamma_{v;\varsigma}} \\ 0 & 0 & 0 \end{pmatrix} \begin{pmatrix} \hat{I}^e_x \\ \hat{I}^e_y \\ \hat{I}^e_z \end{pmatrix} \tilde{G}^{\pm-}_{V;\varsigma} \\ \pm&\begin{pmatrix} -\frac{(ik_y)^2\sigma_\varsigma}{\kappa^2\Gamma_{v;\varsigma}} & \frac{ik_x ik_y\sigma_\varsigma}{\kappa^2\Gamma_{v;\varsigma}} & 0 \\ \frac{ik_x ik_y\sigma_\varsigma}{\kappa^2\Gamma_{v;\varsigma}} & -\frac{(ik_x)^2\sigma_\varsigma}{\kappa^2\Gamma_{v;\varsigma}} & 0 \\ 0 & 0 & 0 \end{pmatrix} \begin{pmatrix} \hat{I}^m_x \\ \hat{I}^m_y \\ \hat{I}^m_z \end{pmatrix} \tilde{G}^{\pm-}_{V;\varsigma},\end{aligned} \tag{B.54}$$

$$\tilde{\mathbf{H}}^{\pm+}_{H;\varsigma} = \begin{pmatrix} \frac{\mathrm{i}k_x \mathrm{i}k_y}{\kappa^2} & -\frac{(\mathrm{i}k_x)^2}{\kappa^2} & 0 \\ \frac{(\mathrm{i}k_y)^2}{\kappa^2} & -\frac{\mathrm{i}k_x \mathrm{i}k_y}{\kappa^2} & 0 \\ \mp\frac{\mathrm{i}k_y}{\Gamma_\varsigma} & \pm\frac{\mathrm{i}k_x}{\Gamma_\varsigma} & 0 \end{pmatrix} \begin{pmatrix} \hat{I}^e_x \\ \hat{I}^e_y \\ \hat{I}^e_z \end{pmatrix} \tilde{G}^{\pm+}_{H;\varsigma}$$

$$+ \begin{pmatrix} \frac{(\mathrm{i}k_x)^2\Gamma_\varsigma}{\kappa^2\zeta_\varsigma} & \frac{\mathrm{i}k_x \mathrm{i}k_y\Gamma_\varsigma}{\kappa^2\zeta_\varsigma} & -\frac{\mathrm{i}k_x}{\zeta_\varsigma} \\ \frac{\mathrm{i}k_x \mathrm{i}k_y\Gamma_\varsigma}{\kappa^2\zeta_\varsigma} & \frac{(\mathrm{i}k_y)^2\Gamma_\varsigma}{\kappa^2\zeta_\varsigma} & -\frac{\mathrm{i}k_y}{\zeta_\varsigma} \\ \mp\frac{\mathrm{i}k_x}{\zeta_\varsigma} & \mp\frac{\mathrm{i}k_y}{\zeta_\varsigma} & \pm\frac{\kappa^2}{\Gamma_\varsigma} \end{pmatrix} \begin{pmatrix} \hat{I}^m_x \\ \hat{I}^m_y \\ \hat{I}^m_z \end{pmatrix} \tilde{G}^{\pm+}_{H;\varsigma}, \tag{B.55}$$

$$\tilde{\mathbf{H}}^{\pm-}_{H;\varsigma} = \begin{pmatrix} -\frac{\mathrm{i}k_x \mathrm{i}k_y}{\kappa^2} & \frac{(\mathrm{i}k_x)^2}{\kappa^2} & 0 \\ -\frac{(\mathrm{i}k_y)^2}{\kappa^2} & \frac{\mathrm{i}k_x \mathrm{i}k_y}{\kappa^2} & 0 \\ \pm\frac{\mathrm{i}k_y}{\Gamma_\varsigma} & \mp\frac{\mathrm{i}k_x}{\Gamma_\varsigma} & 0 \end{pmatrix} \begin{pmatrix} \hat{I}^e_x \\ \hat{I}^e_y \\ \hat{I}^e_z \end{pmatrix} \tilde{G}^{\pm-}_{H;\varsigma}$$

$$+ \begin{pmatrix} \frac{(\mathrm{i}k_x)^2\Gamma_\varsigma}{\kappa^2\zeta_\varsigma} & \frac{\mathrm{i}k_x \mathrm{i}k_y\Gamma_\varsigma}{\kappa^2\zeta_\varsigma} & \frac{\mathrm{i}k_x}{\zeta_\varsigma} \\ \frac{\mathrm{i}k_x \mathrm{i}k_y\Gamma_\varsigma}{\kappa^2\zeta_\varsigma} & \frac{(\mathrm{i}k_y)^2\Gamma_\varsigma}{\kappa^2\zeta_\varsigma} & \frac{\mathrm{i}k_y}{\zeta_\varsigma} \\ \mp\frac{\mathrm{i}k_x}{\zeta_\varsigma} & \mp\frac{\mathrm{i}k_y}{\zeta_\varsigma} & \mp\frac{\kappa^2}{\Gamma_\varsigma} \end{pmatrix} \begin{pmatrix} \hat{I}^m_x \\ \hat{I}^m_y \\ \hat{I}^m_z \end{pmatrix} \tilde{G}^{\pm-}_{H;\varsigma}. \tag{B.56}$$

These expressions define the electromagnetic field in the horizontal wavenumber–frequency domain for source and receiver placed in any layer in the model.

The whole space model. In the whole space no reflections occur and when the receiver is below the source only $\tilde{\mathbf{E}}^{++}$ is non-zero, whereas when the receiver is above the source only $\tilde{\mathbf{E}}^{--}$ is non-zero, which also applies to the same parts of the magnetic field. These results can be combined to yield the electric field TE and TM mode matrix Green's functions given in equations 4.61 and 4.62. The magnetic field generated by an electric dipole source is expressed by the matrix Green's functions given by

$$\tilde{\mathbf{G}}^{me}_H = \begin{pmatrix} \frac{k_x k_y \partial}{\kappa^2 \partial z} & -\frac{k_x^2 \partial}{\kappa^2 \partial z} & 0 \\ \frac{k_y^2 \partial}{\kappa^2 \partial z} & -\frac{k_x k_y \partial}{\kappa^2 \partial z} & 0 \\ -\mathrm{i}k_y & \mathrm{i}k_x & 0 \end{pmatrix} \tilde{G}_H, \tag{B.57}$$

where the third column is zero because the vertical electric dipole generates purely TM mode. The TM mode Green's function is found as

$$\tilde{\mathbf{G}}^{me}_V = \begin{pmatrix} -\frac{k_x k_y \partial}{\kappa^2 \partial z} & -\frac{k_y^2 \partial}{\kappa^2 \partial z} & \frac{\sigma}{\sigma_v}\mathrm{i}k_y \\ \frac{k_x^2 \partial}{\kappa^2 \partial z} & \frac{k_x k_y \partial}{\kappa^2 \partial z} & -\frac{\sigma}{\sigma_v}\mathrm{i}k_x \\ 0 & 0 & 0 \end{pmatrix} \tilde{G}_V, \tag{B.58}$$

where the third row is zero because the vertical magnetic field is purely TE mode. The reciprocity relation $\tilde{\mathbf{G}}^{em}_M = \left(\tilde{\mathbf{G}}^{me}_M\right)^t$ can be used to find the expressions for the Green's matrices of the electric field generated by a magnetic dipole source. The matrix Green's function of the magnetic field generated by a magnetic dipole source can be taken from the electric field generated by an electric dipole source in the following way. Replacing $\tilde{G}_V/\sigma_v$ by $\tilde{G}_H/\zeta$ in $\tilde{\mathbf{G}}_V^{ee}$ results in $\tilde{\mathbf{G}}_H^{mm}$, and replacing $\zeta\tilde{G}_H$ by $\sigma\tilde{G}_V$ in $\tilde{\mathbf{G}}_H^{ee}$ results in $\tilde{\mathbf{G}}_V^{mm}$.

The half-space model. In the half-space model we assume the upper half-space to be non-conductive while the conductive lower half-space contains the source. We are interested in the field at and below the surface between the two half-spaces. The direct field in the half-space is equal to the total field in a whole space and given by equations 4.61 and 4.62. It suffices to indicate that we use the superscript ^{i} to indicate the incident field and the corresponding scalar Green's functions in the incident field are denoted $\tilde{G}_M^-$ for $M = \{V, H\}$. The boundary creates a reflected field and the electric field generated by an electric dipole source reflects as $\tilde{\mathbf{G}}_M^{ee;r}$ and comes from $\tilde{\mathbf{E}}_M^{+-}$ with the scalar Green's function given by

$$\tilde{G}_V^{+-} = \frac{1}{2}r_V^- \exp(-\Gamma_v(z_r + z_s)), \tag{B.59}$$

$$\tilde{G}_H^{+-} = \frac{1}{2}r_H^- \exp(-\Gamma(z_r + z_s)). \tag{B.60}$$

Once we have the final expressions we can take the limit of zero conductivity for σ_0 in the upper half-space and find the approximate expressions for the local reflection coefficients given in equations B.19 and B.48 as

$$r_V^- \approx 1, \tag{B.61}$$

$$r_H^- \approx \frac{\kappa - \Gamma}{\kappa + \Gamma} = -1 - 2\frac{\kappa^2 - \kappa\Gamma}{\gamma^2}. \tag{B.62}$$

Substituting these approximations in the expressions for $\tilde{\mathbf{E}}_V^{+-}$ and $\tilde{\mathbf{E}}_H^{+-}$ of equations B.50 and B.52 leads to the Green's matrix functions given in equations 4.102 and 4.103, respectively. The Green's matrices for the magnetic field generated by an electric dipole are obtained using the above results in equations B.54 and B.56 using electric sources, hence putting $\mathbf{I}^m = 0$. They are given by

$$\tilde{\mathbf{G}}_V^{me;r} = \begin{pmatrix} \frac{ik_x ik_y \partial}{\kappa^2 \partial_z} & \frac{(ik_y)^2 \partial}{\kappa^2 \partial z} & -ik_y\lambda^2 \\ -\frac{(ik_x)^2 \partial}{\kappa^2 \partial z} & -\frac{ik_x ik_y \partial}{\kappa^2 \partial z} & ik_x\lambda^2 \\ 0 & 0 & 0 \end{pmatrix} \tilde{G}_V^+, \tag{B.63}$$

$$\tilde{\mathbf{G}}_H^{me;r} = \begin{pmatrix} -\frac{\mathrm{i}k_x \mathrm{i}k_y \partial}{\kappa^2 \partial_z} & \frac{(\mathrm{i}k_x)^2 \partial}{\kappa^2 \partial z} & 0 \\ -\frac{(\mathrm{i}k_y)^2 \partial}{\kappa^2 \partial z} & \frac{\mathrm{i}k_x \mathrm{i}k_y \partial}{\kappa^2 \partial z} & 0 \\ -\mathrm{i}k_y & \mathrm{i}k_x & 0 \end{pmatrix} \left(1 + 2\frac{\kappa^2}{\gamma^2} + \frac{\kappa \partial}{\gamma^2 \partial z}\right) \tilde{G}_H^+, \tag{B.64}$$

in which the scalar Green's functions are given in equations 4.54 and 4.55.

The Green's matrices for the magnetic field generated by a magnetic dipole are obtained using the above results in equations B.54 and B.56 using magnetic sources, hence putting $\mathbf{I}^e = 0$. They are given by

$$\tilde{\mathbf{G}}_V^{mm;r} = \begin{pmatrix} -\frac{(\mathrm{i}k_y)^2 \sigma}{\kappa^2} & \frac{\mathrm{i}k_x \mathrm{i}k_y \sigma}{\kappa^2} & 0 \\ \frac{\mathrm{i}k_x \mathrm{i}k_y \sigma}{\kappa^2} & -\frac{(\mathrm{i}k_x)^2 \sigma}{\kappa^2} & 0 \\ 0 & 0 & 0 \end{pmatrix} \tilde{G}_V^+, \tag{B.65}$$

$$\tilde{\mathbf{G}}_H^{mm;r} = -\begin{pmatrix} \frac{(\mathrm{i}k_x)^2 \Gamma^2}{\kappa^2 \zeta} & \frac{\mathrm{i}k_x \mathrm{i}k_y \Gamma^2}{\kappa^2 \zeta} & \frac{\mathrm{i}k_x \Gamma}{\zeta} \\ \frac{\mathrm{i}k_x \mathrm{i}k_y \Gamma^2}{\kappa^2 \zeta} & \frac{(\mathrm{i}k_y)^2 \Gamma^2}{\kappa^2 \zeta} & \frac{\mathrm{i}k_y \Gamma}{\zeta} \\ -\frac{\mathrm{i}k_x \Gamma}{\zeta} & -\frac{\mathrm{i}k_y \Gamma}{\zeta} & -\kappa^2 \end{pmatrix} \left(1 + 2\frac{\kappa^2}{\gamma^2} + \frac{\kappa \partial}{\gamma^2 \partial z}\right) \tilde{G}_H^+. \tag{B.66}$$

With these expressions, the electromagnetic field in the conductive half-space is determined.

B.1 Space–frequency domain solutions

The marine application of the CSEM method for hydrocarbon exploration has three main acquisition modes of operation. Two modes use a horizontal electric dipole source towed behind a vessel at a certain depth. In one configuration the receivers are stations positioned at fixed locations on the sea floor, whereas the other configuration uses a streamer containing the electrodes. The seabed stations usually measure all components of the magnetic flux, the two horizontal components and sometimes the vertical component of the electric field, whereas a streamer measures only the in-line electric field component. The third method uses vertical electric dipoles for transmitter and receiver at fixed positions for each measurement. This is a time domain method that measures the earth response to a block pulse after switch-off. In the following we develop expressions to obtain a horizontal electric dipole as source and as receiver.

The usual way to find space–domain solutions is to add all contributions together and then perform the inverse Fourier–Bessel transformation. It is interesting to transform every up- and downgoing contribution back to the space domain separately to investigate its contribution to the total field. Here we write down the solution for the x-component of the electric field generated by an x-directed electric dipole. It is given by

$$\hat{E}_{xx}(\mathbf{r}_r, \mathbf{r}_s, \omega) = \hat{E}^{++}_{xx;V} + \hat{E}^{-+}_{xx;V} + \hat{E}^{+-}_{xx;V} + \hat{E}^{--}_{xx;V} + \hat{E}^{++}_{xx;H} + \hat{E}^{-+}_{xx;H} + \hat{E}^{+-}_{xx;H} + \hat{E}^{--}_{xx;H}. \tag{B.67}$$

The TM mode downgoing electric field from the downgoing part of the electric dipole source is given by

$$\hat{E}^{++}_{xx;V}(\mathbf{r}_r, \mathbf{r}_s, \omega) = \frac{\hat{I}^e_x \partial_x \partial_x}{4\pi\sigma_1} \int_{\kappa=0}^{\infty} \frac{\Gamma_{v;1}}{\kappa M_{V;1}} \exp(-\Gamma_{v;1}|h^-|) \mathrm{J}_0(\kappa r) d\kappa, \tag{B.68}$$

where $ik_x \to \partial_x$ has been used. We now carry out the differentiations and find

$$\hat{E}^{++}_{xx;V}(\mathbf{r}_r, \mathbf{r}_s, \omega) = -\frac{\hat{I}^e_x x^2}{4\pi\sigma_1 r^2} \int_{\kappa=0}^{\infty} \frac{\Gamma_{v;1}}{M_{V;1}} \exp(-\Gamma_{v;1}|h^-|) \mathrm{J}_0(\kappa r) \kappa d\kappa + \frac{\hat{I}^e_x (x^2 - y^2)}{4\pi\sigma_1 r^3} \int_{\kappa=0}^{\infty} \frac{\Gamma_{v;1}}{M_{V;1}} \exp(-\Gamma_{v;1}|h^-|) \mathrm{J}_1(\kappa r) d\kappa. \tag{B.69}$$

In the case $r \to 0$ while $h^- \neq 0$, we use equation C.14 in equation B.68 and find

$$\hat{E}^{++}_{xx;V}(0, 0, z_r, z_s, \omega) = -\frac{\hat{I}^e_x}{8\pi\sigma_1} \int_{\kappa=0}^{\infty} \frac{\Gamma_{v;1}}{M_{V;1}} \exp(-\Gamma_{v;1}|h^-|) \kappa d\kappa, \tag{B.70}$$

which presents no numerical difficulties. For all non-zero offsets we already know from the analysis of the whole space that the second integral in equation B.69 will produce an artefact that we will remove by evaluating

$$\begin{aligned}\hat{E}^{++}_{xx;V}(\mathbf{r}_r, \mathbf{r}_s, \omega) = &-\frac{\hat{I}^e_x x^2}{4\pi\sigma_1 r^2} \int_{\kappa=0}^{\infty} \frac{\Gamma_{v;1}}{M_{V;1}} \exp(-\Gamma_{v;1}|h^-|) \mathrm{J}_0(\kappa r) \kappa d\kappa \\ &+ \frac{\hat{I}^e_x (x^2 - y^2)}{4\pi\sigma_1 r^3} \int_{\kappa=0}^{\infty} \left(\frac{\Gamma_{v;1}}{M_{V;1}} - \frac{\gamma_1^2}{M_{V;1}(\kappa = 0)\Gamma_{v;1}} \right) \exp(-\Gamma_{v;1}|h^-|) \mathrm{J}_1(\kappa r) d\kappa \\ &- \frac{\zeta_1 \hat{I}^e_x (x^2 - y^2)}{4\pi\gamma_1 r^4} \frac{\exp(-\gamma_{v;1} R_v)}{M_{V;1}(\kappa = 0)}.\end{aligned} \tag{B.71}$$

To arrive at this result we have used equation A.12 and removed the part of the exponent that depends only on vertical distance. This makes sense because the same term is found in the TE mode with opposite sign, just as in the homogeneous space, because $M_{V;1}(\kappa = 0) = M_{H;1}(\kappa = 0)$. Now both integrals give physically meaningful results together with the third correction factor. The TE mode equivalent can be written at zero offset as

$$\hat{E}^{++}_{xx;H}(0, 0, z_r, z_s, \omega) = -\frac{\zeta_1 \hat{I}^e_x}{8\pi} \int_{\kappa=0}^{\infty} \frac{\exp(-\Gamma_1|h^-|)}{\Gamma_1 M_{H;1}} \kappa d\kappa, \tag{B.72}$$

and for non-zero offsets as

$$\hat{E}^{++}_{xx;H}(\mathbf{r}_r,\mathbf{r}_s,\omega) = -\frac{\zeta_1\hat{I}^e_x y^2}{4\pi r^2}\int_{\kappa=0}^{\infty}\frac{\exp(-\Gamma_1|h^-|)}{\Gamma_1 M_{H;1}}\mathrm{J}_0(\kappa r)\kappa\, d\kappa$$
$$-\frac{\zeta_1\hat{I}^e_x(x^2-y^2)}{4\pi r^3}\int_{\kappa=0}^{\infty}\left(\frac{1}{M_{H;1}}-\frac{1}{M_{H;1}(\kappa=0)}\right)\frac{\exp(-\Gamma_1|h^-|)}{\Gamma_1}\mathrm{J}_1(\kappa r)d\kappa$$
$$+\frac{\zeta_1\hat{I}^e_x(x^2-y^2)}{4\pi\gamma_1 r^4}\frac{\exp(-\gamma_1 R)}{M_{H;1}(\kappa=0)}. \tag{B.73}$$

These results and the results for the other up/down combinations in both modes are given in equations 4.165–4.181.

The whole space model. In the whole space the electric field generated by an electric dipole source is given in Appendix A. The others are given here and we can write the expressions without having to evaluate new integrals. The magnetic field impulse response to an electric dipole in a VTI whole space is obtained with the aid of equations A.8–A.10 and A.12 in the inverse spatial Fourier transformation of equations B.57 and B.58, and carrying out all derivatives with the aid of Appendix C. For $r > 0$ we remove the exponential term that depends only on vertical distance and find

$$\hat{\mathbf{G}}^{me}_H(\mathbf{r}_r,\mathbf{r}_s,s) = \begin{pmatrix} -\frac{xyh^-}{r^2} & \frac{x^2h^-}{r^2} & 0 \\ -\frac{y^2h^-}{r^2} & \frac{xyh^-}{r^2} & 0 \\ y & -x & 0 \end{pmatrix}(1+\gamma R)\frac{\exp(-\gamma R)}{4\pi R^3}$$
$$-\begin{pmatrix} \frac{2xy}{r^2} & \frac{x^2-y^2}{r^2} & 0 \\ \frac{x^2-y^2}{r^2} & -\frac{2xy}{r^2} & 0 \\ 0 & 0 & 0 \end{pmatrix}\frac{h^-\exp(-\gamma R)}{4\pi r^2 R}, \tag{B.74}$$

and

$$\hat{\mathbf{G}}^{me}_V(\mathbf{r}_r,\mathbf{r}_s,s) = \begin{pmatrix} \frac{xyh^-_v}{r^2} & \frac{y^2h^-_v}{r^2} & -\lambda y \\ -\frac{x^2h^-_v}{r^2} & -\frac{xyh^-_v}{r^2} & \lambda x \\ 0 & 0 & 0 \end{pmatrix}(1+\gamma_v R_v)\frac{\exp(-\gamma_v R_v)}{4\pi R_v^3}$$
$$+\begin{pmatrix} \frac{2xy}{r^2} & \frac{x^2-y^2}{r^2} & 0 \\ \frac{x^2-y^2}{r^2} & -\frac{2xy}{r^2} & 0 \\ 0 & 0 & 0 \end{pmatrix}\frac{h^-_v\exp(-\gamma_v R_v)}{4\pi r^2 R_v}. \tag{B.75}$$

In the whole space we can find the others from $\hat{\mathbf{G}}^{em}_M = (\hat{\mathbf{G}}^{me}_M)^t$, and $\hat{\mathbf{G}}^{mm}_H$ can be obtained from $\hat{\mathbf{G}}^{ee}_V$ by changing σ and σ_v to ζ and changing ζ to σ. Unlike in the wavenumber–frequency domain, it is not possible to obtain $\hat{\mathbf{G}}_V^{mm}$ from $\hat{\mathbf{G}}_H^{ee}$ by simple substitution. The space–frequency domain result for $\hat{\mathbf{G}}_V^{mm}$ is given by

$$\hat{\mathbf{G}}_V^{mm}(\mathbf{r}_r, \mathbf{r}_s, s) = -\sigma \begin{pmatrix} \frac{x^2}{r^2} & \frac{xy}{r^2} & 0 \\ \frac{xy}{r^2} & \frac{y^2}{r^2} & 0 \\ 0 & 0 & 0 \end{pmatrix} \frac{\exp(-\gamma_v R_v)}{4\pi\lambda R_v}$$
$$-\sigma \begin{pmatrix} \frac{x^2-y^2}{r^2} & 2\frac{xy}{r^2} & 0 \\ 2\frac{xy}{r^2} & \frac{y^2-x^2}{r^2} & 0 \\ 0 & 0 & 0 \end{pmatrix} \frac{\exp(-\gamma_v R_v)}{4\pi\gamma r^2}. \tag{B.76}$$

With these expressions the electromagnetic field is determined in the whole space.

The half-space model. In the half-space we can compare all expressions for the electromagnetic field of equations 4.102–4.103 and B.63–B.66. We observe that if we find expressions for the electric field generated by an electric dipole, we can write down the other fields. The TM mode reflection coefficient has the unit value and the reflected electric field is known. Apart from the derivatives in the matrix part of the TE mode reflected electric field Green's function, the new integral that comes from the inverse spatial Fourier transformation is given by

$$\mathcal{I}_H = \frac{1}{2\pi}\int_{\kappa=0}^{\infty} \tilde{G}_H^+(k_x, k_y, z, z_s, s) \mathrm{J}_0(\kappa r) d\kappa. \tag{B.77}$$

This is a known integral, (see Gradshteyn and Ryzhik, 1996: 1098, formula 6.637 1), and is given by

$$\mathcal{I}_H = \frac{1}{4\pi}\mathrm{I}_0(\xi^-)\mathrm{K}_0(\xi^+), \tag{B.78}$$

where I_0 denotes the modified Bessel function of the first kind and order 0 and K_0 denotes the modified Bessel function of the second kind and order 0. The arguments of the Bessel functions are given by

$$\xi^\pm = \frac{1}{2}\gamma(R^+ \pm h^+), \qquad R^+ = \sqrt{r^2 + (h^+)^2}. \tag{B.79}$$

For fixed h^+ and large values of γr, the Bessel function product in equation B.78 has an exponential decay as a function of h^+ only and polynomial decay as an inverse function of horizontal distance and as the ratio of the vertical distance over horizontal distance squared. By taking the limit we find the leading order term as

$$\lim_{\gamma r \to \infty} \mathrm{I}_0(\xi^-)\mathrm{K}_0(\xi^+) = \frac{\exp(-\gamma h^+)}{\gamma r}\left(1 + \mathcal{O}(r^{-2})\right), \tag{B.80}$$

which has the geometrical interpretation of diffusion vertically up in the conductive half-space, then lateral propagation in the air just above the surface and then diffusion vertically down into the conductive half-space. It can now be understood from our large horizontal offset analysis that the diffusive field contribution from the wave that has actually travelled through the air can be called the 'air wave'. The matrix elements in equation 4.103 transform to horizontal derivatives in the space domain from which we conclude that the air wave has a large horizontal offset geometrical spreading factor of r^{-3}. This means it is always a near-field term, which can be understood from the fact that we have taken $\varepsilon_0 = 0$ even in the air. With infinite propagation velocity, all finite distances travelled in the air are in the near-field. The electric field generated by an electric dipole is given in Section 4.3.1.

Appendix C

Green's Functions and Their Derivatives

Green's functions must be differentiated up to two times with respect to the coordinates. For the TM mode Green's function of a whole space they are given by

$$
\begin{aligned}
\frac{\partial}{\partial x}\hat{G}_V(\mathbf{r}-\mathbf{r}_s,s) &= \frac{\partial}{\partial x}\frac{\exp(-\gamma_v R_v)}{4\pi\lambda R_v}, \\
&= -\frac{x-x_s}{R_v}\left(\frac{1}{R_v}+\gamma_v\right)\hat{G}_V(\mathbf{r}-\mathbf{r}_s,s), \qquad (C.1)\\
\frac{\partial}{\partial z}\hat{G}_V(\mathbf{r}-\mathbf{r}_s,s) &= \frac{\partial}{\partial z}\frac{\exp(-\gamma_v R_v)}{4\pi\lambda R_v}, \\
&= -\frac{\lambda h_v^-}{R_v}\left(\frac{1}{R_v}+\gamma_v\right)\hat{G}_V(\mathbf{r}-\mathbf{r}_s,s). \qquad (C.2)
\end{aligned}
$$

The derivative with respect to y can be obtained from equation C.1 by replacing $x-x_s$ by $y-y_s$ in the numerator of the first fraction of that expression. It can also be seen that the vertical derivative is similar to a horizontal derivative but the scale factor λ occurs due to the vertical transverse isotropy (VTI) medium. Hence for second-order derivatives the one with respect to depth suffices to know the others. It is given by

$$
\begin{aligned}
\frac{\partial^2}{\partial z^2}\hat{G}_V(\mathbf{r}-\mathbf{r}_s,s) &= -\frac{\partial}{\partial z}\frac{\lambda h_v^-}{R_v}\left(\frac{1}{R_v}+\gamma_v\right)\hat{G}_V(\mathbf{r}-\mathbf{r}_s,s), \\
&= \left[\left(\frac{3\lambda^2(h_v^-)^2}{R_v^2}-\lambda^2\right)\left(\frac{1}{R_v^2}+\frac{\gamma_v}{R_v}\right)+\gamma_v^2\frac{\lambda^2(h_v^-)^2}{R_v^2}\right]\hat{G}_V(\mathbf{r}-\mathbf{r}_s,s). \qquad (C.3)
\end{aligned}
$$

For the derivatives with two coordinates, we show the one with respect to z and to x, which is given by

$$\frac{\partial^2}{\partial x \partial z}\hat{G}_V(\mathbf{r}-\mathbf{r}_s,s) = -\frac{\partial}{\partial x}\frac{\lambda h_v^-}{R_v}\left(\frac{1}{R_v}+\gamma_v\right)\hat{G}_V(\mathbf{r}-\mathbf{r}_s,s),$$
$$=\frac{\lambda x h_v^-}{R_v^2}\left(\frac{3}{R_v^2}+\frac{3\gamma_v}{R_v}+\gamma_v^2\right)\hat{G}_V(\mathbf{r}-\mathbf{r}_s,s), \tag{C.4}$$

and the others follow in similar ways. Derivatives for TE mode functions are readily obtained from these results by putting $\lambda = 1$ in all expressions. The Bessel function of the first kind and order zero $\mathrm{J}_0(\kappa r)$ is defined as

$$\mathrm{J}_0(\kappa r) = \frac{1}{2\pi}\int_{\psi=0}^{2\pi} \exp[-\mathrm{i}\kappa r\cos(\psi)]d\psi. \tag{C.5}$$

We need up to two orders of derivatives and when taken with respect to the argument of the Bessel function these can be written as

$$\mathrm{J}_0'(\kappa r) = -\mathrm{J}_1(\kappa r), \tag{C.6}$$
$$\mathrm{J}_1'(\kappa r) = \mathrm{J}_0(\kappa r) - \frac{1}{\kappa r}\mathrm{J}_1(\kappa r). \tag{C.7}$$

These two results permit the necessary differentiations, expressing them such that only J_0 and J_1 occur. Derivatives with respect to x and y are given by

$$\frac{\partial \mathrm{J}_0(\kappa r)}{\partial x} = \frac{x}{r}\kappa \mathrm{J}_0'(\kappa r) = -\frac{x}{r}\kappa \mathrm{J}_1(\kappa r), \tag{C.8}$$
$$\frac{\partial^2 \mathrm{J}_0(\kappa r)}{\partial x^2} = -\frac{x^2}{r^2}\kappa^2 \mathrm{J}_1'(\kappa r) - \left(\frac{1}{r}-\frac{x^2}{r^3}\right)\kappa \mathrm{J}_1(\kappa r),$$
$$= -\frac{x^2}{r^2}\kappa^2 \mathrm{J}_0(\kappa r) + \frac{x^2-y^2}{r^3}\kappa \mathrm{J}_1(\kappa r), \tag{C.9}$$
$$\frac{\partial^2 \mathrm{J}_0(\kappa r)}{\partial x \partial y} = -\frac{xy}{r^2}\kappa^2 \mathrm{J}_1'(\kappa r) + \frac{xy}{r^3}\kappa \mathrm{J}_1(\kappa r),$$
$$= -\frac{xy}{r^2}\kappa^2 \mathrm{J}_0(\kappa r) + 2\frac{xy}{r^3}\kappa \mathrm{J}_1(\kappa r). \tag{C.10}$$

The double differentiation with respect to y can be taken from the result from the double differentiation with respect to x by interchanging x and y. We also need the limiting values to be able to compute results for zero horizontal offset while the vertical offset is non-zero. For that we use

$$\mathrm{J}_0(0) = 1, \tag{C.11}$$
$$\lim_{r\to 0}\mathrm{J}_1(\kappa r) = \frac{1}{2}\kappa r, \tag{C.12}$$

and use these results in the right-hand sides of equations C.8–C.10 to find

$$\lim_{r\to 0}\frac{\partial \mathrm{J}_0(\kappa r)}{\partial x} = -\frac{1}{2}\kappa^2 \lim_{r\to 0} x = 0, \tag{C.13}$$

$$\lim_{r\to 0}\frac{\partial^2 \mathrm{J}_0(\kappa r)}{\partial x^2} = -\kappa^2 \lim_{r\to 0}\left(\frac{x^2}{r^2} - \frac{x^2 - y^2}{2r^2}\right) = -\frac{1}{2}\kappa^2, \tag{C.14}$$

$$\lim_{r\to 0}\frac{\partial^2 \mathrm{J}_0(\kappa r)}{\partial x \partial y} = -\kappa^2 \lim_{r\to 0}\left(\frac{xy}{r^2} - \frac{xy}{r^2}\right) = 0. \tag{C.15}$$

The modified Bessel functions occur only in the TE mode expression in equation B.78 and the results of their derivatives in the expression for the air wave in equation 4.113. They have arguments given in equation B.79. The modified Bessel function of the first kind and order zero $\mathrm{I}_0(\xi^-)$ is an almost exponentially growing function and that of the second kind and order zero $\mathrm{K}_0(\xi^+)$ is decreasing faster than exponential with increasing argument. They can be represented in integral form as

$$\mathrm{I}_0(\xi^-) = \frac{1}{\pi}\int_{\psi=0}^{\pi} \exp[\pm\xi^- \cos(\psi)]d\psi, \tag{C.16}$$

$$\mathrm{K}_0(\xi^+) = \frac{1}{\pi}\int_{\alpha=0}^{\infty} \frac{\cos(\alpha\xi^+)}{\sqrt{1+\alpha^2}}d\alpha. \tag{C.17}$$

From this representation it is clear that $\mathrm{I}_0(0) = 1$ and that $\mathrm{K}_0(0)$ is unbounded; $\mathrm{K}_0(\xi^+) \approx -\log(\xi^+)$ for small ξ^+. The rules for differentiation and addition are slightly different from those for the ordinary Bessel function, and are given by

$$\mathrm{I}_n'(\xi^-) = \frac{1}{2}\left[\mathrm{I}_{n-1}(\xi^-) + \mathrm{I}_{n+1}(\xi^-)\right], \tag{C.18}$$

$$\mathrm{I}_n(\xi^-) = \frac{\xi^-}{2n}\left[\mathrm{I}_{n-1}(\xi^-) - \mathrm{I}_{n+1}(\xi^-)\right], \tag{C.19}$$

$$\mathrm{K}_n'(\xi^+) = -\frac{1}{2}\left[\mathrm{K}_{n-1}(\xi^+) + \mathrm{K}_{n+1}(\xi^+)\right], \tag{C.20}$$

$$\mathrm{K}_n(\xi^+) = -\frac{\xi^+}{2n}\left[\mathrm{K}_{n-1}(\xi^+) - \mathrm{K}_{n+1}(\xi^+)\right]. \tag{C.21}$$

We further have

$$\mathrm{I}_0'(\xi^-) = \mathrm{I}_1(\xi^-), \tag{C.22}$$

$$\mathrm{K}_0'(\xi^+) = -\mathrm{K}_1(\xi^+). \tag{C.23}$$

In these expression $\xi^{\pm} = \gamma(R^+ \pm h^+)/2$, with $h^+ = z + z_s$ and R^+ as defined below equation 4.106. The vertical derivative of the product of modified Bessel functions can be written as

$$\partial_z \mathrm{I}_0(\xi^-)\mathrm{K}_0(\xi^+) = -\frac{1}{R^+}\left[\xi^-\mathrm{I}_1(\xi^-)\mathrm{K}_0(\xi^+) + \xi^+\mathrm{I}_0(\xi^-)\mathrm{K}_1(\xi^+)\right]. \tag{C.24}$$

In this equation we have used $\partial_z\xi^\pm = \pm\xi^\pm/R^+$. Then we need the horizontal derivative of the vertical derivative result. Partial results are given as

$$\partial_y\xi^-\mathrm{I}_1(\xi^-) = \frac{\gamma y}{2R^+}\left[\mathrm{I}_1(\xi^-) + \xi^-\mathrm{I}_1'(\xi^-)\right] = \frac{\gamma y}{2(R^+)^2}\xi^-\mathrm{I}_0(\xi^-), \tag{C.25}$$

$$\partial_y\xi^+\mathrm{K}_1(\xi^+) = \frac{\gamma y}{2R^+}\left[\mathrm{K}_1(\xi^+) + \xi^+\mathrm{K}_1'(\xi^+)\right] = -\frac{\gamma y}{2R^+}\xi^+\mathrm{K}_0(\xi^+), \tag{C.26}$$

where equations C.18–C.21 have been used to simplify the expressions. We therefore find

$$\begin{aligned}\partial_y\partial_z\mathrm{I}_0(\xi^-)\mathrm{K}_0(\xi^+) &= -\partial_y\frac{1}{R^+}\left[\xi^-\mathrm{I}_1(\xi^-)\mathrm{K}_0(\xi^+) + \xi^+\mathrm{I}_0(\xi^-)\mathrm{K}_1(\xi^+)\right],\\ &= \frac{y}{(R^+)^3}\left[\xi^-\mathrm{I}_1(\xi^-)\mathrm{K}_0(\xi^+) + \xi^+\mathrm{I}_0(\xi^-)\mathrm{K}_1(\xi^+)\right]\\ &\quad + \frac{\gamma^2 y h^+}{2(R^+)^2}\left[\mathrm{I}_0(\xi^-)\mathrm{K}_0(\xi^+) - \mathrm{I}_1(\xi^-)\mathrm{K}_1(\xi^+)\right],\end{aligned} \tag{C.27}$$

where $\xi^+ - \xi^- = \gamma h^+$ has been used. Now we need to take the second derivative to y, which results in

$$\begin{aligned}&\partial_y\frac{y}{(R^+)^3}\left[\xi^-\mathrm{I}_1(\xi^-)\mathrm{K}_0(\xi^+) + \xi^+\mathrm{I}_0(\xi^-)\mathrm{K}_1(\xi^+)\right]\\ &\quad = \left(1 - \frac{3y^2}{(R^+)^2}\right)\frac{1}{(R^+)^3}\left[\xi^-\mathrm{I}_1(\xi^-)\mathrm{K}_0(\xi^+) + \xi^+\mathrm{I}_0(\xi^-)\mathrm{K}_1(\xi^+)\right]\\ &\quad - \frac{\gamma^2 y^2 h^+}{2(R^+)^4}\left[\mathrm{I}_0(\xi^-)\mathrm{K}_0(\xi^+) - \mathrm{I}_1(\xi^-)\mathrm{K}_1(\xi^+)\right],\end{aligned} \tag{C.28}$$

and

$$\begin{aligned}&\frac{\gamma^2 h^+}{2}\partial_y\frac{y}{(R^+)^2}\left[\mathrm{I}_0(\xi^-)\mathrm{K}_0(\xi^+) - \mathrm{I}_1(\xi^-)\mathrm{K}_1(\xi^+)\right]\\ &\quad = \left(1 - \frac{2y^2}{(R^+)^2}\right)\frac{\gamma^2 h^+}{2(R^+)^2}\left[\mathrm{I}_0(\xi^-)\mathrm{K}_0(\xi^+) - \mathrm{I}_1(\xi^-)\mathrm{K}_1(\xi^+)\right]\\ &\qquad - \frac{\gamma^3 y^2 h^+}{2(R^+)^3}\left[\mathrm{I}_0(\xi^-)\mathrm{K}_1(\xi^+) - \mathrm{I}_1(\xi^-)\mathrm{K}_0(\xi^+)\right.\\ &\qquad \left. + \mathrm{I}_1'(\xi^-)\mathrm{K}_1(\xi^+) + \mathrm{I}_1(\xi^-)\mathrm{K}_1'(\xi^+)\right].\end{aligned} \tag{C.29}$$

In this last expression we use equation C.18 to avoid a numerical singularity for I_1' when $r = 0$, and equation C.20 to simplify the expression. We then obtain,

$$
\begin{aligned}
&\partial_y \frac{\gamma^2 y h^+}{2(R^+)^2} \left[\mathrm{I}_0(\xi^-)\mathrm{K}_0(\xi^+) - \mathrm{I}_1(\xi^-)\mathrm{K}_1(\xi^+)\right] \\
&= \left(1 - \frac{2y^2}{(R^+)^2}\right) \frac{\gamma^2 h^+}{2(R^+)^2} \left[\mathrm{I}_0(\xi^-)\mathrm{K}_0(\xi^+) - \mathrm{I}_1(\xi^-)\mathrm{K}_1(\xi^+)\right] \\
&\quad - \frac{\gamma^3 y^2 h^+}{4(R^+)^3} \left\{[3\mathrm{I}_0(\xi^-) + \mathrm{I}_2(\xi^-)]\mathrm{K}_1(\xi^+) - \mathrm{I}_1(\xi^-)[3\mathrm{K}_0(\xi^+) + \mathrm{K}_2(\xi^+)]\right\}.
\end{aligned} \tag{C.30}
$$

Derivatives with respect to x and y or twice with respect to x can be easily obtained from the above results.

Appendix D

The Final Value Theorem

If a measurement of a transient signal can be described by the causal time function $f(t)$ with $f(t) = 0$ for $t < 0$, then its Laplace transform is given by

$$\hat{f}(s) = \int_{t=0}^{\infty} f(t) \exp(-st) dt. \tag{D.1}$$

Using the rule for integration by parts, the derivative of the time function, $df(t)/dt$, has a Laplace transform which may be written as

$$\begin{aligned} \int_{t=0}^{\infty} \frac{df(t)}{dt} \exp(-st) dt &= \int_{t=0}^{\infty} \frac{df(t) \exp(-st)}{dt} dt + s \int_{t=0}^{\infty} f(t) \exp(-st) dt, \\ &= \lim_{t \to \infty} f(t) \exp(-st) - f(0) + s\hat{f}(s), \end{aligned} \tag{D.2}$$

where the first two terms are the endpoint evaluations of the first integral and last term comes from recognizing the second integral in the right-hand side as the definition of the Laplace transform. This result, under the assumption that

$$\lim_{t \to \infty} f(t) \exp(-st) = 0 \quad \forall \ \Re\{s\} \geq 0, \tag{D.3}$$

reduces to

$$\int_{t=0}^{\infty} \frac{df(t)}{dt} \exp(-st) dt = s\hat{f}(s) - f(0). \tag{D.4}$$

If we take the limit of s to zero, we find

$$\lim_{s \to 0} \int_{t=0}^{\infty} \frac{df(t)}{dt} \exp(-st) dt = \lim_{s \to 0} [s\hat{f}(s)] - f(0). \tag{D.5}$$

We can interchange the integration and the limit to find

$$\int_{t=0}^{\infty} \frac{df(t)}{dt} \lim_{s \to 0} \exp(-st) dt = \int_{t=0}^{\infty} \frac{df(t)}{dt} dt = f(\infty) - f(0). \tag{D.6}$$

If the integral on the left-hand side of equation D.5 exists and the assumption of equation D.3 holds, the limit and integral can be interchanged without changing the outcome. In that case the right-hand sides of equations D.5 and D.6 are the same and we find

$$f(\infty) = \lim_{s \to 0} [s\hat{f}(s)]. \tag{D.7}$$

Equation (D.7) is the Final Value Theorem and holds under the specified conditions. It states that the end value in the time domain step response is equal to the zero s limit of the Laplace transformed impulse response. For example, equation 4.141 gives the time domain expression of the anisotropic half-space step response and equation 4.138 the Laplace transformed domain impulse response. It can be observed that

$$IG_{xx}^{ee}(r, 0, 0, \infty) = \hat{G}_{xx}^{ee}(r, 0, 0, 0) = \frac{\lambda}{\pi \sigma r^3}. \tag{D.8}$$

Under these strict conditions we can also use the Fourier transformation by taking $s = -i\omega$ and writing

$$f(\infty) = \lim_{i\omega \to 0} [-i\omega \hat{f}(\omega)]. \tag{D.9}$$

It is clear the function $\hat{f}(s)$ or $\hat{f}(\omega)$ can have only a single pole at the origin. The strict causality conditions ensure that all non-zero poles or other roots occur with negative real values of s and hence there are no roots in complex-conjugate pairs on the imaginary s or real ω axes.

References

Abramowitz, M., and Stegun, I. A. 1972. *Handbook of mathematical functions*. 10th edn. Washington, DC: National Bureau of Standards. ISBN: 0-486-61272-4.

Abubakar, A., and Habashy, T. M. 2006. A closed-form expression of the electromagnetic tensor Green's functions for a homogeneous TI-anisotropic medium. *IEEE Geoscience and Remote Sensing Letters*, **3**(4), 447–451. DOI: 10.1109/LGRS.2006.874162.

Alumbaugh, D. L., Newman, G. A., Provost, L. and Shadid, J. N. 1996. Three-dimensional wideband electromagnetic modeling on massively parallel computers. *Radio Science*, **31**(1), 1–23. DOI: 10.1029/95RS02815.

Anderson, W. L. 1973. *Fortran IV programs for the determination of the transient tangential electric field and vertical magnetic field about a vertical magnetic dipole for an m-layer stratified earth by numerical integration and digital linear filtering*. Report PB-221240. United States Geological Survey.

Archie, G. E. 1942. The electrical resistivity log as an aid in determining some reservoir characteristics. *Petroleum Transactions of AIME*, **146**, 54–62. DOI: 10.2118/942054-G.

Bannister, P. R. 1968. Determination of the electrical conductivity of the sea bed in shallow waters. *Geophysics*, **33**(6), 995–1003. DOI: 10.1190/1.143999.

Bannister, P. R. 1984. New simplified formulas for ELF subsurface-to-subsurface propagation. *IEEE Journal of Oceanic Engineering*, **OE-9**, 154–163. DOI: 10.1109/JOE.1984.1145620.

Barsukov, P. O., and Fainberg, E. B. 2017. Marine transient electromagnetic sounding of deep buried hydrocarbon reservoirs: Principles, methodologies and limitations. *Geophysical Prospecting*, **65**(3), 840–858. DOI: 10.1111/1365-2478.12416.

Berkhout, A. J. 1984. *Seismic resolution*. London: Geophysical Press. ISBN: 978-0946631124.

Bracewell, R. 1965. *The Fourier transform and its applications*. New York: McGraw-Hill. ISBN 978-0073039381.

Bussian, A. E. 1983. Electric conductance in porous medium. *Geophysics*, **48**, 1258–1268. DOI: 10.1190/1.1441549.

Cagniard, L. 1953. Basic theory of the magnet-telluric method of geophysical prospecting. *Geophysics*, **18**, 605–635. DOI: 10.1190/1.1437915.

Cai, H., Hu, X., Li, J. Endo, M. and Xiong, B. 2017. Parallelized 3D CSEM modeling using edge-based finite element with total field formulation and unstructured mesh. *Computers & Geosciences*, **99**, 125–134. DOI: 10.1016/j.cageo.2016.11.009.

Carcione, J. M., Ursin, B. and Nordskag, J. I. 2007. Cross-property relations between electrical conductivity and the seismic velocity of rocks. *Geophysics*, **72**(5), E193–E204. DOI: 10.1190/1.2762224.

Chave, A. D., Constable, S. C. and Edwards, R. N. 1987. *Electrical exploration methods for the seafloor*. In *Electromagnetic methods and applied geophysics*, ed. Nabhigian, M., pp. 931–966. Tulsa: Society of Exploration Geophysics. ISBN: 978-1-56080-263-1.

Chave, A.D., and Cox, C. S. 1982. Controlled electromagnetic sources for measuring electrical conductivity beneath the oceans: 1: Forward problem and model study. *Journal of Geophysical Research*, **87**(B7), 5327–5338. DOI: 10.1029/JB087iB07p05327.

Chen, J., and Dickens, T. A. 2009. Effects of uncertainty in rock-physics models on reservoir parameter estimation using seismic amplitude variation with angle and controlled-source electromagnetics data. *Geophysical Prospecting*, **57**, 61–74. DOI: 10.1111/j.1365-2478.2008.00721.x

Clemens, M., and Weiland, T. 2001. Discrete electromagnetism with the finite integration technique. *Progress in Electromagnetic Research: PIER*, **32**, 65–87. DOI: 10.2528/PIER00080103.

Constable, S. 2010. Ten years of marine CSEM for hydrocarbon exploration. *Geophysics*, **75**(5), A67–A81. DOI: 10.1190/1.3483451.

Constable, S., and Cox, C. S. 1996. Marine controlled-source electromagnetic sounding 2. The PEGASUS experiment. *Journal of Geophysical Research*, **101**(B3), 5519–5530. DOI: 10.1029/95JB03738.

Constable, S., and Srnka, L. 2007. An introduction to marine controlled-source electromagnetic methods for hydrocarbon exploration. *Geophysics*, **72**(2), WA3–WA12. DOI: 10.1190/1.2432483.

Constable, S., Parker, R. L. and Constable, C. G. 1987. Occam's inversion: A practical algorithm for generating smooth models from electromagnetic data. *Geophysics*, **52**, 289–320. DOI: 10.1190/1.1442303.

Constable, S., Orange, A. S., Hoversten, G. M. and Morrison, H. F. 1998. Marine magnetotellurics for petroleum exploration. Part 1: A sea-floor equipment system. *Geophysics*, **63**(3), 816–825. DOI: 10.1190/1.1444393.

Cox, C. S. 1980. On the electrical conductivity of the oceanic lithosphere. *Physics of the Earth and Planetary Interiors*, **25**, 196–201. DOI: 10.1016/0031-9201(81)90061-3.

Cox, C. S., Deaton, T. K. and Pistek, P. 1981. *An active source EM method for the seafloor*. Technical Report. Scripps Institution of Oceanography.

Cox, C. S., Constable, S. C., Chave, A. D. and Webb, S. C. 1986. Controlled-source electromagnetic sounding of the oceanic lithosphere. *Nature*, **320**(6), 52–54. DOI: 10.1038/320052a0.

D'Arienzo, D., Dell'Aversana, P., Cantarella, G. and Visentin, C. 2010. Multi-transient electromagnetic method in shallow water: A case history in the Mediterranean Sea. EGM2010 International Workshop, Capri, Italy.

da Silva, N. V., Morgan, J. V., MacGregor, L. and Warmer, M. 2012. A finite element multifrontal method for 3D CSEM modeling in the frequency domain. *Geophysics*, **77**(2), E101–E115. DOI: 10.1190/GEO2010-0398.1.

Davydycheva, S., Druskin, V. and Habashy, T. 2003. An efficient finite difference scheme for electromagnetic logging in 3D anisotropic inhomogeneous media. *Geophysics*, **68**, 1525–1536. DOI: 10.1190/1.1620626.

De Witte, A. J. 1957. Saturation and porosity from electric logs in shaly sands. *Oil and Gas Journal*, **55**(9), 89–97.

Djanni, A. T., Ziolkowski, A. and Wright, D. 2016. Electromagnetic induction noise in a towed electromagnetic streamer. *Geophysics*, **81**(3), 1–13. DOI: 10.1190/GEO2014-0597.1.

Druskin, V. I., and Knizhnerman, L. A. 1994. Spectral approach to solving three-dimensional Maxwell's diffusion equations in the time and frequency domains. *Radio Science*, **29**(4), 937–953. DOI: 10.1029/94RS00747.

Duncan, P. M., Hwang, A., Edwards, R. N., Bailey, R. C. and Garland, G. D. 1980. The development and applications of a wide band electromagnetic sounding system using a pseudo-noise source. *Geophysics*, **45**(8), 1276–1296. DOI: 10.1190/1.1441124.

Edwards, R. N. 1988. Two-dimensional modelling of a towed in-line electric dipole–dipole sea-floor electromagnetic system: The optimum time delay or frequency for target resolution. *Geophysics*, **53**(6), 846–853. DOI: 10.1190/1.1442519.

Edwards, R. N. 1997. On the resource evaluation of marine gas hydrate deposits using sea-floor transient electric dipole–dipole methods. *Geophysics*, **62**(1), 63–74. DOI: 10.1190/1.1444146.

Edwards, R. N., and Chave, A. D. 1986. A transient electric dipole–dipole method for mapping the conductivity of the sea floor. *Geophysics*, **51**(4), 984–987. DOI: 10.1190/1.1442156.

Edwards, R. N., Law, L. K. and DeLaurier, J. M. 1981. On measuring the electrical conductivity of the oceanic crust by a modified magnetometric resistivity method. *Journal of Geophysical Research*, **86**(B12), 11609–11615. DOI: 10.1029/JB086iB12p11609.

Eidesmo, T., Ellingsrud, S., MacGregor, L. M., Constable, S., Sinha, M. C., Johanssen, S., Kong, F. N. and Westerdahl, H. 2002. Sea bed logging (SBL), a new method for remote and direct identification of hydrocarbon filled layers in deepwater areas. *First Break*, **20**(3), 144–152.

Ellingsrud, S., Eidesmo, T., Johanssen, S., Sinha, M. C., MacGregor, L. M. and Constable, S. 2002. Remote sensing of hydrocarbon layers by seabed logging (SBL): Results from a cruise offshore Angola. *The Leading Edge*, **21**(10), 972–982. DOI: 10.1190/1.1518433.

Ellis, D. V., and Singer, J. M. 2007. *Well Logging for Earth Scientists*. New York: Springer. ISBN: 978-1-4020-4602-5.

Engelmark, F. 2010. Velocity to resistivity transform via porosity. 80th Annual International Meeting, SEG Expanded Abstracts, 2501–2505. DOI: 10.1190/1.3513358.

Engelmark, F., Mattsson, J., McKay, A. and Du, Z. 2014. Towed streamer EM comes of age. *First Break*, **32**(4), 75–78.

Gassmann, F. 1951. Über die Elastizität poröser Medien. *Vierteljahrsschrift der Naturforschenden Gesellschaft in Zurich*, **96**, 1–23.

Gaver, D. P. 1966. Observing stochastic processes and approximate transform inversion. *Operations Research*, **14**(3), 444–459. DOI: 10.1287/opre.14.3.444.

Ghosh, D. P. 1970. The application of linear filter theory to the direct interpretation of geoelectrical resistivity measurements. Ph.D. Thesis, TU Delft.

Ghosh, D. P. 1971. The application of linear filter theory to the direct interpretation of geoelectrical resistivity sounding measurements. *Geophysical Prospecting*, **19**(2), 192–217. DOI: 10.1111/j.1365-2478.1971.tb00593.x.

Golomb, S. W. 1955. *Sequences with randomness properties*. Baltimore, MD: Glen L. Publishing Company.

Golomb, S. W. 1982. *Shift register sequences*. Walnut Creek, CA: Aegean Park Press. ISBN: 978-981-4632-00-3.

Gradshteyn, I. S., and Ryzhik, I. M. 1996. *Tables of integrals, series, and products*. 5th edn. New York: Academic Press. ISBN: 978-0-12-294756-8.

Gunning, J., Glinsky, M. E. and Hedditch, J. 2010. Resolution and uncertainty in 1D CSEM inversion: A Bayesian approach and open-source implementation. *Geophysics*, **75**(6), F151–F171. DOI: 10.1190/1.3496902.

Hamilton, A. J. 2000. Uncorrelated modes of the non-linear power spectrum. *Monthly Notices of the Royal Astronomical Society*, **312**, 257–284. DOI: 10.1046/j.1365-8711.2000.03071.x.

Hanssen, P., Nguyen, A. K., Fogelin, L. T. T., et al. 2017. The next generation offshore CSEM acquisition system. SEG International Exposition and 87th Annual Meeting, Tulsa, OK, 1194–1198.

Holten, T., Flekkoy, E. G., Singer, B., Blixt, E. M., Hanssen, A. and Maloy, K. J. 2009. Vertical source, vertical receiver, electromagnetic technique for offshore hydrocarbon exploration. *First Break*, **27**(5), 89–93.

Hunziker, J., Thorbecke, J. and Slob, E. 2015. The electromagnetic response in a layered vertical transverse isotropic medium: A new look at an old problem. *Geophysics*, **80**(1), F1–F18. DOI: 10.1190/GEO2013-0411.1.

Hunziker, J., Thorbecke, J., Brackenhoff, J., and Slob, E. 2016. Inversion of controlled-source electromagnetic reflection responses. *Geophysics*, **81**(5), F49–F57. 10.1190/GEO2015-0320.1.

Hursán, G., and Zhdanov, M. S. 2002. Contraction integral equation method in three-dimensional electromagnetic modelling. *Radio Science*, **37**(6), 1–13. DOI: 10.1029/2001RS002513.

Jakosky, J. J., and Hopper, R. H. 1937. The effect of moisture on the direct current resistivities of oil sands and rocks. *Geophysics*, **2**, 33–55. DOI: 10.1190/1.1438064.

Kaufman, A. A., and Keller, G. V. 1983. *Frequency and transient soundings*. New York Elsevier. ISBN 0-444-42032-0.

Kaye, G. W. C., and Laby, T. H. 1971. *Tables of physical and chemical constants*. Harlow: Longman. ISBN 0-582-46326-2.

Key, K. 2009. 1D inversion of multicomponent, multifrequency marine CSEM data: Methodology and synthetic studies for resolving thin resistive layers. *Geophysics*, **74**(2), F9–F20. DOI: 10.1190/geo2011-0237.1.

Key, K. 2012a. Is the fast Hankel transform faster than quadrature? *Geophysics*, **77**(3), F21–F30. DOI: 10.1190/1.3058434.

Key, K. 2012b. Marine EM inversion using unstructured grids: A 2D parallel adaptive finite element algorithm. SEG Technical Program Expanded Abstracts. DOI: 10.1190/segam2012-1294.1.

Key, K., and Lockwood, A. 2010. Determining the orientation of marine CSEM receivers using orthogonal Procrustes rotation analysis. *Geophysics*, **75**, F63–F70. DOI: 10.1190/1.3378765.

Key, K., and Ovall, J. 2011. A parallel goal-oriented adaptive finite element method for 2.5-D modelling. *Geophysical Journal International*, **186**(1), 137–154. DOI: 10.1111/j.1365-246X.2011.05025.x.

Koefoed, O., Ghosh, D. P. and Polman, G. J. 1972. Computation of type curves for electromagnetic depth sounding with a horizontal transmitting coil by means of a digital linear filter. *Geophysical Prospecting*, **20**, 406–420. DOI: 10.1111/j.1365-2478.1972.tb00644.x.

Kong, F. N. 2007. Hankel transform filters for dipole antenna radiation in a conductive medium. *Geophysical Prospecting*, **55**(1), 83–89. DOI: 10.1111/j.1365-2478.2006.00585.x.

Kong, F. N. 2012. Evaluation of Fourier cosine/sine transforms using exponentially positioned samples. *Journal of Applied Geophysics*, **79**, 46–54. DOI: 10.1016/j.jappgeo.2011.12.007.

Kragh, E., and Christie, P. 2002. Seismic repeatability, normalised RMS, and predictability. *The Leading Edge*, **21**(7), 640–647. DOI: 10.1190/1.1497316.

Kruglyakov, M. and Bloshanskaya, Li. 2017. High-performance parallel solver for integral equations of electromagnetics based on Galerkin method. *Mathematical Geosciences*, **49**(6), 751–776. DOI: 10.1007/s11004-017-9677-y.

Leverett, M. C. 1939. Flow of oil–water mixtures through unconsolidated sands. Transactions of the AIME, **7**, 325–345. DOI: 10.2118/939149-G.

Levinson, N. 1946. The Wiener RMS (root mean square) error criterion in filter design and prediction. *Journal of Mathematics and Physics*, **25**(4), 261–278. DOI: 10.1002/sapm1946251261.

Li, G., and Han, B. 2017. Application of the perfectly matched layer in 2.5D marine controlled-source electromagnetic modeling. *Physics of the Earth and Planetary Interiors*, **270**, 157–167. DOI: 10.1016/j.pepi.2017.07.006.

Loke, M. H., and Barker, R. D. 1996. Rapid least-squares inversion of apparent resistivity pseudo-sections using quasi-Newton method. *Geophysical Prospecting*, **48**, 152–181. DOI: 10.1111/j.1365-2478.1996.tb00142.x.

Lu, X., and Srnka, L. J. 2005. Logarithmic spectrum transmitter waveform for controlled-source electromagnetic surveying. U.S. Patent Application WO 2005/117326A2.

Lu, X., and Srnka, L. J. 2009. Logarithmic spectrum transmitter waveform for controlled-source electromagnetic surveying. U.S. Patent 7,539,279 B2.

Maaø, F. A. 2007. Fast finite-difference time-domain modeling for marine-subsurface electromagnetic problems. *Geophysics*, **72**(2), A19–A23. DOI: 10.1190/1.2434781.

Martin, M., Murray, G.H. and Gillingham, W.J. 1938. Determination of the potential productivity of oil-bearing formations by resistivity measurements. *Geophysics*, **3**, 258–272. DOI: 10.1190/1.1439502.

Mattsson, J., Lindqivst, P., Juhasz, R. and Björnemo, E. 2012. Noise reduction and error analysis for a towed EM system. Paper presented at SEG Las Vegas 2012 Annual Meeting. DOI: 10.1190/segam2012-0439.1.

Mittet, R. 2010. High-order finite-difference simulations of marine CSEM surveys using a correspondence principle for wave and diffusion fields. *Geophysics*, **75**(1), F33–F50. DOI: 10.1190/1.3278525.

Mittet, R., and Schaug-Pettersen, T. 2008. Shaping optimal transmitter waveforms for marine CSEM surveys. *Geophysics*, **73**(3), F97–F104. DOI: 10.1190/1.2898410.

Mittet, R., Aakervik, O. M., Jensen, H. R., Ellingsrud, S., and Stovas, A. 2007. On the orientation and absolute phase of marine CSEM receivers. *Geophysics*, **72**(4), F145–F155. DOI: 10.1190/1.2732556.

Mulder, W. A. 2006. A multigrid solver for 3D electromagnetic diffusion. *Geophysical Prospecting*, **54**(5), 633–649. DOI: 10.1111/j.1365-2478.2006.00558.x.

Mulder, W. A., Wirianto, M. and Slob, E. C. 2008. Time-domain modeling of electromagnetic diffusion with a frequency-domain code. *Geophysics*, **73**(1), F1–F8. DOI: 10.1190/1.2799093.

Myer, D., Constable, S. and Key, K. 2011. Broad-band waveforms and robust processing for marine CSEM surveys. *Geophysical Journal International*, **184**, 689–698. DOI: 10.1190/GEO2011-0380.1.

Myer, D., Constable, S., Key, K., Glinsky, M. E. and Liu, G. 2012. Marine CSEM of the Scarborough gas field, part 1: Experimental design and data uncertainty. *Geophysics*, **77**(4), E281–E299. DOI: 10.1190/geo2011-0380.1.

Nabighian, M. N., and Corbett, J. D. 1994. *Electromagnetic methods in applied geophysics – Vol. 1 Theory: investigations in geophysics.* Tulsa, OK: Society of Exploration Geophysicists. ISBN 978-0-93183-051-8.

Nahin, P. 2002. *Oliver Heaviside: The life, work, and times of an electrical genius of the Victorian age.* Baltimore, MD: Johns Hopkins University Press. ISBN 978-0801869099.

Nekut, A. G., and Spies, B. R. 1989. Petroleum exploration using controlled-source electromagnetic methods. *Proceedings of the IEEE*, **77**(2), 338–362. DOI: 10.1109/5.18630.

Newman, G., Commer, M. and Carazzone, J. 2010. Imaging CSEM data in the presence of electrical anisotropy. *Geophysics*, **75**(2), F51–F61. DOI: 10.1190/1.3295883.

Nguyen, A. K., Nordskag, J. I., Wiik, T., Bjørke, A. K., Boman, L., Pedersen, O. M., Ribaudo, J. and Mittet, R. 2016. Comparing large-scale 3D Gauss–Newton and BFGS CSEM inversions. SEG Technical Program Expanded Abstracts. DOI: 10.1190/segam2016-13858633.1

Oberhettinger, F., and Badii, L. 1973. *Tables of Laplace Transforms*. 1st edn. Berlin: Springer-Verlag. ISBN 978-3-642-65645-3

Pankratov, O. V., Avdeyev, D. B. and Kuvshinov, A. V. 1995. Electromagnetic field scattering in a heterogeneous Earth: A solution to the forward problem. *Physics of the Solid Earth*, **31**(3), 201–209.

Parker, R. L. 1994. *Geophysical inverse theory.* Princeton, NJ: Princeton University Press. ISBN 9780691036342.

Plessix, R.-É., and Mulder, W. A. 2008. Resistivity imaging with controlled-source electromagnetic data: Depth and data weighting. *Inverse Problems*, **24**(3), 034012. DOI: 10.1088/0266-5611/24/3/034012.

Rabiner, L. R., and Gold, B. 1975. *Theory and application of digital signal processing.* Englewood Cliffs, NJ: Prentice-Hall. ISBN: 0-13-914101-4

Rider, M. 1996. *The geological interpretation of well logs.* Caithness: Whittles Publishing. ISBN: 1-870325-36-2

Robinson, E. A. 1954. Predictive decomposition of time series with applications to seismic exploration. Ph.D. thesis, Massachusetts Institute of Technology.

Robinson, E. A. 1967. *Statistical communication and detection.* London: Griffin.

Robinson, E. A., and Treitel, S. 1967. Principles of digital Wiener filtering. *Geophysical Prospecting*, **15**(3), 311–333. DOI: 10.1111/j.1365-2478.1967.tb01793.x.

Schwalenberg, K., Willoughby, E., Mir, R. and Edwards, R. N. 2005. Marine gas hydrate electromagnetic signatures in Cascadia and their correlation with seismic blank zones. *First Break*, **23**(4), 57–63.

Schwarzbach, C., Börner, R.-U. and Spitzer, K. 2011. Three-dimensional adaptive higher order finite element simulation for geo-electromagnetics – A marine CSEM example. *Geophysical Journal International*, **187**(1), 63–74. DOI: 10.1111/j.1365-246X.2011.05127.x.

Sen, P. N., Scala, C. and Cohen, M. H. 1981. A self-similar model for sedimentary rocks with application to the dielectric constant of fused glass beads. *Geophysics*, **46**, 781–795. DOI: 10.1190/1.1441215.

Sheriff, R. E. 1973. *Encyclopedic dictionary of exploration geophysics.* Tulsa, OK: Society of Exploration Geophysicists. ISBN 978-1-56080-118-4

Si, H. 2015. TetGen, a Delaunay-based quality tetrahedral mesh generator. *ACM Transactions on Mathematical Software*, **41**, 11.1–11.36. DOI: 10.1145/2629697.

Singer, B. Sh. 1995. Method for solution of Maxwell's equations in non-uniform media. *Geophysical Journal International*, **120**, 590–598. DOI: 10.1111/j.1365-246X.1995.tb01841.x.

Sinha, M. C., Patel, P. D., Unsworth, M. J., Owen, T. R. E. and MacCormack, M. R. G. 1990. An active source electromagnetic sounding system for marine use. *Marine Geophysical Researches*, **12**, 59–68. DOI: 10.1007/BF00310563.

Slob, E., Hunziker, J. and Mulder, W. A. 2010. Green's tensors for the diffusive electric field in a VTI half-space. *Progress in Electromagnetics Research: PIER*, **107**, 1–20. DOI: 10.2528/PIER1005280.

Smith, R. 2010. Airborne electromagnetic methods: Applications to minerals, water and hydrocarbon exploration. *Recorder*, **35**(3), 1–9. DOI: 10.1016/j.cageo.2013.04.004.

Sommer, M., Hlz, S., Moorkamp, M., Swidinsky, A., Heincke, B., Scholl, C. and Jegen, M. 2013. GPU parallelization of a three dimensional marine CSEM code. *Computers & Geosciences*, **58**, 91–99. DOI: 10.1016/j.cageo.2013.04.004

Spies, B. R., and Frischknecht, F. C. 1987. Electromagnetic sounding. In: *Electromagnetic Methods in Applied Geophysics: Volume 2*, ed. Nabighian, M., pp. 285–425. Tulsa, OK: Society of Exploration Geophysicists. ISBN: 978-1-56080-022-4

Srnka, L., Carazzone, J. J., Ephron, M. S. and Eriksen, E. A. 2006. Remote reservoir resistivity mapping. *The Leading Edge*, **25**(8), 972–975. DOI: 10.1190/1.2335169.

Stehfest, H. 1970. Numerical inversion of Laplace transforms. *Communications of the ACM*, **13**(1), 47–49. DOI: 10.1145/361953.361969.

Strack, K.-M. 1992. *Exploration with deep transient electromagnetics*. New York: Elsevier. ISBN: 0-444-89541-8.

Strack, K.-M., Hanstein, T., LeBroq, K., Moss, D. C., Vozoff, K. and Wolfgram, P. A. 1989. Case histories of LOTEM surveys in hydrocarbon prospective areas. *First Break*, **7**(12), 467–477.

Streich, R. 2009. 3D finite-difference frequency-domain modeling of controlled-source electromagnetic data: Direct solution and optimization for high accuracy. *Geophysics*, **74**(5), F95–F105. DOI: 10.1190/1.3196241.

Streich, R. 2016. Controlled-source electromagnetic approaches for hydrocarbon exploration and monitoring on land. *Surveys in Geophysics*, **37**, 47–80. DOI: 10.1007/s10712-015-9336-0.

Summerfield, P. J., Fielding, B. J. and Gale, L. S. 2012. Method to maintain towed dipole source orientation. U.S. Patent No. 8,183,868.

Tarantola, A. 1987. *Inverse problem theory*. Amsterdam: Elsevier. ISBN: 9780444599674.

Tikhonov, A. N., and Arsenin, V. Y. 1977. *Solutions of ill-posed problems*. New York: Winston. ISBN: 0470991240.

Weiglhofer, W. S. 1990. Dyadic Green's functions for general uniaxial media. *IEE Proceedings Part H, Microwaves, Antennas and Propagation*, **137**, 5–10. DOI: 10.1049/ip-h-2.1990.0002.

Weir, G. J. 1980. Transient electromagnetic fields about an infinitesimally long grounded horizontal electric dipole on the surface of a uniform half-space. *Geophysical Journal of the Royal Astronomical Society*, **61**, 41–56. DOI: 10.1111/j.1365-246X.1980.tb04302.x.

Werthmüller, D. 2017. An open-source full 3D electromagnetic modeler for 1D VTI media in Python: Empymod. *Geophysics*, **82**(6), WB9–WB19. DOI: 10.1190/geo2016-0626.1.

Werthmüller, D., Ziolkowski, A. and Wright, D. 2013. Background resistivity model from seismic velocities. *Geophysics*, **78**(4), E213–E223. DOI: 10.1190/geo2012-0445.1.

Werthmüller, D. 2009. Inversion of multi-transient EM data from anisotropic media. M.Sc. Thesis, TU Delft, ETH Zurich, RWTH Aachen.

Wiener, N. 1949. *Time series*. Cambridge, MA: MIT Press. ISBN: 9780262230025.

Wilt, M., and Alumbaugh, D. 1998. Electromagnetic methods for development and production: State of the art. *The Leading Edge*, **17**(4), 487–490. DOI: 10.1190/1.1437997.

Wright, D., Ziolkowski, A. and Hobbs, B. 2002. Hydrocarbon detection and monitoring with a multicomponent transient electromagnetic (MTEM) survey. *The Leading Edge*, **21**(9), 852–864. DOI: 10.1190/1.1508954.

Wright, D., Ziolkowski, A. and Hall, G. 2006. Improving signal-to-noise ratio using pseudo-random binary sequences in multi-transient electromagnetic (MTEM) data. Poster P065 presented at EAGE 68th Conference and Exhibition, Vienna, Austria.

Wright, D. A., Ziolkowski, A. M. and Hobbs, B. A. 2005. Detection of subsurface resistivity contrasts with application to location of fluids. U.S. Patent No. 6,914,433 B2.

Wyckoff, R., and Botset, H. 1936. The flow of gas–liquid mixtures through unconsolidated sands. *Physics: A Journal of General and Applied Physics*, **7**, 325–345. DOI: 10.1063/1.1745402.

Wyllie, M. R. J., and Rose, W. D. 1950. Some theoretical considerations related to the quantitative evaluation of the physical characteristics of reservoir rock from electrical log data. *Journal of Petroleum Technology*, **2**, 105–118. DOI: 10.2118/950105-G.

Yee, K. 1966. Numerical solution of initial boundary value problems involving Maxwell's equations in isotropic media. *IEEE Transactions on Antennas and Propagation*, **16**, 302–307. DOI: 10.1109/TAP.1966.1138693.

Yoon, D., Zhdanov, M. S., Mattsson, J., Cai, H. Z. and Gribenko, A. 2016. A hybrid finite-difference and integral-equation method for modeling and inversion of marine controlled-source electromagnetic data. *Geophysics*, **81**(5), E323–E336. DOI: 10.1190/GEO2015-0513.1.

Zhdanov, M. S. 2002. *Geophysical inverse theory and regularization problems*. New York: Elsevier. ISBN: 9780444510891

Zhdanov, M. S., Lee, S. K., and Yoshioka, K. 2006. Integral equation method for 3D modeling of electromagnetic fields in complex structures with inhomogeneous background conductivity. *Geophysics*, **71**(6), G333–G345. DOI: 10.1190/1.2358403.

Ziolkowski, A. 2007. Developments in the transient electromagnetic method. *First Break*, **25**, 99–106.

Ziolkowski, A., and Carson, R. 2007. Maximising signal-to-noise ratio in transient EM data. Paper E048 presented at EAGE 69th Conference and Exhibition, London.

Ziolkowski, A., Hobbs, B. A. and Wright, D. 2007. Multitransient electromagnetic demonstration survey in France. *Geophysics*, **72**(4), F197–F209. DOI: 10.1190/1.2735802.

Ziolkowski, A., Parr, R., Wright, D., Nockles, V., Limond, C., Morris, E. and Linfoot, J. 2010. Multi-transient electromagnetic repeatability experiment over the North Sea Harding field. *Geophysical Prospecting*, **58**, 1159–1176. DOI: 10.1111/j.1365-2478.2010.00882.x.

Ziolkowski, A., Wright, D. and Mattsson, J. 2011. Comparison of pseudo-random binary sequence and square-wave transient controlled-source electromagnetic data over the Peon gas discovery, Norway. *Geophysical Prospecting*, **59**, 1114–1131. DOI: 10.1111/j.1365-2478.2011.01006.x.

Index